SPONTANEOUS SYMMETRY BREAKDOWN AND RELATED SUBJECTS

XXI WINTER SCHOOL OF THEORETICAL PHYSICS

SPONTANEOUS SYMMETRY BREAKDOWN AND RELATED SUBJECTS

18 February-2 March 1985

Karpacz, Poland

Editors L MICHEL
J MOZRZYMAS
A PĘKALSKI

World Scientific

Published by

World Scientific Publishing Co. Pte. Ltd.
P. O. Box 128, Farrer Road, Singapore 9128

SPONTANEOUS SYMMETRY BREAKDOWN AND RELATED SUBJECTS

ISBN 9971-978-54-7

Printed in Singapore by Singapore National Printers (Pte) Ltd.

XXI Winter School of Theoretical Physics

SPONTANEOUS SYMMETRY BREAKDOWN AND RELATED SUBJECTS

Organized by : Institute of Theoretical Physics
University of Wrocław

ORGANIZING COMMITTEE

Directors : L Michel (IHES, Bures-sur-Yvette)
J Mozrzymas (University of Wrocław)

Members : W Cegła (University of Wrocław)
J Jędrzejewski (University of Wrocław)
A Pękalski (University of Wrocław)
J Stelmach (University of Wrocław)

FOREWORD

The subject of the XXIst Karpacz Winter School was "Spontaneous symmetry breakdown and related subjects". This covers many domains of Physics! The first week was devoted mainly to statistical mechanics, condensed matter physics, more specially second order phase transitions, modulated crystals (an off-schedule lecture was organized on the newly discovered quasicrystals!), topological classification of symmetry defects. In the second week we moved to particle physics and field theory, to supersymmetry and cosmology. Moreover, during the two weeks there were a few lectures on some aspects of group theory.

There were more lectures: those of the "kindergarten" organized on a more spontaneous basis for the students (mainly from Poland) who came to the School. These more general and pedagogical lectures do not appear in the proceedings.

By its interdisciplinary program this Karpacz session may have required more efforts from all participants. One can hope that this will make these Proceedings more attractive to the reader.

To triumph over all possible types of difficulty in Poland for organizing an International School of Physics, our Polish hosts made many miracles. In the name of all participants (who came from fifteen countries) I thank them once more for their cordial hospitality and the excellent organization of the School. So many worked so hard for it. Among them Dr. Janusz Jędrzejewski, the School secretary. It was a happy coincidence that he received formally during this session the Association of Mathematical Physics prize for this beautiful theoretical work on supraconductivity.

I had the pleasure of working as co-director of this school with my friend Jan Mozrzymas, elected a few months before Rector of Wrocław University. This very heavy responsibility forced him to leave the School for several trips to Wrocław on snowy roads.

All of us participants felt that this XXIst session was a genuine success (21 is a lucky number in Poland). I hope this success will extend to these Proceedings; I add my best wishes to the XXIInd and other sessions of the International Karpacz School of Physics.

Niech żyje Polska!

Louis MICHEL
Co-director of the session

CONTENTS

INTRODUCTION TO SPONTANEOUS SYMMETRY BREAKING

SOME EXAMPLES

Louis MICHEL
Institut des Hautes Etudes Scientifiques
35, route de Chartres
91440 Bures-sur-Yvette
FRANCE

0. Introduction.

Let me begin this first lecture of the school by an historical reference [1]. E.P. Wigner was working as an engineer in a Budapest leather factory when at 24 he received an invitation to become assistant of the new theoretical physics professor (Becker) in Berlin. When he arrived, the assistant position was not yet established and he was advised, while waiting, to work with Dr. Weissenberg, a known crystallographer, who told him : "There is a miracle . Why in a crystal atoms are most often on a symmetry axis or on a symmetry plane. Why ?". The day after, Wigner tried to give him an answer : "On a symmetry axis or a symmetry plan the potential is more likely to have an extremum". "Well, well you seem to be right, but one needs an elegant proof" and in his interview Wigner adds : "Then I started with the idea to write a book on group theory".

You will not find this subject treated in the famous Wigner book [2], which appeared soon after (1931), or in most books on physical application of group theory. So I will deal with it in these lectures:§7.

These lectures are nearly self-containing. The first part (Sections 1 to 5) will teach the basic of group actions, illustrated by physical examples, and explain the principal mechanism for the mathematical study of spontaneous symmetry breaking (Section 6). The last three sections give the results of recent, mostly unpublished work to which I am collaborating, in three domains : 1) basic concepts of crystallography (useful for the lectures on modulated crystals), 2) renormalisation group application to Landau theory, 3) a Spin 10 grand unification scheme (sections 8,9,10).

1. Group actions. Orbits and Strata. Examples.

You all know what is a group G . If $H < G$ (< reads "subgroup") it is interesting to consider the cosets g_1H , g_2H , etc. of H . We denote by $[G:H]$ the set of these cosets. If G is finite and has $|G|$ elements, the number of cosets is $|G|/|H|$. If G and H are Lie groups, then $[G:H]$ is a manifold whose dimension is :

$$\dim[G:H] = \dim G - \dim H . \qquad (1)$$

If for every $g \in G$, $gH = Hg$, we note $H \triangleleft G$ and say that H is an invariant subgroup of G . Then there is a natural group law on the set $[G:H]$, given by $g_1H.g_2H = g_1g_2.H$. This group is denoted by G/H and it is called the quotient group of G by H .

All the books on group theory and quantum mechanics study the linear representations of a group G on a vector space E . Such a representation is a homomorphism $G \xrightarrow{f} GL(E)$ of G in the general linear group on E . The group $GL(E)$ is the automorphism group $\mathrm{Aut}\, E$ of E . The set of elements of G which are represented by the identity on E form the kernel of f , $\ker f \triangleleft G$ and $G/\ker f \sim \mathrm{Im}\, f$, the image of f , which is $< GL(E)$. In quantum physics, E is the Hilbert space of state vectors, and we need to consider only unitary G representations, i.e. $\mathrm{Im}\, f < U(E)$, where $U(E) = \mathrm{Aut}\, E$, the automorphism group of the Hilbert space.

But symmetry groups may also enter into physics through an action on a mathematical structure M (e.g. a manifold) which is defined by the group homomorphism $G \xrightarrow{f} \mathrm{Aut}\, M$. The action is effective if $\ker f = [1]$.

Given two G actions G,f,M and Gf'M' , by <u>definition</u> an equivriant map $M \overset{\theta}{\to} M'$, satisfies the commutative diagram 1 for every element of G .

$$\forall y \in G \qquad \begin{array}{ccc} M & \xrightarrow{\theta} & M' \\ \Big\downarrow {\scriptstyle f'(y)} & & \Big\downarrow {\scriptstyle f'(g)} \\ M & \xrightarrow{\theta} & M' \end{array}$$

Diagram 1

<u>Definition</u>. The two actions are equivalent when θ is a bijective map (and therefore an isomorphism between M and M'). When M is a vector space or a Hilbert space, this definition of equivalence coincides with the usual one for G-linear representations.

When a physical system has a symmetry group G , all functions describing physical properties of this system must be G-invariant or G-covariant. But these conditions depend only on the image Im f . So a weaker definition of equivalence is often useful in physics, as J. Mozrzymas and I showed [3].

<u>Definition</u>. Two actions G,f,M and G',f',M' are weakly equivalent if there is an isomorphism $M \overset{\theta}{\to} M'$ such that the corresponding automorphism Aut M $\overset{\theta^*}{\to}$ Aut M' identifies the two images : θ^*(Im f) = (Im f') . For instance when M = M' , Im f and Im f' are conjugate subgroups of Aut M . The two non trivial inequivalent representations of Z_3 (the cyclic group with 3 elements) or the two 3-dimensional inequivalent representations of SU(3) are weakly equivalent. Weak equivalence of actions (which is even defined for two different groups) will appear as a natural and important concept in the study made below of spontaneous symmetry breaking in phase transitions.

To simplify notations, we will often use g.m instead of f(g)m , the transform of m by g . The set of all transforms of m is denoted by G.m and is called the G orbit of m . The little group G_m

(mathematiciens often say the isotropy group) is the set of all elements of G such that $g.m = m$. Note that $G_{g.m} = gGg^{-1}$ so the little groups of an orbit form a conjugation class of G subgroups, that we denote by $[G_m]$. When the actions of G on two orbits are equivalent, it is easy to prove that these orbits have the same conjugate class $[H]$ of little groups. They are said to be of the same type. The sets of cosets $[G:H]$ with the G action $g.xH = gxH$ is a prototype of this type of orbits. By definition, a stratum is the union of all orbits of the same type; equivalently $m' \in S(m)$, the stratum of m , when $G_{m'}$ and G_m are conjugate.

The decomposition of a group action into strata yields a primary important information, very relevant physically. For instance : 1) in the linear representation action of the Lorentz group on Minkowski space there are three other strata outside the origin (unique fixed point) : their elements are respectively the time-like, space-like and light-like vectors. Let us choose four other examples : 2) The symmetry group of an axially symmetric ellipsoid (as a simplified model of the earth) is $D_{\infty h}$, generated by C_∞ , the group of rotations around the axis containing the two poles, the rotation by π around axes in the equatorial plane (with C_∞ , they generate the group D_∞) and finitly the symmetry h through the equatorial plane. (Note that this figures has a symmetry center; when taken as origin $-I \in D_{\infty h}$) . There are three strata : the two poles (i.e. a two-point orbit), the equator (one connected orbit) and the rest, an open dense set in which the orbits are the pair of parallel circles with the same North and South latitude. 3) Consider the n dimensional hypercube, centered at the origin. Its 2^n vertices have coordinates $\varepsilon_1,\varepsilon_2,\ldots,\varepsilon_n$ with $\varepsilon_i^2 = 1$; the center of its $2n$ faces are at the tops of $\pm$ the unit vectors of the coordinate axes. Its symmetry group is generated by the diagonal matrices with ± 1 as elements: (they form an Abelian group $\sim Z_2^n$) and the permutation group S_n of the coordinate axes. This $2^n.n!$ element group is denotes B_n in the classification by Coxeter of the finite groups generated by reflections. Chemists, physicists and crystallographers also use the notations : when $n = 2$, C_{2v} or $2mm$, the element group of symmetry of the square, and when $n = 3$, O_h or m3m , the 48 element group of symmetry of the cube. Outside the origin, the 2 dimensional linear action of C_{2v} containing three strata, the 2 coordinate axes, the two diagonal axes and the rest. The three dimensional representation of O_h contains strata given in table 1. As we see the strata correspond to the symmetry elements.

Table 1. Strata of the 3 dimensional representation of O_h. The little group and its number of elements are given at the end of the lines.

(0)	the origin	O_h , 48
(1)	the 8 axes containing the vertices −(0)	C_{3v}, 6
(2)	the 12 axes containing the center of edges −(0)	C_{2v}, 2
(3)	the 6 axes containing the center of faces −(0)	C_{4v}, 4
(4)	the 3 symmetry planes (coordinate planes) −(0) − (2) −(3)	C_s , 2
(5)	the 6 other symmetry planes −(0) −(1) −(2) −(3)	C_s' , 2
(6)	the rest, open dense	1 , 1

4) In the action of the 230 crystallographic groups on the 3-dimensional space, the strata are tabulated in the international Tables for Crystallography under the name "Wyckoff positions" [4]. For each space group, there is a finite number of them. There is only one of dimension 3, and one can check that it is open dense. 5) In an n dimensional real vector space R^n a lattice is a closed subgroup Z^n generated by n basis (i.e. linearly independent vectors of R^n . The general linear group GL(n,R) transforms any basis into any basis, so the set L of lattice is the orbit [GL(n,R) : GL(n,Z)] . Indeed the little group of a lattice transforms the set of lattice points (i.e. vectors with integral coordinates) into itself. Note that $GL(n,Z) = \mathrm{Aut}\, Z^n$. The orthogonal subgroup $O(n) < GL(n,R)$ respects the space metric. The strata of its action on L correspond to crystallographic systems, the corresponding little groups P_H are called the holohedries of the lattices. For $n = 3$ there are (*) 7 crystallographic systems :

(*) These crystallographic systems were listed by Weiss in 1815. This stratum definition corresponds to the "French systems" in ref [4]. Strangely enough the International Tables have adopted an unnatural definition.

Systems	Triclinic	Monoclinic	Orthorhombic	Tetragonal	Trigonal
	Tri	Mon	Ort	Tet	Trg
P_H	$T = C_i$	$2/m = C_{2n}$	$mmm = D_{2h}$	$4/mmm = D_{4h}$	$\overline{3}m = D_{3d}$

Hexagonal	Cubic
Hex	Cub
$6/mmm = D_{6h}$	$m3m = O_h$

For $n = 2$ there are 4 crystallographic systems. We will prove it from a natural description of the space of 2 dimensional lattices as an orbit space. (See below).

In all these examples, the number of strata is finite. This will be the case in most physics problems.

2. Orbit Space. Examples.

In the action of G on M the set of orbits we denoted by $M|G$ the set of orbits and by $\overline{M|G}$ the set of strata and by π, σ the canonical surjective maps

$$M \xrightarrow{\pi} M|G \xrightarrow{\sigma} \overline{M|G} \qquad (2)$$

In the five examples of the preceding section we have studied $\overline{M|G}$. Let us now study the orbit space.

1) The scalar product $S = (a,a)$ is a real number and it is an invariant of the Lorentz group. To any value of (a,a) corresponds a unique orbit except $(a,a) = 0$ which is both the length of light-like vectors and the $\vec{0}$ vector. So if we consider the Minkowski space minus the origin, the orbit space is R and the three strata are defined by $S > 0$, $S = 0$, $S < 0$.

2) The orbit space is $0 \leq \theta \leq \pi/2$ where θ is the absolute value of the latitude; the three strata are $\theta = 0$, $\theta = \frac{\pi}{2}$, $0 < \theta < \frac{\pi}{2}$.

3) The orbit space is one of the convex connected cones formed by the symmetry hyperplanes ex : for $n = 2$, $x_1 \geq 0$, $x_2 - x_1 > 0$, for $n > 2$, $x_1 \geq 0$, $x_2 - x_1 \geq 0$, $x_3 - x_2 > 0$,..., $x_n - x_{n-1} > 0$. It is called a Weyl chamber.

4) The orbit space for the translation group Z^n est $R^n \mid Z^n$; it has the topology of a torus $(S_1)^n$ (where S_k is the k dimensional sphere). It is what the crystallographer calls a Wigner Satz cell with its opposite faces identified. The action of G on R^n defines an action of the point group $P = G/Z^n$ on the torus $R^n \mid Z^n$. We leave to the reader the determination of the orbit space in general. It is not a manifold except in the case where there is one stratum only : this occurs for respectively 2 and 13 crystallographic groups for $n = 2$ and 3 . Then $R^n \mid G$ is a flat Riemann manifold with G as first homotopy group.

5) We will study the action of the group $O(n) \times R^x$, including the dilations for the case $n = 2$. Indeed the symmetry of a lattice is independent from its scale. We choose as first generator $\vec{a}$ of the lattice, one of the shortest vectors (*); by a dilation and rotation we

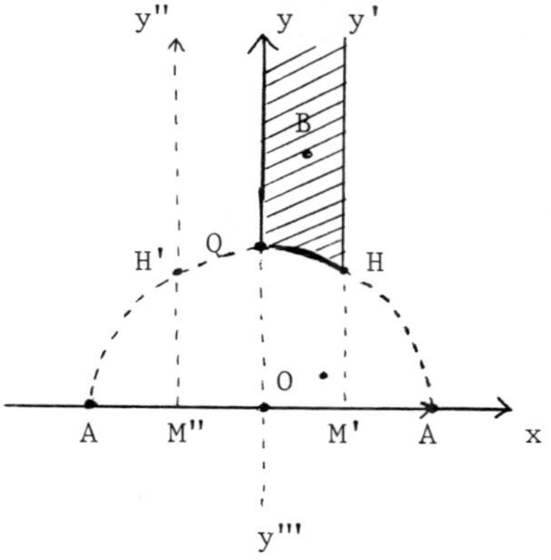

Figure 1

(*) Beware that this is not always possible for $n \geq 5$. I am grateful to Henry Bacry for advice and discussions on Fig. 1.

bring it to the unit vector of the x axis. Then we can look for the second generator $\vec{b}$ in a vertical band of width 1, for instance $-\frac{1}{2} < x \leq \frac{1}{2}$ and also require its component y to be > 0 (since $\vec{b} \in Z^2 \Rightarrow -\vec{b} \in Z^2$, $\vec{b}+m\vec{a} \in Z^2$), moreover since $|\vec{b}| \geq |\vec{a}|$, the point B, top of the vector $\vec{OB}$, must be outside the open unit circle centered in O. Finally, if the abscissa of B is negative, by the reflection through the y-axis, we change its sign. So the orbit space is the hatched domain $yQHy'$. The (open dense) inside represents the diclinic system, $P_H = \{I,-I\} = C_i$ or $\bar{1}$. The boundary minus Q, H represents the orthorhombic system $P_H = C_{2v}$ or $2mm$, Q the quadratic system $P_H = C_{4v} = 4mm$ and H the hexagonal system $P_H = C_{v}$ or $6mm$. The corresponding study for $n = 3$ has been done in [5].

Let us add another example : 6) Consider the decay of a particle of energy momentum $\underline{P}$ into three particles of energy momenta p_i, $i = 1,2,3$. The phase space M is defined by the relations :

$$(\underline{P},\underline{P}) = M^2 , \quad (p_i,p_i) = m_i^2 , \quad \underline{P} = \Sigma_i \underline{p}_i . \tag{3}$$

The little group G of $\underline{P}$ in the Lorentz group is isomorphic to $O(3)$ (generally the initial particle is considered at rest; it is not relevant). It acts on M. The orbit space $M|G$ is the Dalitz plot. There are two strata, the interior, when the $3p_i$'s span a 2-plane, and the boundary, when the $3p_i$'s are colinear.

3. Action on subsets and substructures. Examples.

The action $G \xrightarrow{f} \mathrm{Aut}\, M$ of G on M defines an action of G on the set $\mathcal{P}(M)$ of subsets of M. Given such a subset $X \subset M$ one defines the centralizer in G of X as :

$$C_G(X) = \bigcap_{x \in X} G_x \quad ; \tag{4}$$

it is the largest G-subgroup which leaves fixed every element of X. Similarly one defines the stabilizer in G of X as the largest G subgroup which transforms X in itself. We denote it by $S_G(X)$. It is easy to prove that $C_G(X) \triangleleft S_G(X)$. We shall denote the quotient

$$W_G(X) = S_G(X)/C_G(X) \ ;$$

it acts effectively on X . If G is a compact semi-simple group, its Lie algebra $\mathcal{G}$ has an orthogonal scalar product, the Cartan Killing form. The adjoint representation of G is the natural linear representation on $\mathcal{G}$(as a vector space). If X is a Cartan subalgebra, i.e. a maximal Abelian subalgebra (they are all conjugate by G) , then $W_G(X)$ is the Weyl group; it is a Coxeter group.

Note a fundamental relation satisfies by centralizers in any group action

$$C_G(\cup_i M_i) = \cap\, C_G(M_i) \quad .$$

<u>Exercise</u>. Prove that for linear representations of finite groups or enumerable groups (i.e. crystallographic groups), interactions of little groups are little groups. (This is not true for other groups in general!). The proof is an appendix A of [6] . It is also useful to introduce the traditional notation M^g for the set of elements of M invariant by G . Similarly

$$H < G \quad , \quad M^H = \bigcap_{g \in H} M^g \quad .$$

4. <u>Partial ordering of the strata. Compact group actions.</u>

There is a partial ordering, by inclusion, on $\{< G\}$, the set of subgroups of G . When G is compact (this includes finite), it defines a partial order (by inclusion up to a conjugation) on the set of conjugation classes of closed subgroups of G . This is also true for crystallographic groups or for the conjugate classes of finite subgroups of an arbitrary group. This induces a partial ordering on the set of strata.

If the action of G is continuous, M^g is closed, so is M^H as an intersection of closed sets. $\bigcup_{H \in [H]} M^H$ is the union of all strata with little group conjugate class $\geq [H]$. For finite groups, as a finite union of closed sets, it is closed, so the strata for maximal isotropy groups are closed. This extends to compact group action. Similarly, it is easy to prove [3a] that for a finite group action, there is a unique minimal conjugate class of little groups (Kerf itself); the corresponding stratum is open dense. We call it "generic". This is also true for smooth compact group actions [7]. We verify these two properties on examples 2,3,5,6; they are also true for 4, but not true for 1 : The Lorentz group is not compact. For smooth compact group action G on a finite dimensional differentiable manifold M, with a finite number of strata (e.g. this is the case when M is compact), Mostow [8] proved that there exists a smooth injective equivariant Map $M \xrightarrow{\theta} E$ into a real vector space E of finite dimension, carrying a linear orthogonal representation of G. So the case of linear action is pretty general !

5. Action on a group. Action of G on itself.

We consider the actions on G preserving its group law. They are defined by $K \xrightarrow{f} \mathrm{Aut}\, G$ and most properties of the action depend only on $\mathrm{Im}\, f < \mathrm{Aut}\, G$. Consider the particular case $G = K$; then, for the "natural" action of G on itself, $\mathrm{Im}\, f = \mathrm{In\,Aut}\, G$, the group of inner automorphisms. One proves that $\mathrm{In\,Aut}\, G \triangle \mathrm{Aut}\, G$ and one defines $\mathrm{Out}\, G = \mathrm{Aut}\, G/\mathrm{In\,Aut}\, G$. Obviously $\mathrm{Ker}\, f = C(G)$, the center of G. The orbit $G\cdot x = \{gxy^{-1}, \forall g \in G\}$ is called the conjugation class and the isotropy group G_x is the centraliser of x. The corresponding action of G on the set $\{< G\}$ of its subgroups, defines for each $H < G$, the centralizer $C_G(H)$ and the stabilizer $N_G(H)$; the latter is also called the normalizer : it is the largest subgroup of G which has H as invariant subgroup. Since both H and $C_G(H)$ are invariant subgroups of $N_G(H)$, this is also the case of $H.C_G(H)$ and one finds (proof left to the reader) that there is an injective homomorphism

$$\frac{N_G(H)}{H.C_G(H)} \rightarrowtail \mathrm{Out}\, H \qquad (7)$$

The G orbit of H is the conjugation class of subgroups $[H]_G$ that we have already studied. There is also a natural action of $\mathrm{Aut}\, G$ on

G . The action of G on the orbits $[G:H_1]$ and $[G:H_2]$ when the subgroups H_1 and H_2 belong to the same orbit of Aut G are quasi-equivalent but may be non equivalent.

Given an action $Q \xrightarrow{f} \text{Aut } G$ one forms the semi-direct product $G \rtimes Q$ defined by the group law

$$(g_1,q_1)(g_2,q_2) = (g_1 \cdot f(q_1)[g_2], q_1q_2) \tag{8}$$

When f is the trivial homomorphism, the law (8) is that of the direct product. More generally one calls extension E of Q by G a group such that $G \triangleleft E$ and the action by E inner automorphisms $E \xrightarrow{\tilde{f}} \text{Aut } G$ factorizes : $\tilde{f} = f \circ s$ where s is defined by $E \xrightarrow{s} E/G = Q$. Two extensions E and E' are equivalent if there is a commutative diagram

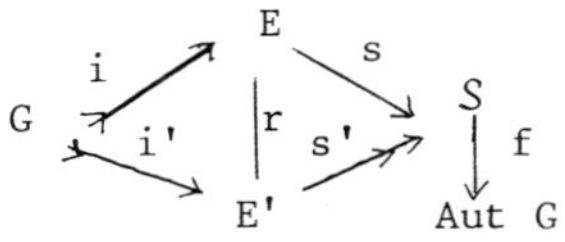

Diagram 2

(then r is isomorphism) and there is a natural group law on the set of equivalence classes of extension. This group is denoted by $H^2_f(Q,G)$ and it is called the second cohomology group of Q with value in G . (For lectures in physics Summer School, see e.g. [9] where the original mathematical literature is quoted and explained). The semi-direct product represents the unit of the cohomology group. When G is not Abelian, there is a natural action $G \xrightarrow{\phi} \text{Aut } C(G)$ on its center, and the set of equivalence classes of extension of Q by G is isomorphic to $H^2_{\phi\circ f}(Q,C(G))$.

Examples of semi-direct products are the Euclidean groups $E(n) = R^n \rtimes O(n)$ and the affine group $Aff(n) = R^n \rtimes GL(n,R)$.

6. Spontaneous symmetry breaking. Example of Landau.

When a physical problem has a symmetry group G , a solution is not necessarily G-invariant. If H is its isotropy group, then one can build an orbit $[G:H]$ of solutions. The set S of solutions of the problem is G invariant. As a general example assume that S is a set of stable states of the system and they depend on one (or several) G invariant parameters λ (e.g. temperature, time...). For some value of λ the state representative function $s(\lambda)$ may change of strata on S ; then the symmetry of the physical system changes. If it decreases(*), one says the symmetry is spontaneously broken.

There are half a dozen of mathematical schemes used for describing symmetry breaking in physics. We describe here the most common one, and the only one used in the other lectures. A system of equations for a G-symmetric physical problem is a smooth G-equivariant map ϕ between two functional spaces F_1, F_2, carrying a linear representation of G. Moreover, we assume that it depends differentiably on parameters λ. Assume that at λ_o we know a unique solution $u_o \in F_1$ of the problem, i.e.

$$\phi(u_o,\lambda_o) = 0 \tag{9}$$

If the Fréchet derivative

$$\forall v \in F_1 \quad \frac{d\phi_{u_o}}{du}(v) = \lim_{\varepsilon \to o} \frac{1}{\varepsilon}(\phi(u_o + v,\lambda_o) - \phi(u_o,\lambda_o)) \tag{10}$$

is inversible in a neighbourhood of λ_o, by the implicit function theorem, we can compute a G-invariant solution $u(\lambda)$ satisfying $\phi(u,\lambda) = 0$. If, for the value λ_i of λ, the linear operator $\frac{d\phi_{u_c}}{du}$ has a non trivial kernel, there is a bifurcation and Ker $\frac{d\phi_{u_c}}{du}$ is the tangent plane to the set of possible solutions which appear; it is stable by G and carries a linear representation of this group. It is in general irreducible (otherwise we have an "accidental degeneracy"). At λ_c, the symmetry G will be broken into one of the isotropic groups of the representation on Ker$\frac{d\phi_{u_c}}{du}$. Which one? This will

(*) If the transformation is reversible, e.g. phase transitions in thermodynamics, it increases one way and decreases the other way.

be seen in the example of the Landau theory of second order phase transition which will be treated by several lecturers, mainly Prof. P. Toledano. In this theory $\mathcal{F}_1$ is a space of physical functions (e.g. electron density, etc...) defined on the crystal and ϕ is the set of derivative of a thermodynamic potential, e.g. Gibbs Free energy V, when the parameters λ are the temperature, the pressure etc. The orthogonal irreducible (on the real) representation of a crystallographic group G are finite dimensional. Let $\mathcal{E}$ be the orthogonal vector space $\mathrm{Ker}\,\frac{d\phi_{u_c}}{du}$, carrier of an irrep of G. The simplest modelization of the restriction $V|_{\mathcal{E}}$ on this kernel has been proposed by Landau nearly fifty years ago : it is a G-invariant degree four polynomial bounded below :

$$x \in \mathcal{E} \qquad V(x) = p_4(x) + a(T)(x,x) \tag{11}$$

where p_4 is a strictly positive homogeneous G-invariant quartic polynomial, (x,x) is the invariant orthogonal scalar product and $a(T)$ is a function of temperature whose value has the same sign as $T-T_c$. When the physical signification of x is clear, it is called the order parameter; its number of component is $\dim \mathcal{E}$. The physically stable states are represented by the minima of V as a function of T. The symmetry is broken into their isotropy group. The potential $V(x)$ in (11) has no third degree term. As Landau pointed out, this is necessary for avoiding a first order transition (with a jump at $T = T_c$ from one minimum to another one). However, such generalized potential is useful for the study of the symmetry change in weak 1st order transitions. As we will see later (see also lecture of Prof. Ruegg) Higgs potentials are of a similar nature.

7. Minima of a G-invariant potential.

It is time to prove Prof. Wigner's answer. We first recall that the Euclidean group $E(3)$ is the semi-direct product of the orthogonal group $O(3)$ by the translation group R^3 :

$$E(3) = R^3 \rtimes O(3) .$$

A crystallographic group G is a discrete closed subgroup of E(3) containing a translation lattice $\sim Z^3$. The little group G_x of any point x is finite. Indeed $E(3)_x \sim O(3)$ so $G_x = O(3) \cap G$ and the intersection of a compact and a discrete closed subgroup of E(3) is a finite subgroup. This is also true of the point group G/Z^3. But P is a subgroup of G, and then an isotropy group P_x only when G is a semi-direct product (crystallographers say "symmorphic").

The gradient of a G-invariant potential at x is invariant by G_x. So it vanishes at a symmetry center; it is along a symmetry axis or in a symmetry plane. Consider a symmetry axis which does not carry symmetry centers or a symmetry plane which does not contain higher symmetry elements. Due to its periodicity, a continuous function reaches maxima and minima on these symmetry elements; on these points its gradient vanishes, so they are extrema for the full function. As we will see later, Morse theory can give some conditions for the localisation of the minima.

In the case of a smooth compact (and in particularly finite) group action on a manifold M (linear representations are a special case) there are classical theorems easy to prove (see e.g. [6], appendix C and references there). For any $x \in M$, there is a neighbourhood V_x such that for all $y \in V_x$, $[G_y] \leq [G_x]$. As a corollary a G equivariant differentiable tangent vector field is tangent to its stratum.

In a linear representation, smooth G-invariant functions are smooth functions of G-invariant polynomials whose ring is finitely generated. So one can establish invariant equations for localizing the zero of G equivariant vector fields. This has been done in [10] for the representations of all closed subgroups of O(3); one sees easily that it is "easier" to have zero on symmetry elements (i.e. non generic strata).

A degree four polynomial on $\mathcal{E}$ (such as a Landau or Higgs polynomial) bounded below and maximum at the origin 0 , has two radial minima on each straight line containing 0 . Let M be the set of these radial minima. It is a smooth manifold, homotopic to a sphere (and with 0 as symmetry center if there are no 3rd degree terms). It has extrema on every closed strata on M (see e.g. [11] and earlier quoted references); hence every maximal conjugation class of entropy groups is that of extrema. To obtain conditions on minima one can apply The Morse theory [3a], [11]. One can also prove that for irreducible representations there are no extrema in the generic stratum [11], [12]. What can be said about the lowest minima. It has been conjectured that they occur only with maximal isotropy groups [13]. This has no meaning. Indeed, one has to make the following distinctions. Given the image G of the symmetry group, one chooses a Landau polynomial with a quartic term $p_4 \in P_4^G$ where P_4 is the vector space of quartic polynomials in n variables; dim $P_4 = \binom{n+3}{4}$ and we define $\nu_G = \dim P_4^G$. One has to consider the centralizer in O(n) , $C_{O(n)}(P_4^G) \geq G$. Finally any mathematical theorem or conjecture can be formulated only in term of the exact isotropy group $O(n)_{p_4} = \tilde{G} \gneqq G$ (see [11]) . However, counter-examples have recently been found [14], [15], to the conjecture that the isotropy group of the absolute minimum of a Landau potential on E , is a maximal isotropy group on E of the isotropy group in O(n) , (i.e. $\tilde{G} = O(n)_{p_4}$) of the Landau potential. Professor Ruegg will give in his lectures a similar counter-example for a Higgs potential published in [16] and another example has been found in [17].

The Wigner problem for a crystal can be transformed into a problem of action of the finite point group P on the torus $R^3|Z^3$, the orbit space of the translation group (indeed, a triply periodic function is defined by its values on this torus) and the Morse theory is also applicable.

8. Basic Concepts in Crystallography.

This section is based on an unpublished manuscript with Prof. Jan Mozrzymas. It is a direct application of the concepts of group action.

It will help the participants of these school not acquainted with crystallography to follow some of the lectures. In §1, example 5 we defined the lattices and classed them into crystallographic systems. All other definitions we will give here are independent of the dimension n ; we give them here for $n = 3$. The isotropy groups of the lattices belonging to a crystallographic system from a conjugation class $[P_H]$ of $O(n)$. The conjugation classes $\leq$ that of the holohedries P_H are called geometric classes. There are 32 in $n = 3$ dimensions. Most macroscopic properties of the crystal are classified according to these classes. A group P of one of these classes is called the point group of the crystal. Its action in the lattice is an injective homomorphism $P \overset{f}{\rightarrowtail} GL(3,Z) = \mathrm{Aut}\, Z^3$. However, the equivalence chosen is neither the usual one, nor the weak one; indeed the point group is given as an $O(3)$ subgroup (up to a conjugation). For instance the holohedry group $O_n = m3m$ of the cubic system has an automorphism which exchanges the conjugation class of the 6 plane symmetry with that of the rotations by π around axes forming the middle of the edges. Such "non geometric" automorphisms are not considered. So for crystallographers, two actions f, f' of the point group P are equivalent if the images $\mathrm{Im}\, f$ and $\mathrm{Im}\, f'$ are conjugated in $GL(3,Z)$. So the 73 conjugation classes of finite subgroups of $GL(3,Z)$ correspond to all possible actions of the 32 geometric classes. They are called arithmetic classes. To the seven holohedries correspond 14 arithmetic classes; they are exactly the Bravais classes (the 1850 definition of Bravais was different!). For each of the 73 arithmetic classes one has to solve an extension problem. The equivalence of extensions defined by diagram 2 is too fine for the crystallographers. The normalizer $N_{GL(3,Z)}(\mathrm{Im}\, f)$ acts on P and also on the lattice Z^3 , so it acts on the cohomology group $H^2_f(P,Z^3)$. To each orbit of the normalizer is corresponding one crystallographic class. There are 219 of them for $n = 3$. One shows that this equivalence corresponds exactly to the following : two crystallographic space groups (i.e. closed discrete subgroups of $E(3)$ containing a lattice Z^3) belong to the same mathematical crystallographic class if they are conjugated by an element of the affine group $\mathrm{Aff}(3)$. It is a remarkable theorem of

Bieberbach [18] that isomorphic space groups (in n dimensions) are conjugated in Aff(n) . The equivalence definition used in crystallography is slightly stricter. Indeed, although the interatomic distances in a crystal phase change with temperature, the symmetry is considered the same. However, since temperature changes are continuous, two crystallographic space groups belong to the same physical crystallographic class if they are conjugated by an element of the connected affine group $\mathrm{Aff}_+(n) = R^n \rtimes GL_+(n,R)$, where $GL_+(n,R)$ is the group of linear transformation with positive determinant. For $n = 3$, 11 mathematical classes split into a pair of "enantiomorphic" physical classes, so there are 230 of the latter. Other basic references to n dimensional crystallography are [19] and [20].

9. <u>Renormalization of the Landau theory of second order phase transition</u>.

As you will hear in other lectures, Landau theory of second order phase transition in crystals is rather successful for explaining symmetry changes. However, it fails completely for giving the critical exponents. So as soon as the Wilson renormalisation with $\varepsilon = 4-d$ (d is the space dimension) expansion was proposed, it was used for Landau theory by adding to the potential V of equation (11) a kinetic energy term. Here I will simply give a general but abstract formulation of the so-called renormalisation group technique and explain the main results of some of my recent papers [21], [22] and in collaboration with J.C. Toledano [23] and also with P. Toledano and Brézin [24].

The right hand side of the renormalization group equation

$$\frac{dg(\lambda)}{\lambda d\lambda} = \beta(g(\lambda)) \qquad (12)$$

is a vector field β defined on the vector space P_4 of quartic polynomials. A fixed point g satisfies $\beta(\tilde{g}) = 0$. As we have noted in §7, the quartic part p_4 of the Landau potential depends only on the image $G < O(n)$ of the representation of the symmetry group. Hence weakly equivalent representations yield the same potential. The expression of p_4 depends on the orthonormal basis chosen for the n dimensional representation space. However, physics must not depend on this choice of basis, so the vector field β must be $O(n)$ equivariant, and the critical exponents are $O(n)$ invariants. As a consequence, the trajectory by equation (12) of any $g \in P_4^G$ (the subspace of G-invariant quartic polynomials) stays in this subspace and stops at a fixed point $\tilde{g} \in P_4^G$. It is stable if

$$\frac{d\beta}{dg}(\tilde{g})\Big|_{P_4^G} > 0 \qquad (13)$$

If this condition is satisfied, and $\tilde{g} > 0$ then the effective Landau potential is obtained from (11) by replacing p_4 by $\tilde{g}$. One interprets the non satisfaction of (13) by a lack of second order phase transition. The vector field β has been computed in [25]. Some successful predictions based on this scheme were made.
(P. Toledano's lectures are more critical). The recent results were obtained

(i) At the approximation computed in [25], the vector field β is a gradient.

(ii) If it exists, the stable fixed point $\tilde{g}$ is unique ([21] completed in [24]) and

(iii) its isotropy group $O(n)_{\tilde{g}}$ is equal to its normalizer in $O(n)$.

(iv) If the stabilizer $S_{O(n)}(P_4^G)$ does not leave invariant a quartic polynomial, there are no stable fixed points.

Property (i) depends essentially on the approximation. One can hope it is not the case for (ii). Property (iii) is a consequence of (ii) and (iv) is a simple corollary of (iii). Remark that (iii) is very restrictive. For $n = 2,3,4$ the number of closed strict subgroups of $O(n)$ which satisfies it are respectively $0,1,3$.

In [22] there are some remarks for arbitrary n.

10. A Spin 10 grand unification theory (*)

Let me first remind a few facts about simple compact Lie algebras $\mathcal{G}$ and groups G and their representation. A Cartan subalgebra of such a $\mathcal{G}$ is a maximal Abelian subalgebra (it corresponds phycically to a complete system of commuting observables); they are all conjugated. Their common dimension ℓ is the rank of the algebra. For $\ell = 5$ or $\ell \geq 9$ there are 4 such algebras (**), labelled by their Dynkin diagrams. For instance for $\ell = 5$:

A_5	B_5	C_5	D_5	
0-0-0-0-0	0-0-0-0=0	0-0-0-0=0	0-0-0<(0, 0)	(14)
SU(6)	Spin(11)	Sp(10)	Spin(10)	

(Sp is for symplectic).

To a simple compact Lie algebra $\mathcal{G}$ corresponds a unique simply connected compact Lie group $\tilde{G}$. Its irreducible representations are labelled by a set of ℓ non negative integers placed at the vertices of the Dynkin diagram. The center of the group $\tilde{G}$ is :

$$Z_{\ell+1} \text{ for } SU(\ell+1) = A_\ell \;,\quad Z_2 \text{ for } Spin(2\ell+1) = B_\ell \tag{15}$$

$$1 \text{ for } Sp(2\ell) = C_\ell \;,\; Z_4 \text{ for } Spin(4k+2) = D_{2k+1},\; Z_2^2 \text{ for } Spin(4k)=D_{2k}$$

The other groups G with the same Lie algebra $\mathcal{G}$ are quotient $\tilde{G}/F$ of $\tilde{G}$ by a finite subgroup of the center. So its irreducible representations form a subset of those of $\tilde{G}$. More generally a compact Lie group H is of the form :

(*) See also O'Raifeartaigh and Ruegg lectures.

(**) For other dimensions one has to add the five exceptional Lie algebras : G_2, F_4, E_6, E_7, E_8 . The series B,C are defined for $\ell \geq 2$ and D for $\ell \geq 4$. Moreover, we have the group isomorphisms : $SU(2) = Spin(3)$, $SU(2) \times SU(2) = Spin(4)$, $Spin(5) = Sp(4)$, $SU(4)=Spin(6)$.

$$H = N/F \qquad N = U(1)^k \times (\times_{i=1}^{j} \tilde{K}_i) \ , \quad F \text{ finite } < C(N) \tag{16}$$

i.e. N is the direct product of k Abelian $U(1)$ and j simply connected compact simple groups and F is a finite subgroup of the center of N. The irreps of H are the tensor products of those of the factors of N for which F is represented trivially. Example : (e.g. see [6], [9a]).

$$U(n) = \frac{SU(n) \times U(1)}{Z_n} \quad , \quad Z_n = \{(e^{2\pi ik/n}I_n, e^{-\pi ik/n}) \ , \ 0 \leq k < n\} \tag{17}$$

so the irreps of $U(n)$ are labelled by the integers

$$a_i \geq 0 \ , \ a_1, a_2, \ldots, a_{n-1}, m \quad \text{with} \quad \sum_{k=1}^{n-1} k a_k + m \equiv 0 \text{ mod } n \tag{18}$$

For $SU(2)$ the tradition is to use the spin $t = \frac{1}{2} a$, where a is the Dynkin label and $2t+1 = a+1$ the dimension of the irreducible representation. It is also interesting to replace (17) by

$$U(n) = \frac{SU(n) \times R}{Z} \quad , \quad Z \text{ generated by } (e^{2\pi i/n}I_n, \alpha) \tag{19}$$

because $SU(n) \times R$ is the universal covering of $U(n)$ and we wish to emphasize that α is an arbitrary real number, so there is no natural scale for the value of the real parameter; there is the relation (18) for quantum numbers. This relation is satisfied for the standard unified electroweak theory and for what we believe are exactly preserved gauge symmetry interactions, i.e. electrochromodynamics, see table 2. The gauge groups of these theories are respectively $U(2)$ and $U(3)$. Any symmetry group G of a grand unified theory (GUT) must contain $U(2)$ and $U(3)$ as subgroups; since the electromagnetic gauge is common to both, these subgroups have an intersection $U(1)$ and they generate a subgroup

$$S(U(3) \times U(2)) \qquad \frac{SU(3) \times SU(2) \times U(1)}{Z_6(\xi)} \tag{20}$$

where ξ , the generator of Z_6 is

$$\xi = (e^{2\pi i/3} I_3, -I_2, e^{-2\pi i(5(3))}) \qquad (20')$$

where $H = S(U(3) \times U(2))$ is the group of matrices :

$$H = \{ \left(\begin{array}{c|c} u & o \\ \hline o & v \end{array}\right) , \; u \in U(3) \; , \; v \in U(2) \; , \; (\det u)(\det v) = 1 \} \qquad (21)$$

This group H of rank 4 is a maximal subgroup of $SU(5)$ and it is an isotropy group of the $SU(5)$ adjoint representation $\overset{1\ 0\ 0\ 1}{0\text{-}0\text{-}0\text{-}0}$ of dimension 24. So $SU(5)$ is the smallest possible GUT symmetry group and this model was proposed ten years ago [26]. It has very good features, but one rather inelegant : the 15 fermion states of one horizontal family (12 = 3 colors × 2 spin states × 2 quark states u,d + 2 for electron +1 for neutrinos) are in a reducible representation $\overset{0\ 0\ 0\ 1}{0\text{-}0\text{-}0\text{-}0} \oplus \overset{0\ 1\ 0\ 0}{0\text{-}0\text{-}0\text{-}0}$ of dimension 5+10 of $SU(5)$.

Table 2. Multiplet of particles; q is the electric charge.

U(2) symmetry: representation t,y with $2t+y = 0 \bmod 2$; $q = t_z + \frac{1}{2} y$

irrep	1,0 + 0,0	$\frac{1}{2}$,−1	$\frac{1}{2}$, 1	0,2	0,−2	$\frac{1}{2}$,1
particle	$\gamma W^+ W^- Z^o$	$\nu_L \varepsilon^-_L$	ν_R, ε^+_R	ε^+_L	ε^-_R	Higgs.

U(3) symmetry: representation c_1, c_2, x ; $c_1 + 2c_2 + x \equiv 0$ mod ; $q = \frac{x}{3}$

1,0,2	1,0,−1	0,1,−2	0,1,1
u	d	$\bar{u}$	$\bar{d}$

SU(5) symmetry, irrep $a_1 a_2 a_3 a_4$ reduces on H into $c_1, c_2; 2t, m$ with $c_1 + 2c_2 \equiv m \bmod 3$, $2t \equiv m \bmod 2$, $q = t_z + \frac{m}{6}$

irrep $\bar{5}$: (0,0,0,1) = (0,1;0,2) + (0,0;1,−3)

$(\bar{d}_L)$ ν_L, ε^-_L

irrep 10 : (0,1,0,0) = (0,1;0,−4) + (0,0;0,6) + (1,0;1,1)

$(\bar{u})_L$ ε^+_L $u_L + d_L$

However, if the recent possible observations of neutrino oscillations (in Bugey and CERN) or of a neutrino mass (in Moscow) are confirmed, the neutrino must also have two states, so each fermion family require a 16 dimensional representation. It is a hard problem to find a G which contains the exact number of families (presently believed to be three); provisorily the simplet extension of SU(5) symmetry is up to Spin(10) , the covering of SO(10), which has two complex conjugate irreducible spinor representations $\overset{0}{\circ}-\overset{0}{\circ}-\overset{0}{\circ}<\begin{smallmatrix}\circ\ 1\\ \circ\ 0\end{smallmatrix}$ and $\overset{0}{\circ}-\overset{0}{\circ}-\overset{0}{\circ}<\begin{smallmatrix}\circ\ 0\\ \circ\ 1\end{smallmatrix}$ of dimension 16. This is well known among high energy physicists. The adjoint representation $\overset{0}{\circ}-\overset{1}{\circ}-\overset{0}{\circ}<\begin{smallmatrix}\circ\ 0\\ \circ\ 0\end{smallmatrix}$ is of dimension 45; so this is the number of gauge bosons. In which representation should the Higgs scalar be in order to break the symmetry on the subgroup H of equation (21) ? I am working on this unsolved problem with Ömer Kaymakcalan(*), K.C. Wali, L. O'Raifeartaigh, W.D. McGlinn. I sketch here the method for solving it. On the ground of physical elegance we consider only representations of small dimension d (say d < 100). They are

irrep	(00000),	(10000),	(20000),	(01000),	(00010),	(00001)
d	1	10	54	45	16	$\overline{16}$

where $(a_1 a_2, a_3, a_4, a_5)$ labels the representation $\overset{a_1}{\circ}-\overset{a_2}{\circ}-\overset{a_3}{\circ}<\begin{smallmatrix}\circ\ a_4\\ \circ\ a_5\end{smallmatrix}$

To obtain the observed breaking on H , one has to choose a reducible representation. The isotropy subgroups of a direct sum of inequivalent irreps are the intersections of the isotropy groups of the irreps. A more efficient method to compute the isotropy subgroups of the direct sum of irreps carried by the space $E = E_1 + E_2$ is to look for the isotropy subgroups of E_1 and study their action and corresponding isotropy subgroups on E_2 .

Results added in April when these notes have been written : H is an isotropy subgroup of the representation 45 + 54 , and corres-

(*) This gifted young physicist died of illness in Syracuse, N.Y. (USA) two days after I was giving this lecture.

ponding to the absolute minimal of a Higgs potential (depending on 11 parameters). However, there are two distinct conjugate classes of Spin 10 subgroups isomorphic to H . One obtains the wrong class : this [H] is not < [SU(5)] . The right class [H] appears as isotropy subgroups of the representations : a) $54 + 16 + \overline{16}$, b) $45 + 16 + \overline{16}$. In case a), the minimum of the Higgs potential covers an infinity of orbits and there are pseudo Goldstone bosons. The case b) gives a good solution of the problem. This solution has already been found in [27], [28]. It is quite elegant. Indeed it uses only two types of representations. The adjoint one for the spin 1 gauge bosons and spin 0 Higgs bosons. The spinor representations for the Fermions (quarks and leptons) and the rest of the Higgs bosons. So this model presents some remnants of a supersymmetry.

REFERENCES

[1] Doncel, M.G., Michel, L., Six, J., Interview de Eugen P. Wigner sur sa vie scientifique, Archives internationales d'histoire des sciences, 34 (1984) 177-217 (n°112, juin 1984).

[2] Wigner, E.P., English translation "Group Theory and its application to the quantum mechanics of atomic spectra", Academic Press, 1959, New-York.

[3] Michel, L., Mozrzymas, J., a) Application of Morse Theory to the symmetry breaking in the Landau Theory of second order phase transitions, VIth International Colloquium "Group Theoretical Methods in Physics", Tübingen 1977, Lecture Notes in Physics 79, 447-461, Springer (1978). b) Weak equivalence of irreducible representations of little space groups, Match (= Communications in Mathematical Chemistry) 10 (1980) 223-226.

[4] International Tables for Crystallography, Reidel 1983, Dordrecht.

[5] Schwarzenberger, R.L.E., Proc. Camb. Phil. Soc. 72 (1972) 325-349.

[6] Michel, L., Symmetry Defects and broken Symmetry Configurations. Hidden Symmetry, Rev. Mod. Phys. 52 (1980) 617.

[7] Montgommery, D., Yang, C.T., Trans. Am. Math. Soc. 87 (1958) 284-297.

[8] Mostow, G., Ann. Math. (1957) 432 and 513.

[9] Michel, L., a) Invariance in Quantum Mechanics and Group Extensions, Istambul Summer School, 16 juillet - 4 août 1962, Gordon and Breach (New-York) 1964. b) Relativistic invariance and internal symmetries, 1965 Brandeis Summer Institute in Theoretical Physics, Vol.I, p.247, Gordon and Breach, New-York 1966. c) Relations entre symétries internes et Invariance relativiste, Cargèse 1965 lectures in theoretical Physics, p.409, Gordon and Breach, New-York 1966.

[10] Jarić, M., Michel, L., Sharp, R.T., Invariant formulation for the zeros of covariant vector fields, Proceedings, Group Theoretical Methods in Physics XI, Lecture Notes Phys. 180, 317-318, Springer (1983) and J. Physique 45 (1984) 1.

[11] Michel, L., Minima of Higgs-Landau Polynomials, p.157-203 in Regards sur la Physique Contemporaine, Edition CNRS (Paris 1980).

[12] Jaric, M., in Group Theoretical Methods in Physics (IX), Lecture Notes in Physics 135, 12.

[13] Ascher, E., J. Phys. 10 (1977) 1365.

[14] Mukamel, D., Jaric, M.V., Phys. Rev. B 29 (1984) 1465.

[15] Jaric, M.V., Phys. Rev. Lett. 51 (1983) 2073.

[16] Abud, M., Anastaze, G., Eckert, P., Ruegg, H., Phys. Lett. 142 B 1984, p.371.

[17] Burzlaff, M., O'Raifeartaigh, L., To appear in Phys. Lett. B.

[18] Bieberbach, L., Math. Ann. 72 (1912) 400.

[19] Brown, H., Bülow, R., Neubüser, J., Wondratschek, H., Zassenhaus, H., Crystallographic groups of four dimensional space, John Wiley, New-York, 1978.

[20] Schwarzenberger, R.L.E., N-dimensional crystallography, Pitman Publ., London, 1980.

[21] Michel, L., Phys. Rev. B 29 (1984) 2777.

[22] Michel, L., in Group Theoretical Methods in Physics, p.162-184, World Scientific, Singapore 1984.

[23] Michel, L., Toledano, J.C., Phys. Rev. Lett.

[24] Toledano, J.C., Michel, L., Toledano, P., Brézin, E., Fixed points and stability for anisotropic systems with 4-component order parameters, Phys. Rev. B (to appear).

[25] Brézin, E., Le Guillou, J., Zinn-Justin, J., Phys. Rev. B, 10 (1974) 892.

[26] Georgi, H., Glashow, S.L., Phys. Rev. Lett. 32 (1974) 438.

[27] Bucella, F., Ruegg, H., Savoy, I.A., Phys. Lett. 94 B (1980) 491.

[28] Yasue, M., Phys. Rev. D 24 (1981) 1005.

INTRODUCTION TO SPONTANEOUS SYMMETRY BREAKDOWN IN CLASSICAL LATTICE SYSTEMS

Christian Gruber* and Charles Ed. Pfister**

*Département de Physique
**Département de Mathématique
Ecole Polytechnique Fédérale, CH-1015 Lausanne (Switzerland)

1. INTRODUCTION

Gibbs States - also called Gibbs Random Fields - are introduced in statistical mechanics to describe the equilibrium properties of macroscopic systems. Using the well-known Gibbs formula (1878) these states are defined by means of the hamiltonian, function on the configuration space which represents the interaction energy between the subsystems which constitute the macroscopic system. The Symmetry Group of the system $\tilde{G}$ is a group of transformations on the configuration space which leave the hamiltonian invariant. There is a Spontaneous Symmetry Breakdown (SSB) whenever there exist states which are not invariant under this symmetry group, i.e. there is a change in the symmetry of the state without changing the symmetry of the system (this justifies the term "spontaneous"); since $\tilde{G}$ acts on the states, there exists more than one Gibbs states.

The fundamental problem in this domain is to prove the existence - or the absence - of S.S.B. and to obtain a description of all possible Gibbs states which may exist at a given temperature. Let us note that under general conditions the set of Gibbs states forms a Choquet simplex and thus we can limit ourselves to a discussion of the extremal states.

From a physical point of view the existence of S.S.B. is associated with "Phase Transitions" and "Coexistence of phases". In this context, we recall that there are two types of phase transitions associated with SSB: either the phase transition is such that the symmetry groups of the high and low temperature phases are in the relation from group to subgroup, or this condition is not satisfied. In the second type (which appear especially in metallurgy) the transition is generally discontinuous. The first type of phase transitions are usually described by means of "local order parameters", and may be either continuous or discontinuous; this first type of phase transition has a long history; we shall only mention "Curie Principle (1895)" which states that the symmetry groups of the ordered phase is the intersection of the symmetry group of the disordered phase with the symmetry group of the order parameter.

Among the many questions one would like to answer in this domain, we mention the following ones for which partial results exist:

- How many phases can coexist at a given temperature?
- What are the properties of the different phases, in particular their symmetries? What are the possible subgroups which can appear in S.S.B.?
- What are the local order parameters? How many non equivalent parameters are necessary to describe the S.S.B.? Do they vary continuously or not with respect to external parameters such as temperature, magnetic field, and so on?
- What is the relation between Local Order Parameter and Long range order?
- Existence of Non Local Order parameter such as the surface tension? In what cases does coexistence of phases imply a non vanishing surface tension? What are the properties of the interface between two pure phases?
- Decreasing the temperature from infinity to zero how many phase transitions occur? At a critical temperature do we have coexistence of the disordered phase with the ordered phase or not?

As we shall see, for ferromagnetic systems there exists a unique $\tilde{G}$-invariant state for almost all temperatures (and, generally, for all but a finite number of temperatures); furthermore, in this case, the group we need to consider is not the full symmetry group of the system but only an abelian subgroup. Hence the following problems are of interest:

- What is the smallest internal symmetry group which is relevant for the analysis of Gibbs states and S.S.B.? What is the smallest group S with respect to which there exists a unique S-invariant state? For what systems is it possible to prove the unicity of the invariant state?

To conclude this general discussion we recall that the physical origin of phase transition with S.S.B. can often be found in the existence of ground states which are not invariant under the symmetry group of the system. In fact under some natural assumptions it is possible to show that the phase diagram depends weakly on T around T = 0. Finally it is clear that any S.S.B. is associated with a phase transition but one should insist on the fact that the converse is not true: there exist phase transitions without S.S.B., without coexistence of phases and without local order parameter (Gauge models, Clock models, rotators).

In Sec. 2, we introduce the classical lattice systems and their symmetry groups; we then define the notions of Gibbs states and spontaneous symmetry breakdown. Sec. 3 contains the analysis of S.S.B. for abelian internal symmetry group and in Sec. 4 we discuss S.S.B. in the case where the configuration space itself is a compact metrizable abelian group. In this case the most general results are obtained for ferromagnetic systems. Finally the analysis of Sec. 2 - 4 is illustrated on specific examples in Sec. 5.

2. CLASSICAL LATTICE SYSTEMS [1, 2, 3]

2.1. Lattice Systems*

A lattice L is a countable set of points in $\mathbb{R}^\nu$, called "sites", denoted by x, y, z, ...

With each site x is associated a random variable θ_x with value in some measure space $\mathcal{G}_x$ with probability measure $d\nu_x$. For the sake of simplicity we assume that $\mathcal{G}_x$ is a Hausdorff - Metrizable - Compact space. This is however not necessary and the discussion below is clearly valid under more general circumstances. The space $\mathcal{G}_x$ describes the configurations of the subsystem located at the site x. We shall assume that all subsystems are identical, i.e. for all x in L, $\mathcal{G}_x$ is isomorphic to some $\mathcal{G}_o$, with measure $d\nu_o$.

The configuration space of the system (without constraints) is then defined by the diret product:

$$\mathcal{G}_L = \prod_{x \in L} \mathcal{G}_x = \{\underline{\theta}\}$$

(endowed with the topology and measure induced by the product). Therefore a configuration $\underline{\theta}$ is a function defined on L with value in $\mathcal{G}_o$:

$$\begin{array}{rccc} \underline{\theta} : & L & \to & \mathcal{G}_o \\ & \cup & & \cup \\ & x & \mapsto & \theta_x \end{array}$$

For any subset Λ of L, we introduce $\Lambda^c = L/\Lambda$, $\mathcal{G}_\Lambda = \prod_{x \in \Lambda} \mathcal{G}_x$ and the measure $d\nu_\Lambda = \prod_{x \in \Lambda} d\nu_x$. For any $\underline{\theta} \in \mathcal{G}_L$, $\underline{\theta}_\Lambda$ represents the projection of $\underline{\theta}$ on $\mathcal{G}_\Lambda$:

---

* The reader should look at the models discussed in Sec.5 as illustration of the definitions.

$$\begin{array}{rcl} \underline{\theta}_\Lambda : \Lambda & \to & \mathcal{G}_0 \\ \psi & & \psi \\ x & \mapsto & \theta_x \end{array}$$

and we identify $\underline{\theta}$ and $(\underline{\theta}_\Lambda, \underline{\theta}_{\Lambda^c})$.

The algebra $\mathcal{O}\!\mathcal{l} = \mathcal{C}(\mathcal{G}_L)$ of complex continuous functions on L defines the algebra of <u>observables</u>; for any finite subset Λ of L, $\mathcal{O}\!\mathcal{l}_\Lambda$ is the algebra of observables in Λ, i.e.

$$\mathcal{O}\!\mathcal{l}_\Lambda = \{ A \in \mathcal{O}\!\mathcal{l} \; ; \; A(\underline{\theta}) = A(\underline{\theta}_\Lambda) \}$$

The <u>states</u> are finally defined as probability measures on $\mathcal{G}_L$ or equivalently as normed, positive linear functionals on $\mathcal{O}\!\mathcal{l}$.

The interaction among the subsystems is described by a <u>potential</u> Φ which is a real function on $\bigcup_{B \in \mathcal{P}_f(L)} \mathcal{G}_B$ (*):

$$\forall B \in \mathcal{P}_f(L), \quad \begin{array}{rcl} \phi_B : \mathcal{G}_B & \to & \mathbb{R} \\ \psi & & \psi \\ \underline{\theta}_B & \mapsto & \phi_B(\underline{\theta}_B) \end{array}$$

We consider ϕ_B to be defined on all $\mathcal{G}_L$ by

$$\phi_B(\underline{\theta}) = \phi_B(\underline{\theta}_B)$$

For any finite $\Lambda \subset L$ we introduce

$$H_\Lambda = \sum_{\substack{B \in \mathcal{P}_f(L) \\ B \cap \Lambda \neq \phi}} \phi_B$$

The <u>hamiltonian of the finite system Λ with boundary condition</u> $\underline{\theta}_{\Lambda^c}$ is the function defined on $\mathcal{G}_\Lambda$ by

$$H_{\Lambda, \underline{\theta}_{\Lambda^c}}(\underline{\theta}_\Lambda) = H_\Lambda(\underline{\theta}_\Lambda, \underline{\theta}_{\Lambda^c})$$

---

* $\mathcal{P}_f(L) = \{ X \subset L \; ; \; |X| < \infty \}$ where $|X|$ denotes the cardinality of the subset X.

It represents the energy of the finite system Λ in the configuration $\underline{\theta}_\Lambda$, while the rest of the system outside Λ is in the configuration $\underline{\theta}_{\Lambda^c}$.

We also introduce the hamiltonian of the finite system Λ with free boundary conditions as

$$H_\Lambda^{(free)} = \sum_{B \subset \Lambda} \phi_B$$

and the formal hamiltonian for the infinite system

$$H = \sum_{B} \phi_B$$

Finally we assume that the potential Φ satisfy the following condition:

$$\sup_x \sum_{B \ni x} \int_{\mathcal{G}_B} d\nu_B \, |\phi_B| < \infty$$

Systems with constraints

Let $\mathfrak{F}$ be a family of finite subsets of L; for each $\Lambda \in \mathfrak{F}$ the admissible configurations of the finite system Λ are defined by $\mathcal{G}_\Lambda^{adm} \subset \mathcal{G}_\Lambda$. The space of admissible configurations is then defined as :

$$\mathcal{G}_L^{adm} = \{ \underline{\theta} \in \mathcal{G}_L \; ; \; \forall \Lambda \in \mathfrak{F} \; , \; \underline{\theta}_\Lambda \in \mathcal{G}_\Lambda^{adm} \}$$

2.2. Symmetry Group of the System

2.2.1. Euclidean Symmetry Group $\mathcal{E}$

It is usually assumed that there is an action of $\mathbb{Z}^\nu$ which leaves L and Φ invariant, i.e.

i) $\forall a \in \mathbb{Z}^\nu$, $\quad x \in L$ then $\quad x + a \in L$

which yields an action of $\mathbb{Z}^\nu$ on $\mathcal{G}_L$ defined by

$$(a.\underline{\theta})_x = \theta_{x-a}$$

and

$$\text{ii)} \quad \phi_{B+a}\,(a\cdot\underline{\theta}) = \phi_B\,(\underline{\theta})$$

More generally let $\mathcal{E}$ be a subgroup of the Euclidean group which leaves L and Φ invariant, i.e.

i) $\forall e \in \mathcal{E}\,,\ x \in L$ then $e\cdot x \in L$ which yields an action of $\mathcal{E}$ on $\mathcal{G}_L$ defined by $(e\circ\underline{\theta})_x = \theta_{e^{-1}\cdot x}$

and

$$\text{ii)} \quad \phi_{e\cdot B}\,(e\cdot\underline{\theta}) = \phi_B\,(\underline{\theta})$$

2.2.2. Internal Symmetry Group G^{int}

Let G_0 be a topological group acting on $\mathcal{G}_0$ and leaving the measure $d\nu_0$ invariant.

We define $\quad G = \prod_{x\in L} G_x \quad$ with $\quad G_x \cong G_0$

i.e.

$$\forall g \in G \qquad g: \begin{array}{ccc} L & \rightarrow & G_0 \\ \text{\rotatebox{90}{$\in$}} & & \text{\rotatebox{90}{$\in$}} \\ x & \mapsto & g(x) \end{array}$$

The group G acts on $\mathcal{G}_L$ as: $\qquad (g\circ\underline{\theta})_x = g(x)\,\theta_x$

and the internal symmetry group is the subgroup of G such that:

$$\phi_B\,(g\cdot\underline{\theta}) = \phi_B\,(\underline{\theta}) \qquad \forall B$$

Let us note that G^{int} contains as subgroup the "Gauge Group" G_f^{int} defined by :

$$G_f^{int} = \{ g \in G^{int}\ ;\ g(x) \neq \mathbb{1}_x \text{ for finitely many } x \}$$

which may of course be trivial (see examples sec.5)

2.2.3. Symmetry Group $\tilde{G}$

$\mathcal{E}$ acts as a group of automorphisms on G, as

$$(T_e g)(x) = g(e^{-1}x)$$

which satisfies

$$g \cdot (e\underline{\theta}) = e \cdot (T_{e^{-1}} g \cdot \underline{\theta})$$

and $\mathcal{E}$ leaves G^{int} invariant.

We can thus consider the symmetry group $\tilde{G}$ defined by the semi-direct product $\tilde{G} = G^{int} \circledS \mathcal{E} = \{\tilde{g} = (g, e)\}$

$$\tilde{g}_1 \tilde{g}_2 = (g_1 \cdot T_{e_1} g_2, e_1 e_2) \qquad \tilde{g} = g \cdot e$$

$$e\, g = (T_e g) \cdot e$$

Let us note that $\tilde{G}$ is <u>not</u> the largest symmetry group; one could for example consider the subgroup of $G \circledS \mathcal{E}$ such that

$$\phi_{eB}(\tilde{g}\underline{\theta}) = \phi_B(\underline{\theta})$$

but as we mention in the introduction it is not clear what is the relevant symmetry group and for the following discussion we will need in fact a subgroup of $\tilde{G}$ taken with $\mathcal{E} = \mathbb{Z}^{\nu}$

The symmetry group $\tilde{G}$ acts on the configuration space

$$\tilde{g} \circ \underline{\theta} = g \circ (e\underline{\theta})$$

and thus it acts as a group of automorphims on the observables

$$(\tau_{\tilde{g}} A)(\underline{\theta}) = A(\tilde{g}^{-1} \circ \underline{\theta})$$

which induces a group of linear transformations on the states:

$$(\tau^*_{\tilde{g}}\, \omega)[A] = \omega[\tau_{\tilde{g}^{-1}} A]$$

Formally the symmetry group $\tilde{G}$ is a group of transformations such $\tau_{\tilde{g}} H = H$.

2.3. Gibbs States or Gibbs Random Fields

Let $\omega_\Lambda^{\underline{\theta}}$ be the state on $\mathcal{O}\!\mathcal{l}_\Lambda$ defined by the probability measure

$$P_\Lambda^{\underline{\theta}}[d\underline{\theta}_\Lambda] = Z_{\Lambda,\underline{\theta}}^{-1} \quad e^{-\beta H_\Lambda(\underline{\theta}_\Lambda, \underline{\theta}_{\Lambda^c})} \quad d\nu_\Lambda(\underline{\theta}_\Lambda)$$

$$Z_{\Lambda,\underline{\theta}} = \int_{\mathcal{G}_\Lambda} d\nu_\Lambda(\underline{\theta}_\Lambda)\, e^{-\beta H_\Lambda(\underline{\theta}_\Lambda, \underline{\theta}_{\Lambda^c})}$$

We consider all possible limits

$$\omega^{\underline{\theta}} = \lim_{\Lambda \to L} \omega_\Lambda^{\underline{\theta}}$$

Definition

The closed, convex, hull of all states obtained from the above definition in the limit $\Lambda \to L$ defines the set of all Gibbs States $\Delta(\beta, \Phi)$ associated with the interaction Φ and the inverse temperature β. The Gibbs State defined by the free boundary condition is denoted ω_f.

Theorem 1

$\Delta(\beta,\phi)$ is a non empty, convex set, which is a Choquet simplex. The set $\Delta_0(\beta,\phi)$ of all $\mathbb{Z}^\nu$-invariant Gibbs states has the same property.

Definition

The extremal points of $\Delta_0(\beta, \phi)$ are called "Pure Phases".

2.4. Action of $\tilde{G}$ on Gibbs state

Theorem 2

1) For any $\tilde{g} \in \tilde{G}$ and any $\omega \in \Delta(\beta, \Phi)$ then $\tau'_{\tilde{g}}\, \omega \in \Delta(\beta, \Phi)$;

furthermore if ω is extremal in $\Delta(\beta, \Phi)$ then $\tau'_{\tilde{g}}\,\omega$ is also extremal in $\Delta(\beta, \Phi)$.

2) For any $g \in G_f^{int}$, $\tau'_g\, \omega = \omega$

3) For any $g \in G^{int}$, $\tau'_g\, \omega = \omega$ on $\mathcal{O}\!\!\mathcal{L}^{sym}$

where $\mathcal{O}\!\!\mathcal{L}^{sym} = \{A \in \mathcal{O}\!\!\mathcal{L};\ \tau_g A = A\}$,

furthermore $\tau'_g\, \omega_f = \omega_f$ and $\tau'_g\, \omega^{\underline{\theta}} = \omega^{g \circ \underline{\theta}}$.

Let us remark that the above theorem implies that the whole set of extremal states of $\Delta(\beta,\Phi)$ can be decomposed into orbits of the group $\tilde{G}$.

Definition

The Gibbs state ω is "invariant" if $\tau'_{\tilde{g}}\, \omega = \omega$ for all $\tilde{g} \in \tilde{G}$; the family of invariant Gibbs states is a compact, convex set.

We say that there exist a "Spontaneous Symmetry Breakdown" if there exist states which are not invariant.

Remarks

1/ Usually we can show that at high temperature the Gibbs state is unique and thus invariant.

2/ For ferromagnetic systems (to be defined in Sec. 4) one can usually show that there exists a unique invariant state at low temperatures.

3/ It is impossible to have a spontaneous breaking of the Gauge Group.

Definition

For any Gibbs state ω, let $\mathcal{E}_\omega$ and G_ω^{int} denote the subgroup of $\mathcal{E}$ and G^{int} which leave ω invariant: $\mathcal{E}_\omega$ and G_ω^{int} are the symmetry groups of the state ω (or little groups).

Lemma 1

1) G_ω^{int} is stable under $\mathcal{E}_\omega$, i.e. $\forall g \in G_\omega^{int}$, $e \in \mathcal{E}_\omega$

then $T_e g \in G_\omega^{int}$

(since : $\tau_e^{\prime}(\tau_g^{\prime}\,\omega) = \tau_{T_e g}^{\bullet}\ \tau_e^{\bullet}\,\omega$)

2) $G^{int}_{\tau_g^{\bullet}\omega} = g\ G^{int}_\omega\ g^{-1}$

(since : $\tau_{g_1}^{\cdot}(\tau_g\,\omega) = \tau_g\,\omega \Leftrightarrow \tau_{g^{-1}g_1 g}\,\omega = \omega$)

3) $G^{int}_{\tau_e^{\bullet}\omega} = T_e\ G^{int}_\omega$

3. $G^{int} = \mathcal{S}$ = "COMPACT ABELIAN GROUP"

In this section we consider the case where the internal symmetry group, which we denote now by $\mathcal{S} = \{s\}$, is compact and abelian. We repeat once more that $\mathcal{S}$ is not necessarily the largest internal symmetry group of the system, but only some compact abelian subgroup of the full internal symmetry group.

A Gibbs state is said to be "symmetric" if it is invariant under $\mathcal{S}$.

Let $\Delta_{\mathcal{S}}(\beta) = \{\omega \in \Delta(\beta, \Phi)\,;\ \tau_s^* \omega = \omega \quad \forall s \in \mathcal{S}\}$ denote the set of symmetric states; it is a non empty, compact, convex set, which is a Choquet simplex.

For any Gibbs state ω

$$\mathcal{S}_\omega = \{s \in \mathcal{S};\ \tau_s^* \omega = \omega\}$$

denote the internal symmetry group of ω (little group).

As before the symmetric algebra $\mathfrak{A}^{sym}$ is the algebra of observable invariant under $\mathcal{S}$; any (extremal) state on $\mathfrak{A}^{sym}$ has a unique extension to a symmetric (extremal) state on $\mathfrak{A}$.

The discussion of Sec. 2 yields immediately for abelian internal symmetry group the following:

Theorem 3

1) $\forall s \in \mathcal{S}, \quad \omega \in \Delta(\beta, \Phi)$ then

$$\mathcal{S}_{\tau_s^* \omega} = \mathcal{S}_\omega$$

and thus all states on the orbit of ω under $\mathcal{S}$ have the same internal symmetry group.

2) The only subgroups $\mathcal{S}_{\omega_0}$ of $\mathcal{S}$ which can appear as internal symmetry group of a pure phase ω_0 are those subgroups which are stable under the action of $\mathbb{Z}^{\nu}$ (the same result holds for any $\mathbb{Z}^{\nu}$- invariant state).

Indeed : $\mathcal{E}_0 = \mathbb{Z}^{\nu} \Rightarrow T_e s \in \mathcal{S}_{\omega_0} \quad \forall s \in \mathcal{S}$

Remark

1) For any extremal state ω_0 of $\Delta(\beta,\phi)$ and any s in $\mathcal{S}$ $\tau_s^* \omega_0$ is extremal and has the same internal symmetry group but not necessarily the same euclidean symmetry group. Furthermore any state ω in the convex set generated by the orbit of ω_0 under $\mathcal{S}$ has $\mathcal{S}_{\omega} \supset \mathcal{S}_{\omega_0}$.

2) Since $\mathcal{G}_L$ is metrizable, then $\mathfrak{A}$ is separable; we assume that $(\tau_s^* \omega)[A]$ is a continuous function on $\mathcal{S}$ for all $A \in \mathfrak{A}$ and $\omega \in \Delta(\beta, \Phi)$.

Lemma 2

1) Every symmetric state is the resultant of a unique measure carried by the extremal states of $\Delta_{\mathcal{S}}(\beta)$.

2) For any $\omega \in \Delta(\beta,\Phi)$

then
$$\bar{\omega} \doteq \int_{\mathcal{S}/\mathcal{S}_\omega} ds\, \tau_s^* \omega \qquad \text{is symmetric} \qquad (*)$$

where ds is the normalized Haar measure on the group.

3) For any extremal state ω_0 in $\Delta(\beta,\Phi)$, $\overline{\omega_0}$ is extremal in $\Delta_{\mathcal{S}}(\beta)$ and (*) is the extremal decomposition of ω_0 in $\Delta(\beta,\phi)$.

4) For any ω_1, ω_2 extremal in $\Delta(\beta,\phi)$ such that $\overline{\omega_1} = \overline{\omega_2}$, there exists an element s in $\mathcal{S}$ such that $\omega_2 = \tau_s^* \omega_1$.

5) For any ω_s extremal in $\Delta_{\mathcal{S}}(\beta)$, there exists ω_0 extremal in $\Delta(\beta,\Phi)$ such that $\bar{\omega}_0 = \omega_s$.

Indeed:

1) see [1] prop. 6.4.3.

2) By definition $\bar{\omega}[A] = \int_{\mathcal{S}/\mathcal{S}_\omega} ds \quad \omega[\tau_s^{-1} A]$

implies $(\tau'_{s_1} \bar{\omega})[A] = \bar{\omega}[\tau_{s_1}^{-1} A] = \int_{\mathcal{S}/\mathcal{S}_\omega} ds \quad \omega[\tau_{s_1+s}^{-1} A]$

which concludes the proof by the translation invariance of the Haar measure.

3) Let $\bar{\omega}_0 = \alpha \omega_1 + (1-\alpha)\omega_2 \qquad \omega_1, \omega_2 \in \Delta_{\mathcal{S}}(\beta)$

$$\Rightarrow \bar{\omega}_0 = \int_{\Delta(\beta,\Phi)} [\alpha \, d\lambda_1(\omega) + (1-\alpha)\, d\lambda_2(\omega)]\, \omega = \int_{\Delta(\beta,\Phi)} \omega \, d\lambda(\omega)$$

$$= \int_{\mathcal{S}/\mathcal{S}_\omega} ds \; \tau'_s \omega_0$$

with $\tau'_s \omega_0$ extremal in $\Delta(\beta,\Phi)$.

Therefore the measure $\alpha \, d\lambda_1 + (1-\alpha)\, d\lambda_2$ is carried by the extremal states on the orbit of ω_0 under $\mathcal{S}$. But for all $A \in \mathfrak{A}^{sym}$, $(\tau'_s \omega)[A] = \omega[A]$ and thus

$\omega[A] = \omega_0[A]$, λ a.s. which implies

$\omega_1[A] = \omega_2[A] = \omega_0[A] = \bar{\omega}_0[A]$

i.e. ω_0 is extremal. Finally $\tau'_{s_1}\omega_0 = \tau'_{s_2}\omega_0 \Leftrightarrow s_1 - s_2 \in \mathcal{S}_{\omega_0}$.

4) Follows from the fact that $\tau'_s \omega_1$, $\tau'_s \omega_2$ are extremal and the unicity of the extremal decomposition in $\Delta(\beta,\Phi)$.

5) Let $\omega_s = \int_\Delta \omega \; d\lambda(\omega)$ be the extremal decomposition of ω_s in $\Delta(\beta,\Phi)$; then for all $A \in \mathfrak{A}^{sym}$

$\omega_s[A] = \int_\Delta \omega[A] \; d\lambda(\omega) = \int_\Delta \bar{\omega}[A] \; d\lambda(\omega)$.

Since ω_s is extremal in $\Delta_{\mathcal{S}}(\beta)$, we must have $\omega_s = \bar{\omega}$ λ a.s. therefore there exists ω_0 extremal in $\Delta(\beta,\Phi)$ such that $\omega_s = \overline{\omega_0}$.

<u>Conclusion</u>

The set of extremal Gibbs states decomposes into orbits with respect to $\mathcal{S}$; each orbit is mapped exactly on one extremal symmetric state; two different orbits are mapped onto two different extremal symmetric states; every extremal state on a given orbit has the same internal symmetry group.

We thus have the following picture:

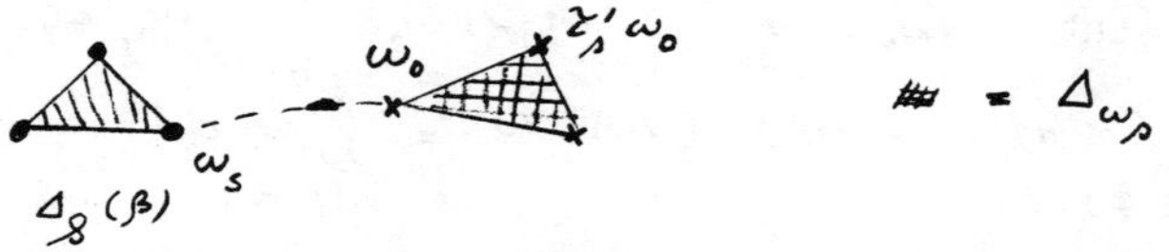

With ω_s an extremal state of $\Delta_{\mathcal{S}}(\beta)$ we associate the convex set

$$\Delta_{\omega_s} = \{\omega \in \Delta(\beta,\Phi)\ ;\ \bar{\omega} = \omega_s\}$$

By definition Δ_{ω_s} is invariant under $\mathcal{S}$ and

$$\omega \in \Delta_{\omega_s} \quad \Leftrightarrow \quad \omega[A] = \omega_s[A] \qquad \forall\ A \in \mathcal{O}\!\!\!\iota^{sym}.$$

<u>Lemma 3</u>

For any state $\omega_1 \in \Delta_{\omega_s}$ whose extremal decomposition in $\Delta(\beta,\Phi)$ is $\omega_1 = \int_\Delta \omega \, d\lambda(\omega)$, we have $\omega \in \Delta_{\omega_s}$.

Indeed: $\bar{\omega}_1 = \omega_s$ is extremal in $\Delta_{\mathcal{S}}$; thus for all $A \in \mathcal{O}\!\!\!\iota^{sym}$

$\omega[A] = \bar{\omega}[A] = \omega_s[A]$ λ a.p. i.e. $\omega \in \Delta_{\omega_s}$ λ a.p.

We thus obtain:

<u>Theorem 4</u>

Let ω_0 be an extremal state in $\Delta(\beta,\Phi)$ such that $\bar{\omega}_0 = \omega_s$ [extremal of $\Delta_{\mathcal{S}}(\beta)$]. Then the extremal points of

Δ_{ω_S} are precisely the Gibbs states $\tau'_\rho \omega_0$ with $\rho \in \mathcal{S}/\mathcal{S}_{\omega_0}$; moreover for any ω in Δ_{ω_S} there is a unique probability measure λ on $\mathcal{S}/\mathcal{S}_{\omega_0}$ such that

$$\omega = \int_{\mathcal{S}/\mathcal{S}_{\omega_0}} d\lambda(\rho) \quad \tau'_\rho \omega_0$$

and any ω in Δ_{ω_S} is invariant under $\mathcal{S}_{\omega_0}$.

The discussion of Sec. 1 leads us to introduce the following definition:

β is said "regular" if there exists a unique extremal state ω_I of $\Delta_{\mathcal{S}}(\beta)$ which is $\mathbb{Z}^\nu$-invariant

Theorem 5

Let β be regular and ω_0 extremal in $\Delta(\beta,\Phi)$ such that $\overline{\omega}_0 = \omega_I$.

1) For all $a \in \mathbb{Z}^\nu$, there exists $\rho \in \mathcal{S}$ such that

$$\tau'_a \omega_0 = \tau'_\rho \omega_0$$

2) The internal symmetry group of ω_0, $\mathcal{S}_{\omega_0}$, is stable under $\mathbb{Z}^\nu$.

(although ω_0 is not necessarily $\mathbb{Z}^\nu$-invariant).

3) Every periodic Gibbs state is in Δ_{ω_I}; in particular $\Delta_0(\beta,\phi) \subset \Delta_{\omega_I}$.

4) Every extremal Gibbs state of Δ_{ω_I} is of the form $\tau'_\rho\omega_0$, $\rho \in \mathcal{S}$, and has the same internal symmetry group as ω_0.

5) Any state ω in Δ_{ω_I} is of the form

$$\omega = \int_{\mathcal{S}/\mathcal{S}_{\omega_0}} d\lambda(\omega) \quad \tau'_\rho \omega_0$$

and

$$\mathcal{S}_\omega \supset \mathcal{S}_{\omega_0}.$$

Proof

1) $\forall a \in \mathbb{Z}^\nu$, $\overline{\tau'_a \omega_0} = \tau'_a \omega_0$ on $\mathfrak{A}^{sym}$

but $\omega_0 = \omega_I$ "

and $\tau'_a \omega_I = \omega_I$ "

implies $\overline{\tau'_a \omega_0} = \omega_I$ "

and thus $\tau'_a \omega_0 \in \Delta_{\omega_I}$.

Since $\tau'_a \omega_0$ is extremal (Th. 2) the proof is concluded using Th. 5.

2) $\forall \rho_1 \in \mathcal{S}_{\omega_0}$ $\tau'_a \omega_0 = \tau'_\rho \omega_0 = \tau'_a \tau'_{\rho_1} \omega_0 = \tau'_{T_a \rho_1} \tau'_a \omega_0 = $
$= \tau'_{T_a \rho_1} (\tau'_\rho \omega_0) \therefore T_a \rho_1 \in \mathcal{S}_{\tau'_\rho \omega_0} = \mathcal{S}_{\omega_0}$.

3) $\forall \omega \in \Delta_0(\beta, \Phi)$, $\bar{\omega} = \omega_I$ implies $\omega \in \Delta_{\omega_I}$.

For any periodic Gibbs state ω, $\frac{1}{N} \sum_a \tau'_a \omega = \omega_I$ on $\mathfrak{A}^{sym}$

Since ω_I is extremal, $\tau'_a \omega = \omega_I$ on $\mathfrak{A}^{sym}$ and $\omega \in \Delta_{\omega_I}$.

4) 5) follows from Th. 5.

Remark

Theorem 2 and 5 imply that on $\mathfrak{A}^{sym}$
$\tau'_\rho \omega_0 = \omega_0$ and $\tau'_a \omega_0 = \omega_0$ $\forall \rho \in \mathcal{S}, a \in \mathbb{Z}^\nu$.

CONCLUSION

We thus have obtain a description of all periodic states in the case where β is regular. The only Gibbs states which are not covered by the general discussion are thus those which are non periodic (see examples).

Finally we shall see in the next section that regular β is the generic case for ferromagnetic systems.

4. GROUP STRUCTURE FOR LATTICE SYSTEMS AND FERROMAGNETIC SYSTEMS

4.1. Definition and results

In the situations usually discussed in the literature there is a natural group G_0 acting transitively on the configuration space $\mathcal{G}_0$ as measure preserving group of transformations, and one is led to consider $\mathcal{G}_0$ itself as a group.

In this section, we thus consider the case where $\mathcal{G}_0$ is a compact, metrizable, abelian group[1] and $d\nu_0 = d\theta$ is the normalised Haar measure; (however under suitable conditions the results will remain valid for locally compact groups).

Let us recall that with any Locally Compact Abelian group $\mathcal{G}_0 = \{\theta\}$ one can associate its dual group $\hat{\mathcal{G}}_0 = \{\chi\}$ which is the group of all continuous complex functions χ on $\mathcal{G}_0$ such that

$$|\chi(\theta)| = 1, \qquad \chi(\theta_1)\cdot\chi(\theta_2) = \chi(\theta_1+\theta_2)$$

together with the composition law $(\chi_1\cdot\chi_2)(\theta) = \chi_1(\theta)\cdot\chi_2(\theta)$ [4].

Furthermore for any subset H of $\mathcal{G}_0$ the annihilator $H^\perp$ is the closed subgroups of $\hat{\mathcal{G}}_0$ defined by

$$H^\perp = \{\chi \in \hat{\mathcal{G}}_0 ;\ \chi(h) = 1 \quad \forall h \in H\}$$

with the properties:

$$(H^\perp)^\perp = \overline{\text{group generated by } H}$$

$$(\mathcal{G}_0/\bar{H})^\wedge \cong H^\perp \qquad \hat{\mathcal{G}}_0/H^\perp \cong \hat{\bar{H}}$$

---

1) We use the sign "+" for the composition law.

We then introduce the following groups[1]):

$$\mathcal{G}_L = \prod_{x \in L} \mathcal{G}_x = \{\underline{\theta}\} \qquad \mathcal{G}_L^\wedge = \prod_{x \in L} \mathcal{G}_x^\wedge = \{\underline{\chi}\}$$

$$\mathcal{G}_{L,f} = \prod_{x \in L}^{(f)} \mathcal{G}_x \qquad \mathcal{G}_{L,f}^\wedge = \prod_{x \in L}^{(f)} \mathcal{G}_x^\wedge$$

with the usual notation: $\underline{\chi}(\underline{\theta}) = \langle \underline{\chi}; \underline{\theta} \rangle = \langle \underline{\theta}; \underline{\chi} \rangle = \prod_{x \in L} \chi_x(\theta_x)$.

Let us also recall the Fourier Decomposition on L.C.A. which yields:

$$\forall A \in \mathcal{O}_\Lambda \qquad \tilde{A}(\underline{\chi}) = \int_{\mathcal{G}_\Lambda} d\underline{\theta}_\Lambda \, A(\underline{\theta}) \langle \underline{\chi}^{-1}; \underline{\theta} \rangle$$

$$A(\underline{\theta}) = \sum_{\chi \in \mathcal{G}_\Lambda^\wedge} \tilde{A}(\underline{\chi}) \langle \underline{\chi}; \underline{\theta} \rangle \qquad 2)$$

Applying the Fourier decomposition to the interaction ϕ_B we can write:

$$\beta H = - \sum_{\underline{\chi} \in \mathcal{G}_{L,f}^\wedge} K(\underline{\chi}) \, \underline{\chi} = - \sum_{b \in \mathcal{B}} [K(b) \chi_b + \overline{K(b)} \chi_b^{-1}]$$

where $\mathcal{B}$ is a countable set of indices, called "bonds" which parametrise those characters $\underline{\chi} \in \mathcal{G}_{L,f}^\wedge$ such that $K(\underline{\chi}) \neq 0$.

---

1) $\prod_{x \in L}$ denotes the "complete direct sum",

$\prod_{x \in L}^{(f)}$ the direct sum: $\mathcal{G}_{L,f}^\wedge = [\mathcal{G}_L]^\wedge$, $\mathcal{G}_L^\wedge = [\mathcal{G}_{L,f}]^\wedge$.

2) Since G_0 is compact its dual $G_0^\wedge$ is discrete.

The characters $\chi \in \mathcal{G}_\Lambda^\wedge$ yield a basis for $\mathfrak{A}_\Lambda$ and thus any state ω will be characterised by its values $\omega[\chi]$ with $\chi \in \mathcal{G}^\wedge_{L,f} \subset \mathfrak{A}$.

Let us apply the general discussion of Sec. 2 to the present case.

$\mathbb{Z}^\nu$-invariance

There is an action of $\mathbb{Z}^\nu$ on $\mathcal{B}$ such that for all $a \in \mathbb{Z}^\nu$, $b + a \in \mathcal{B}$ where $\tau_a \chi_b = \chi_{b+a}$ and $K(b+a) = K(b)$. (More generally the invariance of the system under $\mathcal{E}$ implies that there is an action of $\mathcal{E}$ on $\mathcal{B}$ such that $\tau_e \chi_b = \chi_{eb}$ and $K(eb) = K(b)$.

Internal Symmetry Group

The group $\mathcal{G}_L$ acts:

- as a group of transformations on $\mathcal{G}_L$ defined by

$$\forall \underline{\varphi} \in \mathcal{G}_L \qquad \underline{\varphi}: \begin{array}{ccc} \mathcal{G}_L & \longrightarrow & \mathcal{G}_L \\ \underline{\theta} & \mapsto & \underline{\theta} + \underline{\varphi} \end{array}$$

- as a group of automorphisms on $\mathfrak{A}$:

$$\tau_{\underline{\varphi}} \chi = \chi^{-1}(\underline{\varphi})\, \chi$$

- as a group of transformations on states:

$$(\tau^\bullet_\varphi \omega)^\bullet[\chi] = \chi(\varphi)\, \omega[\chi]$$

It is clear that the action of $\mathcal{G}_L$ is continuous.

Theorem 6

1) The subgroup $\mathcal{S}$ of $\mathcal{G}_L$ defined by

$$\mathcal{S} = \{ \rho \in \mathcal{G}_L \; ; \; \chi_b(\rho) = 1 \; \forall \; b \in \mathcal{B} \} \doteq \{ \chi_b \}^{\perp}$$

is the internal subgroup of the system; it is stable under $\mathbb{Z}^{\nu}$ and theorems 3,4,5 are applicable; the subgroup $\mathcal{S}_f = \mathcal{S} \cap \mathcal{G}_{L,f}$ is the gauge group.

2) The state ω is invariant under the subgroup $\mathcal{S}_\omega$ of $\mathcal{S}$ iff

$$\omega[\chi] = 0 \qquad \forall \; \chi \notin \mathcal{S}_\omega^{\perp}$$

In particular ω is symmetric (i.e. invariant under $\mathcal{S}$) iff

$$\omega[\chi] = 0 \qquad \forall \; \chi \notin \mathcal{S}^{\perp}$$

and $\mathcal{S}^{\perp}$ generates the symmetric algebra.

3) The Gibbs state ω_f defined by the free boundary conditions is symmetric. Any Gibbs state is invariant under $\bar{\mathcal{S}}_f$ (closure in $\mathcal{G}_L$), i.e. $\omega[\chi] = 0 \quad \forall \; \chi \notin \mathcal{S}_f^{\perp}$.

Indeed:

1) $\forall \rho \in \mathcal{S} \qquad \tau_\rho(K(b)\chi_b) = K(b)\chi_b \Rightarrow$

$$\phi_B(\theta + \rho) = \phi_B(\theta) \qquad \forall B$$

2) ω is invariant under $\mathcal{S}_\omega \subset \mathcal{S} \Leftrightarrow \tau'_\rho \omega = \omega \quad \forall \rho \in \mathcal{S}_\omega$

$$\Leftrightarrow [\chi(\rho) - 1]\, \omega[\chi] = 0 \quad \forall \rho \in \mathcal{S}_\omega \Rightarrow \omega[\chi] = 0 \;\; \forall \; \chi \notin \mathcal{S}_\omega^{\perp}.$$

3) By continuity.

Theorem 7

If $K(b) > 0 \quad \forall b \in \mathcal{B}$ (ferromagnetic systems), then

1) The configurations $\underline{s}$ in $\mathcal{S}$ are "ground states". The ground state $\underline{s} \in \mathcal{S}$ are isolated [2] whenever the gauge group is trivial.

2) The state ω_0 defined by the boundary conditions $\theta_x^{\circ} = 0 \quad \forall x \in L$ where o is the identity in $\mathcal{G}_0$, is extremal, $\mathbb{Z}^{\nu}$-invariant and satisfies

$$\omega_0 [\chi_b] \geq \omega_f [\chi_b]$$

$$\omega_0 [\chi] \geq |\omega [\mathrm{Re}\, \chi]| \qquad \forall \omega \in \Delta(\beta, \Phi).$$

Furthermore, if $\omega_0 [\chi_1] \neq 0$ then $\omega_0 [\chi_1 \chi] \neq 0 \quad \forall \chi \in \mathcal{S}^{\perp}$.

3) Any state is invariant under $\mathcal{S}_{\omega_0}$ and $\beta_1 > \beta_2$ implies $\mathcal{S}_{\omega_0(\beta_1)} \subset \mathcal{S}_{\omega_0(\beta_2)}$ i.e. all transitions with S.S.B. are in the relation from group to subgroup.

4) The set of non-regular β is at most countable.

5) The following assertions are equivalent

i) β_0 is regular

ii) $\omega_0 [\chi_b] = \omega_f [\chi_b] \quad \forall b \in \mathcal{B}$

iii) the free energy is differentiable in β at β_0

iv) $\omega_0 [\chi_b]$ is continuous in β at β_0 for all $b \in \mathcal{B}$.

Part (1) is immediate; indeed if $\underline{\theta} = \underline{s}$ (a.s.) i.e. $\underline{\theta} - \underline{s} \in \mathcal{G}_{L,f}$ then:

$$\beta \left(H(\underline{s}) - H(\underline{\theta}) \right) = - \sum_b K(b) \left[2 - (\chi_b + \chi_b^{-1}) \right] \geq 0$$

and

$$H(\underline{s}) - H(\underline{\theta}) = 0 \iff \chi_b (\underline{\theta} - \underline{s}) = 1 \quad \forall b \in \mathcal{B}$$

$$\text{i.e.} \iff \underline{\theta} - \underline{s} \in \mathcal{S}_f$$

the other assertions have been established for the spin 1/2 case in [5,6]; for the general case considered here, see [7].

Remark

It follows from the above theorem that the distinct possible order parameters are given by the elements of

$$\hat{\mathcal{G}}_{L,f} \cap \mathcal{S}_f^{\perp} / \mathcal{S}^{\perp}$$

in particular if $\bar{\mathcal{S}}_f = \mathcal{S}$ there is no S.S.B. and no local order parameter.

4.2. Group structure [8]

In the situation discussed above we have two lattice structures which are similar:

I) $L = \{x\}$ = family of sites; $\forall a \in \mathbb{Z}^\nu$, $x + a \in L$

$\forall x \in L$ $\quad \mathcal{G}_x = \{\theta_x\}$, $\mathcal{G}_L = \prod_{x \in L} \mathcal{G}_x = \{\underline{\theta}\}$ "Group of configurations"

$\hat{\mathcal{G}}_x = \{\chi_x\}$, $\hat{\mathcal{G}}_L = \prod_{x \in L} \hat{\mathcal{G}}_x = \{\underline{\chi}\}$

II) $\mathcal{B} = \{b\}$ = family of bonds; $\forall a \in \mathbb{Z}^\nu$, $b + a \in \mathcal{B}$

$\forall b \in \mathcal{B}$ $\quad \mathcal{G}_b = \mathbb{Z}_q = \{\ell_b\}$, $\mathcal{G}_{\mathcal{B}} = \prod_{b \in \mathcal{B}} \mathcal{G}_b = \{\underline{\ell}\}$ "Group of bond configurations"

$q = |\mathcal{G}_x|$

$\hat{\mathcal{G}}_b = \{\vartheta_b\}$, $\hat{\mathcal{G}}_{\mathcal{B}} = \prod_{b \in \mathcal{B}} \hat{\mathcal{G}}_b = \{\underline{\vartheta}\}$ $\qquad \hat{\mathcal{G}}_b = \begin{cases} \mathbb{Z}_q & q < \infty \\ U(1) & q = \infty \end{cases}$

These two structures are related as follows :

1) $\forall b \in \mathcal{B}$, $\exists \chi_b \in \hat{\mathcal{G}}_{L,f}$

2) $\forall b \in \mathcal{B}$, $X_b = \{x \in L;\ \chi_{b,x} \neq 1_x\}$ = set of sites interacting by means of b.

3) $\forall x \in L$, $\beta_x = \{b \in \mathcal{B};\ \chi_{b,x} \neq 1_x\}$ = set of bonds containing the site x,

4) The interaction on L defines real functions on $\hat{\mathcal{G}}_b$ and $\mathcal{G}_b$ given by:

$$V_b : \quad \hat{\mathcal{G}}_b \longrightarrow \mathbb{R}$$
$$\vartheta_b \mapsto V_b[\vartheta_b] = K(b)\, e^{i\vartheta_b} + \bar{K}(b)\, e^{-i\vartheta_b}$$

$$f_b : \quad \mathcal{G}_b \longrightarrow \mathbb{R}$$
$$\ell_b \mapsto f_b(\ell_b) = \int d\vartheta\, e^{i\ell_b \vartheta}\, e^{V_b[\vartheta]}$$

Furthermore the groups which are useful in the analysis of lattice systems appear as kernels and images of <u>two homomorphisms</u> between the group of configurations and the group of bond configurations:

$$\pi : \quad \mathcal{G}_{\mathcal{B}} \longrightarrow \hat{\mathcal{G}}_L$$
$$\underline{\ell} \mapsto \pi(\underline{\ell}) = \prod_{b \in \mathcal{B}} \chi_b^{\ell_b}$$

$$\gamma : \quad \mathcal{G}_L \longrightarrow \hat{\mathcal{G}}_{\mathcal{B}}$$
$$\underline{\theta} \longmapsto \gamma(\underline{\theta}) \qquad \langle \gamma(\underline{\theta}); \underline{\ell} \rangle \doteq \langle \underline{\theta}; \pi(\underline{\ell}) \rangle =$$
$$= \prod_{b \in \mathcal{B}} \chi_b^{\ell_b}(\underline{\theta})$$

and we have

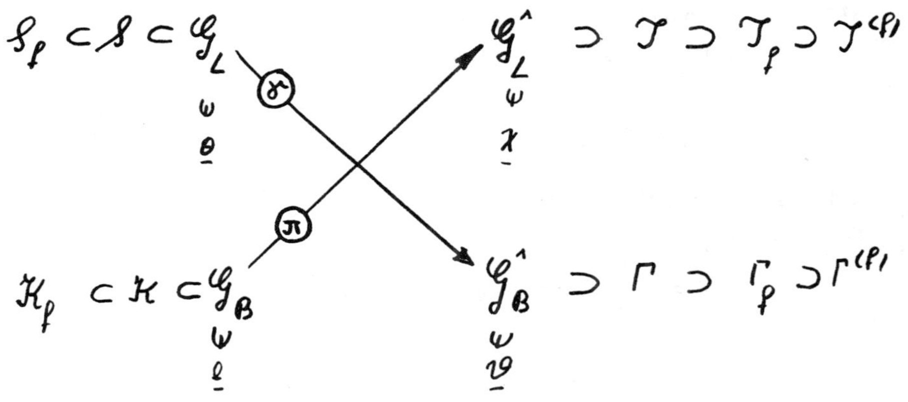

where : $\mathcal{S}$ and $\mathcal{K}$ are the kernels of γ and $\hat{\pi}$

$$\mathcal{S}_f = \mathcal{S} \cap \mathcal{G}_{L,f} \quad , \mathcal{K}_f = \mathcal{K} \cap \mathcal{G}_{B,f} \quad \text{are}$$

the kernels of γ and π restricted to $\mathcal{G}_{L,f}$ and $\mathcal{G}_{B,f}$.
Γ and $\mathcal{T}$ are the images of γ and π
$\Gamma^{(f)}$ and $\mathcal{T}^{(f)}$ the images of γ and π restricted to $\mathcal{G}_{L,f}$, $\mathcal{G}_{B,f}$,

and $\quad \Gamma_f = \Gamma \cap \hat{\mathcal{G}}_{B,f} \qquad \mathcal{T}_f = \mathcal{T} \cap \hat{\mathcal{G}}_{L,f}$

which are related as follow:

$$\mathcal{S}^{\perp} = \mathcal{T}^{(f)} \cong \Gamma^{\wedge} \qquad \mathcal{K}^{\perp} = \Gamma^{(f)}$$
$$\mathcal{S}_f^{\perp} = \mathcal{T} \cong \Gamma^{(f)\wedge} \qquad \mathcal{K}_f^{\perp} = \Gamma$$

The interest of the group $\mathcal{S}$, $\mathcal{S}_f$, $\mathcal{T}$, $\mathcal{T}^{(f)}$ has been illustrated in theorems 6,7; $\Gamma_f = \Gamma^{(f)}$ is a necessary condition in the proof of the existence of a phase transition with S.S.B., or associated with surface tension [8, 22]; the groups Γ and $\mathcal{K}$ are those which appear in connection with duality transformations. For finite systems Λ the group $\mathcal{K}_\Lambda$ is associated with the high-temperature expansion, while the group Γ_Λ is associated with low temperature expansion (for ferromagnetic system); these expansions are also the starting points to establish the unicity and analycity properties of the invariant state by the Asano-Ruelle Method. For example we have:

$$\begin{cases} Z_\Lambda = \sum_{k \in \mathcal{K}_\Lambda} \prod_{b \in \mathcal{B}_\Lambda} f_b(k_b) \\ \omega_\Lambda[\prod_b \chi_b^{\ell_b}] = Z_\Lambda^{-1} \sum_{k \in \mathcal{K}_\Lambda} \prod_b f_b(k_b - \ell_b) \end{cases}$$

which shows immediately that $\quad \omega[\prod_b \chi_b^{\ell_b}] > 0$
for ferromagnetic systems ,

and $$Z_\Lambda = |\mathcal{S}| \int_{\Gamma_\Lambda} d\underline{\gamma} \prod_{b \in \mathcal{B}_\Lambda} e^{V_b[\gamma_b]}$$

It is immediately seen that these two expressions for the partition functions are related by Poisson formula [8] :

Let H be a subgroup of G, then

$$\sum_{\beta \in H} f(\beta) = \sum_{\bar{\beta} \in H^{\perp}} \tilde{f}(\bar{\beta})$$

where

$$\tilde{f}(\bar{\beta}) = |G| \sum_{\beta \in G} f(\beta)\, \chi_{\bar{\beta}}(\beta) .$$

5. EXAMPLES

5.1. Spin 1/2

$$\mathcal{G}_0 = \{-1,+1\} = \{\sigma\} \quad ; \quad \mathcal{G}_L \cong \mathcal{P}(L) \doteq \{X ;\ X \subset L\}$$

$$\phi_B(\underline{\sigma}) = -J(B) \prod_{x \in B} \sigma_x \qquad \mathcal{B} = \{B \subset L ;\ J(B) \neq 0\}$$

$$G_0 = \{-1,+1\} \ ; \ (g \cdot \underline{\sigma})_x = g(x)\sigma_x \ ; \ G \cong \mathcal{P}(L) \text{ with } X \cdot Y = X \cup Y / X \cap Y$$

$$G^{int} = \{\underline{\lambda} \in G_L ;\ \prod_{x \in B} \lambda_x = 1 \quad \forall B \in \mathcal{B}\} \cong \mathcal{S} = \{S \subset L ;\ S \cdot B = \phi \quad \forall B \in \mathcal{B}\}$$

5.1.1. Ising Model $\nu \geq 2$

$$L = \mathbb{Z}^{\nu} , \quad \mathcal{B} = \{(xy) ;\ \|x-y\| = 1\} \ \Rightarrow \mathcal{S} = \{\phi, L\}$$

The only subgroup of $\mathcal{S}$ is the trivial subgroup; by a simple change of variables we can always take J(B) > 0. Theorem 7 implies then the existence of a unique β_c such that for $\beta < \beta_c$ $\mathcal{S}_{\omega_0} = \mathcal{S}$ and for $\beta > \beta_c$, $\mathcal{S}_{\omega_0} = \{\phi\}$. Therefore for $\beta > \beta_c$ there exist at least two pure phases which are also extremal in $\Delta(\beta,\Phi)$ and the only relevant order parameter is $m = \omega[\sigma_x]$. If $\nu = 2$ all β are regular, and thus there are exactly two pure phases for $\beta > \beta_c$ [9], and $\Delta_{\bar{\omega}_0}(\beta) = \Delta(\beta,\Phi)$ [10]. On the other hand for $\nu \geq 3$ there exist Gibbs states which are not in $\Delta_{\bar{\omega}_0}$ [2].

5.1.2. Triangular Model $\nu = 2$

L = triangular lattice

$\mathcal{B} = \{\Delta$ = elementary triangular cell of $L\}$ $\mapsto$

$\mathcal{S} = \{\phi, S_1, S_2, S_3\}$ with $S_i^2 = \phi$ $S_1 S_2 = S_3$, ...

$J(B) > 0$ $\tau_a S_i \neq S_i$

Again the only subgroup of $\mathcal{S}$ stable under $\mathbb{Z}^2$ is the trivial subgroup and there exists a unique β_C such that for $\beta < \beta_C$ $\mathcal{S}_{\omega_0} = \mathcal{S}$ and for $\beta > \beta_C$ $\mathcal{S}_{\omega_0} = \{\phi\}$. For $\beta > \beta_C$ the set $\Delta_{\bar{\omega}_0}$ contains 4 extremal states which give 2 pure phases, namely ω_0 and $\omega' = \frac{1}{3} \sum_{i=1}^{3} \tau'_{S_i} \omega_0$.

The relevant order parameter is $m = \omega[\sigma_x]$.

5.1.3. Baxter Model $\nu = 2$

$L = \mathbb{Z}^2$, $\mathcal{B} = \{(xy), \square$; (xy) = Next Nearest Neighbour

$J(B) > 0$ $\square$ = elementary cell of $L\}$

$\mathcal{S} = \{\phi, S_1, S_2, S_3\}$ is isomorphic to the group of 5.1.2.

But now there is a non-trivial subgroup $\mathcal{S}_1 = \{\phi, S_1\}$ which is stable under $\mathbb{Z}^2$. According to our general discussion we could have two phase transitions with SSB; however, as was shown by Baxter [11], the critical temperature is unique:

for $\beta < \beta_C$ $\mathcal{S}_{\omega_0} = \mathcal{S}$ while for $\beta > \beta_C$ $\mathcal{S}_{\omega_0} = \{\phi\}$.

For $\beta > \beta_C$, $\Delta_{\bar{\omega}_0}$ contains again 4 extremal states but they yield 3 pure phases ω_0, $\omega' = \tau'_{S_1} \omega_0$ and $\omega'' = \frac{1}{2}(\tau'_{S_2}\omega_0 + \tau'_{S_3}\omega_0)$.

The relevant order parameters are in this model $m = \omega[\sigma_x]$ and $p = \omega[\sigma_x \sigma_y]$ with $\|x-y\| = 1$.

5.1.4. Ashkin-Teller Model $\nu = 2$

$L = \mathbb{Z}_A^2 \times \mathbb{Z}_B^2$ $\mathcal{B} = \{ (x,y),$ A—A / B—B $;\ \|x-y\| = 1,$

$J(B) > 0$

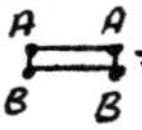
= elementary rectangle $\}$

$\mathcal{S} = \{\phi, S_1, S_2, S_3\}$ is again isomorphic to the group $\mathcal{S}$ of 5.1.2.

Now there are 3 non trivial subgroups stable under $\mathbb{Z}^2$. For suitable values of J(B) it is possible to show [12] that there are two critical temperatures β_1 and β_2 such that for $\beta < \beta_1$ $\mathcal{S}_{\omega_0} = \mathcal{S}$, for $\beta_1 < \beta < \beta_2$ $\mathcal{S}_{\omega_0} = \{\phi, S_i\}$, and for $\beta > \beta_2$ $\mathcal{S}_{\omega_0} = \{\phi\}$. The set $\Delta_{\bar{\omega}_0}$ contains respectively 1, 2, and 4 extremal states which are also pure phases. The relevant order parameters are $m_A = \omega[\sigma_x]$ $m_B = \omega[\sigma_z]$ and $p = \omega[\sigma_x \sigma_z]$ where $x \in \mathbb{Z}_A^2$ $z \in \mathbb{Z}_B^2$ and $(x\,z)$ are nearest neighbours.

5.1.4. Plaquettes models $\nu = 3$

$L = \{x\,;\,x$ = center of an edge of on a cubic lattice $\}$

$\mathcal{B} = \{$ ◇ = elementary square on a face of the cubic lattice$\}$

$J(B) > 0$

Let S_0 denotes the 6 sites nearest to a given vertex of the cubic lattice. The Gauge Group $\mathcal{S}_f$ is then generated by the finite product of elements $\{T_a S_0\}$ $a \in \mathbb{Z}^3$ and $\mathcal{S}$ is the closure of $\mathcal{S}_f$ in $\mathcal{P}(L)$. Therefore, although this model (dual of the Ising model) has a phase transition, there is no spontaneous symmetry breakdown, i.e. for all β, $\mathcal{S}_{\omega_0} = \mathcal{S}$ and for all regular β there is a unique translation invariant Gibbs state (and no other periodic state).

5.1.5. Reducible system $\nu = 2$ $\quad \Gamma_f \neq \Gamma^{(f)}$

$L = \mathbb{Z}^2$ $\quad \mathcal{B} = \{$ 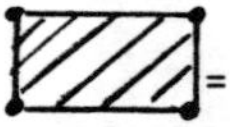 = set of 4 sites on a rectangle with sides of length 1 and 2 $\}$

$\mathcal{S} = \mathcal{S}_0 \cup \mathcal{S}_1$ with $\mathcal{S}_0 = \{s; |s \cap X|$ is even, $\forall X$ unit squares$\}$

$\mathcal{S}_1 = \{s; |s \cap X|$ is odd, $\forall X$ unit squares$\}$

As in 5.1.4. the order of $\mathcal{S}$ is ∞ but now $\mathcal{S}_f = \{\Phi\}$; this model is a HT-HT dual of the Ising model and there exists a unique β_C such that $\mathcal{S}_{\omega_0} = \mathcal{S}$ for $\beta < \beta_C$ and $\mathcal{S}_{\omega_0} = \mathcal{S}_0$ for $\beta > \beta_c$. $\Delta_{\bar{\omega}_0}$ contains exactly two extremal states and they are $\mathbb{Z}^\nu$-invariant [8, 23].

5.2. $\mathcal{G}_0$ finite $\quad |\mathcal{G}_0| = q$

We can always take

$$\mathcal{G}_0 = \{ n \tfrac{2\pi}{q} ; \ n = 0, 1, \dots, q-1 \} \qquad \hat{\mathcal{G}}_0 \cong \mathcal{G}_0$$

$$G_0 = \mathcal{G}_0 \qquad G = \mathcal{G}_L \quad \text{with} \quad (g \cdot \underline{\theta})_x = g_x + \theta_x$$

5.2.1. Generalized Potts Models ($\nu \geq 2$)

$$L = \mathbb{Z}^2 \qquad \mathcal{B} = \{ (xy) ; \ \|x - y\| = 1 \}$$

$$\Phi_B(\underline{\theta}) = -\sum_{\ell=0}^{q-1} J(\ell) \cos \ell(\theta_x - \theta_y) \qquad J(\ell) \geq 0$$

we thus have $\mathcal{S} \cong G_0$ with $\underline{\theta} \in \mathcal{S} \Leftrightarrow \theta_x = \theta \quad \forall x$

As we have remarked in Sec. 2, $\mathcal{S}$ is not the largest internal symmetry group; whatever the values of $J(\ell)$ the system is always invariant under the group generated by $\mathcal{S}$ and the inversion $\theta \mapsto q - \theta$.

5.2.2. Standard Potts Models (also called: Potts model)

It is obtained from 5.2.1. with $J(\ell) = \frac{J}{q}$ i.e.

$$\phi_B = -J\,\delta_{\theta_x,\theta_y}$$

It is known that for q large enough (q > 4 if $\nu = 2$ [13]) there is a $\beta_c(q)$ which is non regular; for $\beta \gtrsim \beta_c(q)$, $\mathcal{S}_{\omega_0}$ is trivial and $\Delta_{\bar{\omega}_0}$ contains q extremal states which are also invariant; at β_c there are at least two distinct $\mathbb{Z}^\nu$-invariant symmetric states $\omega' = \frac{1}{q}\sum_{\rho\in\mathcal{S}} \tau'_\rho\, \omega_0$ and ω_f [14].

Furthermore for $\nu = 2$, $q^{1/4} > 6$, $\beta_c(q) = \log(1+\sqrt{q})$ is the only non regular temperature and for $\beta < \beta_c(q)$ there is a unique Gibbs state [15]. Let us note once more that $\mathcal{S}$ is not the largest internal symmetry group, which in this case is isomorphic to the group of permutations of n elements.

5.2.3. Vector Potts Model (also called "Planar", or "$\mathbb{Z}_q$-model", or "Clock model")

$$\phi_B = -J\cos(\theta_x - \theta_y)$$

In this case there exists β_1 such that for $\beta < \beta_1$ $\mathcal{S}_{\omega_0} = \mathcal{S}$ and for $\beta > \beta_1$ $\mathcal{S}_{\omega_0}$ is trivial; furthermore there exists $\beta_2 < \beta_1$ such that any β in $[\beta_2, \beta_1]$ is a critical temperature:
We have thus an example of phase transitions without SSB [16].

5.3. Continuous Group

Let $L = \mathbb{Z}^{\nu}$, $\mathcal{G}_0 = S^{(1)} = \{\theta;\ \theta \in [0, 2\pi[\}$; $G_0 = \mathcal{G}_0$, $(\underline{g} \circ \underline{\theta})_x = g_x + \theta_x$

$\mathcal{B} = \{(xy);\ \|x-y\| = 1\}$

and $\phi_B(\underline{\theta}) = \phi(\theta_x - \theta_y)$

Then $\mathcal{S} \cong \mathcal{G}_0$, $\underline{s} \in \mathcal{S} \Leftrightarrow s_x = s \quad \forall x$

5.3.1. Rotators

$$\phi_B(\underline{\theta}) = -J\cos(\theta_x - \theta_y)$$

For $\nu = 2$ $\mathcal{S}_{\omega_0} = \mathcal{S}$ for all β and all β are regular [17,19]. However there exist a phase transition without SSB [16]. For $\nu \geq 3$ at low enough temperature $\mathcal{S}_{\omega_0}$ is trivial [18] and it is known that either $\mathcal{S}_{\omega_0} = \mathcal{S}$ or $\mathcal{S}_{\omega_0}$ is trivial [19]. Therefore there exists a unique β_c such that $\mathcal{S}_{\omega_0} = \mathcal{S}$ for $\beta < \beta_c$ and $\mathcal{S}_{\omega_0}$ is trivial for $\beta > \beta_c$. It is expected that all β are regular; furthermore it is expected that $\Delta_{\bar{\omega}_0} = \Delta(\beta, \Phi)$ if $\nu = 2,3$ and $\Delta_{\bar{\omega}_0} \neq \Delta(\beta, \Phi)$ if $\nu \geq 4$ [20].

5.3.2. Villain model

$$\beta\phi_B(\underline{\theta}) = -\ln\Big[\sum_{n=-\infty}^{+\infty} e^{-\frac{\beta}{2}(\theta_x - \theta_y - 2n\pi)^2}\Big]$$

Again for $\nu = 2$ it is known that $\mathcal{S}_{\omega_0} = \mathcal{S}$ for all β and all β are regular; however there exists a phase transition without SSB [16], which is called Kosterlitz-Thouless transition.

5.4. Discrete model

$\mathcal{G}_0 = \mathbb{Z} = \{n\}$, $G_0 = \mathcal{G}_0$, $L = \mathbb{Z}^2$,

It is known [16] that the following models have a Kosterlitz-Thouless transition.

5.4.1. Gaussian models

$$\phi_B(\underline{n}) = c\,|n_x - n_y|^2 \qquad \|x-y\| = 1$$

The Gaussian model with o boundary condition is the dual of the Villain model with free boundary condition [2,8].

5.4.2. Solid on solid

$$\phi_B(\underline{n}) = |n_x - n_y| \qquad \|x-y\| = 1$$

5.4.3. Neutral Coulomb Systems L

$$\mathcal{B} = \{x, (xy)\} \qquad \phi_x = \pi e\, n_x^2 \qquad \phi_{xy} = -\pi\, n_x n_y \ln|x-y|$$

with the constraint $\Sigma\, n_x = 0$.

REFERENCES

[1] Ruelle D., Statistical Mechanics, 1969.
[2] Sinai Ya G. : Theory of Phase Transitions : Rigorous Results Pergamon Press, 1982.
[3] Bratteli o., Robinson D, Operator Algebra and Quantum Statistical Mechanics, Springer, 1979.
[4] Rudin W. : Fourrier Analysis on Groups, Interscience Publishers (1967).
[5] Lebowitz J.L., J. Stat. Phys., 16, 3 (1977).
[6] Gruber C., Lebowitz J.L., Comm. Math. Phys. 59, 97 (1978).
[7] Pfister C.E., 1) Comm. Math. Phys. 86, 375 (1982).
2) Proceeding of the Sixth International Symposium on Information theory, Tashkent, 1984, part. III, p. 259.
3) Infinite dimensional analysis and stochastic processes, p. 98, Ed. S. Abeverio, Res. Notes, Pitman, Maths. 1985.
[8] Gruber C., Hintermann A., Merlin D., "Group Structure Analysis of Classical Lattice Systems", Lecture notes in Physics, Springer (1977).
Gruber C., "Les Systèmes Réticulaires en Mécanique Statistique et théorie des Champs", Cours Louvain-la-Neuve, Belgique (1977).
[9] Messager A., Miracle-Sole S., Comm. Math. Phys., 40, 187 (1975).
[10] Higuchi Y., Colloquia Mathematica Society Janos Bolyai, 27, Random Fields Esztergom (1979).
Aizenman M., Phys. Rev. Lett., 43, 407 (1979).
Comm. Math. Phys., 73, 83 (1980).
[11] Baxter R.J., Ann. Phys. N.Y., 70, 193 (1972).
[12] Pfister C.E., J. Stat. Phys., 29, 113 (1982).
[13] Baxter R.J., J. Phys. A, 15, 3329 (1982).
[14] Kodecky R. and Schlossman S.B., Comm. Math. Phys., 83, 495 (1982).
[15] Laanait L., Messager A., Ruiz J., "Phases coexistence and surface termins for the Potts model", Preprint Marseille (1984).
[16] Fröhlich J., Spencer T., Comm. Math. Phys., 81, 527 (1981).
[17] Bricmont J., Fontaine J.R., Landau L.J., Comm. Math. Phys., 56, 281 (1977).
[18] Fröhlich J., Simon B., Spencer T., Comm. Math. Phys., 50, 79 (1976).
[19] Messager A., Miracle-Sole S., Pfister C.E., Comm. Math. Phys., 58. 19 (1978).
[20] Fröhlich J., Pfister C.E., Communication Math. Phys., 89, 303 (1983).
[21] Dobrushin R.L., Theory Probab. Applic. A, 582 (1972); 18, 253 (1973).
[22] Fontaine J.R., Gruber C., Comm. Math. Phys., 70, 243 (1979).
[23] Gruber C., Wiskott B., Helv. Phys. Act., 52, 597 (1979).

TOPOLOGICAL DEFECTS IN SYSTEMS OF BROKEN SYMMETRY

H.-R. Trebin

Institut für Theoretische Physik der Universität
Regensburg, D-8400 Regensburg, F.R.G. *

ABSTRACT

After a short introduction to the principles two aspects of the topological theory of defects are discussed: the notion of semidefects, and the problem of classifying defects in crystals and layered structures.

1. PRELIMINARIES AND PRINCIPLES

1.1 Growth of defects in phase transitions

Landau theory is the standard method to describe the symmetry breaking of condensed matter systems in a phase transition (see, for example, the lectures of P. Toledano). To investigate the transition from an isotropic liquid of elongated molecules to the nematic state, where the molecules align to form a long-range orientational order, one expands the nonequilibrium free-energy density $f(T,\underline{\underline{Q}})$ into an O(3)-invariant polynomial of $\underline{\underline{Q}}$[1)]. The order parameter $\underline{\underline{Q}}$ is a symmetric, traceless tensor, which characterizes the second moment of the molecules' angular distribution function or, macroscopically, the deviation of the dielectric tensor $\underline{\underline{\varepsilon}}$ from isotropy (the birefringence). For temperatures T above the critical one, T_N, the absolute minimum of f is found at the origin $\underline{\underline{Q}}^o=0$ of the five-dimensional vector space $X=\{\underline{\underline{Q}}\}$. The tensor $\underline{\underline{Q}}^o=0$ is invariant under the full unbroken symmetry group O(3). For $T<T_N$, the

minimum is taken by a quadrupole tensor $\underline{\underline{Q}}^o=\sqrt{\tfrac{3}{2}}Q(T)\{\hat{n}\otimes\hat{n}-\underline{\underline{1}}\}$ of cylinder symmetry $D_{\infty h}$. The symmetry axis is represented by a unit vector $\hat{n}$, the director. Now the symmetry of the equilibrium order parameter $\underline{\underline{Q}}^o$ has been broken from G=O(3) to $H=D_{\infty h}$, but as a remnant of the unbroken symmetry, f is minimized on an entire submanifold, since the direction $\hat{n}$ is arbitrary. The set V of minima is denoted "space of degeneracy" and is identical to the projective plane $P^2=S^2/\mathbb{Z}_2$ (fig. 1). $S^2=\{\hat{n}\}$ is the two-sphere. The factorization of S^2 by $\mathbb{Z}_2$ means that antipodal points have to be identified, since $\hat{n}$ and $-\hat{n}$ produce the same quadrupole tensor $\underline{\underline{Q}}^o$.

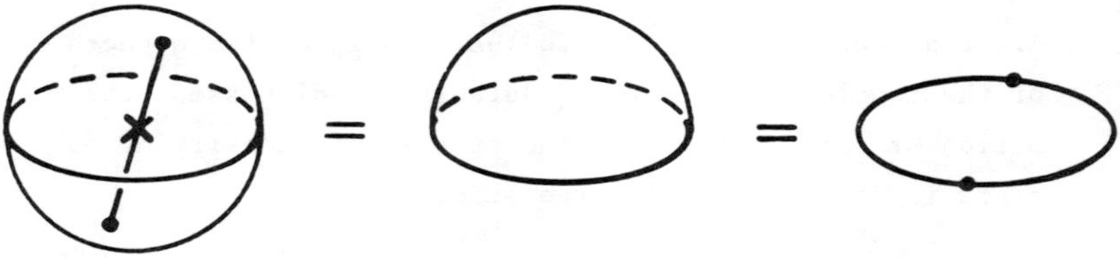

Fig. 1 Space of degeneracy for a uniaxial nematic liquid crystal. The lower hemisphere is redundant. The upper hemisphere can (by parallel projection) be pressed into a disk, where antipodal points on the boundary have to be interpreted as one point.

Ideally, the broken symmetry phase is uniform, corresponding to a single element $x\in V$. Growth conditions, boundaries, or external fields, force the liquid crystal into a nonuniform state: the order parameter varies in space. As long as the distortions are weak (long-wave or hydrodynamic limit) the system looks uniform locally. The order parameter takes values only in the minimum set V, and the liquid crystal is described by a mapping from physical space $\mathbb{R}^d\setminus\Delta$ to V. Here Δ is the defect set. It denotes those points, lines, or walls in space, where the degeneracy parameter is singular and the order parameter must change the stratum.

Symmetry defects strongly influence the physical and chemical properties of matter. Methods of algebraic topology have been applied since 1976 to classify them. In these lectures, first a short introduction to the topological defect classification is presented. Then some recently developed aspects of the theory are sketched: the notion of "semidefects", the classification of point solitons in crystals, and a novel classifying scheme for point singularities in three-dimensional layered structures.

1.2 The classification

1.2.1 Line and point singularities. Several reviews exist on the topological defect classification[2-4]. Therefore the principles are described only briefly. A point singularity in two-dimensional space, and a line singularity in three-dimensional space, are tested by a closed loop ("Burgers circuit"). The degeneracy parameters along the circuit yield a closed loop in V. This loop, and all those which result from it by continuous deformation, form a homotopy class. The homotopy classes label the singularities. They correspond to the conjugacy classes of the fundamental group $\pi_1(V,x_o)$, whose elements stand in one-to-one correspondence with the homotopy classes of loops tied to a fixed point $x_o \in V$. The fundamental group of the projective plane P^2 is $\pi_1(P^2,x_o)=\mathbf{Z}_2$. There is the trivial class of those loops which can be deformed to the constant one, and the nontrivial class of noncontractable loops, all of which are homotopic to their inverse loops. The respective singularity is a 180°-disclination (fig. 2).

A point defect in three-dimensional space is tested by a Burgers sphere S^2. The degeneracy parameters on this sphere compose a mapping $S^2 \rightarrow V$. The singularity is labeled by the homotopy class thereof. The homotopy classes of based mappings $S^2 \rightarrow V$ (northpole of S^2 going into base point x_o) form the second homotopy group $\pi_2(V,x_o)$. The free homotopy classes are in one-to-one correspondence with the orbits of $\pi_2(V,x_o)$ under a certain action of $\pi_1(V,x_o)$ (consult the reviews for details). The group $\pi_2(P^2,x_o)$ is isomorphic to the integers $\mathbf{Z}$. Each point singu-

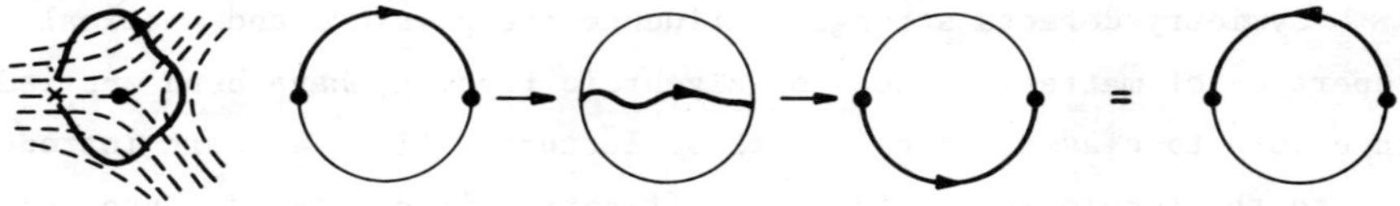

Fig. 2 180°-disclination of a uniaxial nematic liquid crystal, and deformation of the corresponding loop in P^2 into its inverse.

larity is labeled by an orbit $\{n,-n\}$, $n \in \mathbb{Z}$. Hedgehog and hyperbolic point belong to the class $\{\pm 1\}$. Whenever two singularities are associated with the same class, one can exchange the core of one by the core of the other without introducing additional defects (fig. 3).

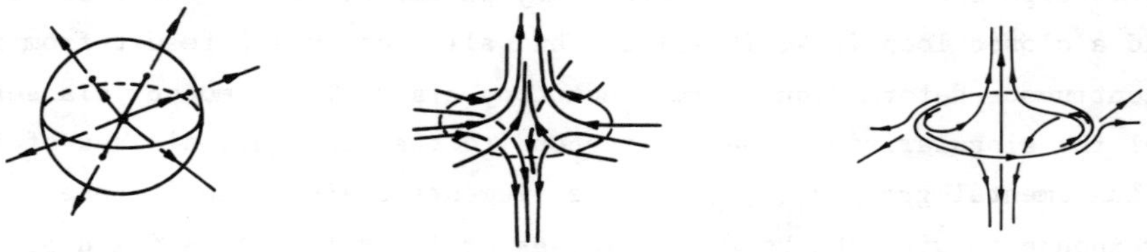

Fig. 3 Hedgehog, hyperbolic point, and exchange of cores.

1.2.2 Topological solitons. These are nonsingular mappings from a d-dimensional disk $D^d \subset \mathbb{R}^d$ into V, where on the boundary $S^{d-1} = \partial D^d$ the range of values is restricted to a subset $A \subset V$. The homotopy classes of based mappings $(D^d, S^{d-1}, p \in S^{d-1}) \to (V, A, x_o \in A)$ are elements of the relative homotopy group $\pi_d(V, A, x_o)$. In case of constant boundary conditions, $A = \{x_o\}$ consists of one point, and the labeling is done by the absolute homotopy groups $\pi_d(V, x_o)$. If V is replaced by the contractable order parameter space X, and A by $V \subset X$, then singularities can be viewed as topological solitons in X due to the isomorphism $\pi_d(X,V,x_o)=\pi_{d-1}(V,x_o)$.

Relative homotopy groups are required: for the classification of surface singularities[5], where A is the subset of degeneracy parameters admitted by the surface anchoring conditions; in studies of defect cores[6]; and for the description of new types of singularities denoted semidefects[7] (section 2).

1.3 Spaces of degeneracy as coset spaces

From the introductory discussion of symmetry breaking in a phase transition follows that the space of degeneracy is the orbit $V=Gx_o$ of an order parameter x_o under the action of the unbroken symmetry group G, whenever x_o minimizes the free-energy. Denote by H the little group of x_o. Then any point $gx_o \in V$, $g \in G$, can be placed in one-to-one correspondence with the left coset gH. This coset is comprised of all elements of G which turn x_o into gx_o. V is homeomorphic to the set of cosets G/H. Note that G/H is a group only if H is a normal divisor of G.

2. SEMIDEFECTS

2.1 Origin and definition

If a uniaxial nematic liquid crystal is cooled, this already broken symmetry state can still further be deprived of symmetry. One case (observed[8] until now only in lyotropic systems, where concentration is the controlling thermodynamic variable rather than temperature, but cultivated here as model case due to its simplicity), is the transition to a biaxial nematic phase. Thereby the two equal eigenvalues ("biaxes") of the quadrupole tensor $\underline{\underline{Q}}^o$ become different and reduce the symmetry from $H_1=D_{\infty h}$ to $H_2=D_{2h}$. If the high temperature uniaxial phase was uniform, of aligned main axes, the space of degeneracy for the new degree of freedom - the biaxes - is now $H_1/H_2=P^1$. Since $P^1=S^1/\mathbb{Z}_2$ is homeomorphic to S^1, the phase transition closely resembles the ordering of planar spins (XY-model).

In this second transition, defects can grow only in the field of the biaxes (fig. 4a). They are denoted semidefects, since they are singularities in only part of the degeneracy parameter. The classifying sets are the homotopy groups $\pi_r(H_1/H_2)$ (the base points are omitted in the following). In our example $\pi_1(H_1/H_2)=\frac{1}{2}\mathbf{Z}$. Even if the field of the main axes is nonuniform, but without singularity in the core of the biaxial defect, the notion of semidefect is reasonable and the classifying element of $\pi_r(H_1/H_2)$ well-defined (up to a group action introduced in section 2.5), because there is a unique procedure to align the main axes[7].

Defects usually are created in a phase transition. When a phase is attained in two steps, also its singularities develop in two steps. The investigation of these sequences has been the main motivation for the notion of semidefects. From a more general viewpoint semidefects are defined in the following way: given a degeneracy parameter, that can be divided into two coupled components, a "rigid" and a "soft" one, a semidefect is a singularity in the field of the soft component, while the field of the rigid component remains continuous. For biaxial nematic liquid crystals, the two components are the main axis and the biaxes, coupled by the requirement to stand perpendicular. For the semidefect in a smectic-A liquid crystal of fig. 4b the components are the nematic director and the mass density wave coupled such that directors and layer normals are locked parallel.

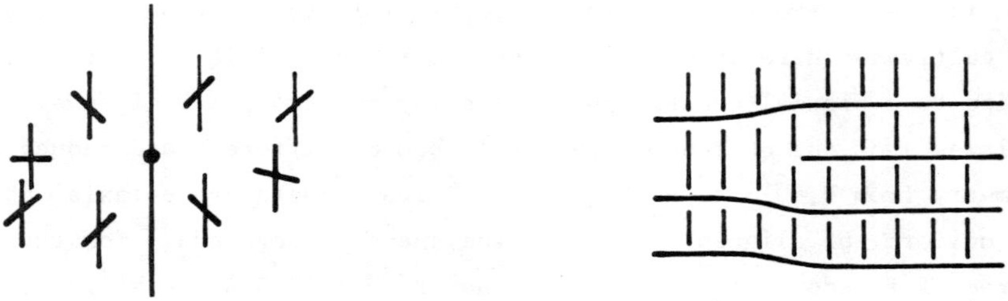

Fig. 4 Semidefects a) of a biaxial nematic liquid crystal, and b) of a smectic-A liquid crystal.

2.2 Properties of semidefects

Contrary to a topologically stable defect in the full degeneracy parameter, a semidefect can terminate or change its type in the bulk of an ordered medium, but only if a singularity in the rigid component bounds it or forms an interface. The source singularity in the main axes (fig. 5) either bounds an n=2 biaxial semidefect (of winding angle 4π), or transforms a semidefect of index n=-1 into one of index n=+1 (the winding numbers are measured with respect to a fixed orientation on the line). For, on each sphere about the source the tangential field of the biaxes must form singularities of total winding angle 4π (with respect to the outer normals of the sphere).

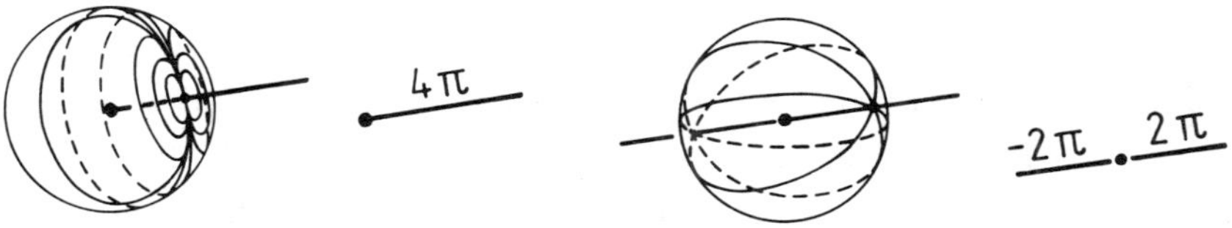

Fig. 5 A source singularitiy in the main axes is boundary of a semi-defect line or interface between two semidefects.

2.3 Topological description

We now study cases, where the coupling can be expressed by a subgroup relation $H_2<H_1<G$ between the little group H_2 of the full degeneracy parameter, and the little group H_1 of its rigid component. The three spaces of degeneracy G/H_2, G/H_1, and H_1/H_2 can be incorporated into a fiber bundle: G/H_2 is the bundle space, G/H_1 the base, and over each point of the base a copy is placed of the fiber H_1/H_2. According to Steenrod[9], the homotopy groups of the three topological spaces are related by a sequence of homomorphisms, of which the following section is being interpreted now:

$$\ldots \rightarrow \pi_2(G/H_2) \xrightarrow{j_2} \pi_2(G/H_1) \xrightarrow{\partial_2} \pi_1(H_1/H_2) \xrightarrow{i_1} \pi_1(G/H_2) \xrightarrow{j_1} \pi_1(G/H_1) \rightarrow \ldots$$

The sequence is exact: the kernel of a homomorphism is equal to the image of the preceding homomorphism (fig. 6).

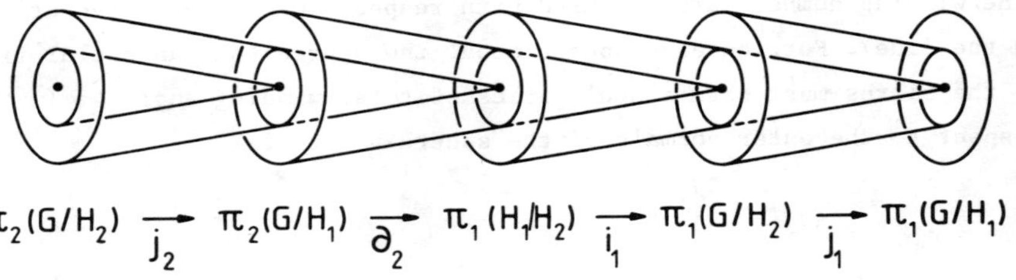

Fig. 6 This illustration of an exact sequence is taken from ref. 3. Each disk represents a group, with the identity element at its center. The inscribed disk represents the kernel of the following and the image of the preceding homomorphism.

a) Homomorphism j_2 describes the fate of a point singularity, if for example in a phase transition the full order parameter is reduced to the rigid component. Defects of ker j_2 disappear in the transition.

b) Homomorphism ∂_2 yields those linear semidefects which emanate from a point singularity in the rigid component of the degeneracy parameter. $\partial_2^{-1}(\alpha)$ characterizes the possible boundaries of a semidefect α, $\partial_2^{-1}(\alpha\beta^{-1})$ the allowed interfaces between semidefects of types α and β. Point singularities of $\ker\partial_2$ do not bound stable semidefects. Since $\ker\partial_2 = \mathrm{im}\, j_2$, they possess an inverse image, into which they return, when in a phase transition the soft component is complementing the rigid component to form the full degeneracy parameter.

c) Homomorphism i_1 relates semidefects to full defects. The semidefects belonging to $\ker i_1$ are unstable as full defects. Because $\ker i_1 = \mathrm{im}\,\partial_2$, they correspond exactly to those which can terminate in the bulk. The full defects of $\mathrm{im}\, i_1 \setminus \{0\}$ are equivalent to semidefects and hence are supposed to be of lowest energy.

d) Moreover, the singularities of $\mathrm{im}\, i_1 \setminus \{0\}$ are included in $\ker j_1$ and belong to those linear defects, which grow spontaneously out of nothing, when the soft component of the degeneracy parameter is developing in a phase transition.

2.4 Relaxation modes

A semidefect line of type $\alpha \in \mathrm{im}\,\partial_2 \subset \pi_1(H_1/H_2)$ can break by forming a pair of point singularities of types ρ and ρ^{-1}, $\rho \in \partial_2^{-1}(\alpha) \subset \pi_2(G/H_1)$, which then move apart. When $\rho \in \partial_2^{-1}(\alpha\beta^{-1})$, in between the two points the line has changed from type α to type β.

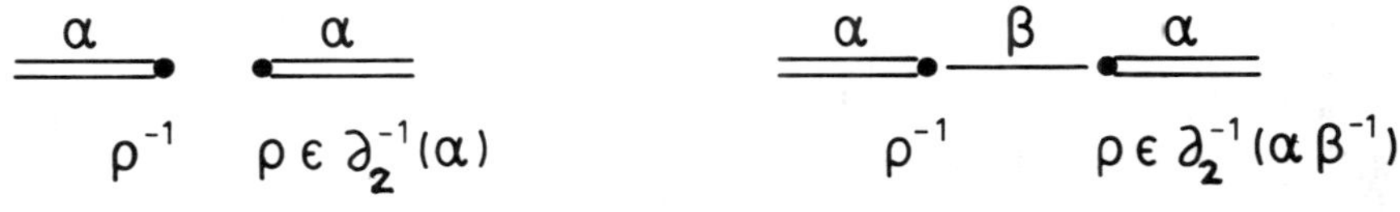

Fig. 7 Relaxation modes of a semidefect line.

2.5 Action of a full line singularity on a semidefect line

When representing a space of degeneracy by a coset space, one can describe an element $\alpha \in \pi_r(H_1/H_2)$ by $\alpha = [h(u)H_2]$, where $h\colon S^r \to H_1$, $u \mapsto h(u)$ is a continuous mapping, and the brackets designate the respective based homotopy class. Once G is chosen a simply connected Lie group, a full line singularity κ corresponds uniquely to a left coset kH_2^o of the connected component H_2^o of H_2 owing to the isomorphism $\pi_1(G/H_2) = H_2/H_2^o$.[2)] If a semidefect α is guided around κ, its based class turns into $\kappa(\alpha) := [k^{-1}h(u)kH_2] \in \pi_r(H_1/H_2)$. This action of κ on $\pi_r(H_1/H_2)$, presented here without proof, is very similar to the action of an element of $\pi_1(G/H)=H/H^o$ on $\pi_2(G/H)=\pi_1(H)$ (see the reviews refs. 2-4 and ref. 7).

2.6 Crossing of a semidefect line with a full line singularity

The semidefect line of fig. 8 can cross the full line singularity κ only, if the double line in between belongs to the trivial semidefect class. The testloop measuring the semidefect on the other side of κ yields a based class $\kappa(\alpha)$, so that the two parallel semidefect lines combine to a type $\kappa(\alpha)\alpha^{-1}$. There are three possibilities:

i) $\kappa(\alpha)\alpha^{-1}=0$: crossing is not obstructed.

ii) $\kappa(\alpha)\alpha^{-1}\notin \mathrm{im}\,\partial_2$: after crossing, the two singularities remain connected by a full line defect of $\mathrm{im}\, i_1$.

iii) $\kappa(\alpha)\alpha^{-1}\in \mathrm{im}\,\partial_2\setminus\{0\}$: crossing is allowed but only by production of a pair of point singularities (fig. 8).

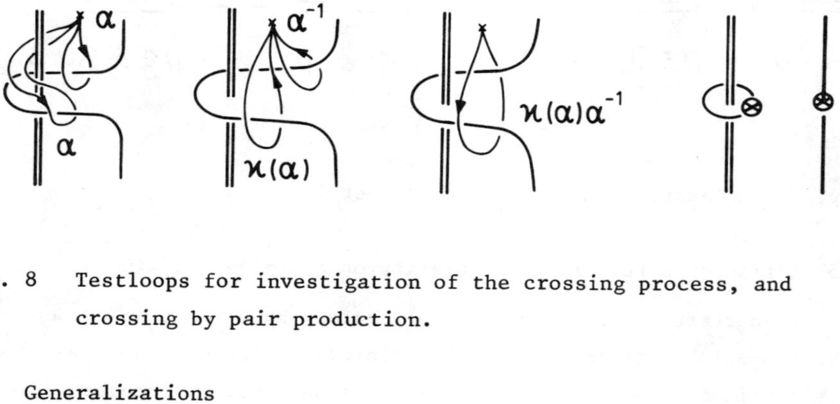

Fig. 8 Testloops for investigation of the crossing process, and crossing by pair production.

2.7 Generalizations

Cholesteric liquid crystals can be viewed as twisted nematics. The rigid component of the degeneracy parameter is the director, the soft component is the pair of twist axis and binormal. There exists, however, no subgroup relation between the little group H_a of the director, and H_b of the cholesteric state[7]. For such a situation one chooses $H_1=H_a$ as the group of the rigid component, and $H_2=H_a\cap H_b$ as the maximal common subgroup of the two phases. Then one arrives at the same semidefect structure as for the biaxial nematic liquid crystals; twist axis and binormal play the part of the biaxes.

In the cases discussed above, $A=H_1/H_2$ denotes the subset $A\subset V$ of energetically favored positions within the space of degeneracy $V=G/H_2$. Since $\pi_r(G/H_1)=\pi_r(G/H_2,H_1/H_2)=\pi_r(V,A)$, one can write the exact sequence as

$$\ldots \rightarrow \pi_r(V) \rightarrow \pi_r(V,A) \rightarrow \pi_{r-1}(A) \rightarrow \pi_{r-1}(V) \rightarrow \ldots$$

This form of the sequence allows to define semidefects without introducing the rigid component or representing V and A as coset spaces. In ref. 7, out of the space of degeneracy V of a crystal a set of "easy" positions A has been selected by energy considerations. The labeling set $\pi_o(A)$ for semidefect walls contains classes, which describe stacking faults. Their boundaries are partial dislocations, labeled by elements of $\pi_1(V,A)$.

3. DEFECT CLASSIFICATION FOR SYSTEMS OF BROKEN TRANSLATIONAL SYMMETRY

3.1 Space of degeneracy

It is not straightforward to deduce the space of degeneracy for crystals, rod lattices, or layered systems from a Landau theory of melting and crystallization. In the homotopic defect classification[2-4] a short-cut was taken, and the space of degeneracy was at once represented by a coset space V=G/H, in analogy to the case of anisotropic fluids. As unbroken symmetry G the Euclidean group E(d) in d-dimensions was chosen, as broken symmetry the space group H of the system. To determine the degeneracy parameter at a point p of the distorted crystal, one has to adjust a uniform crystal tangentially to the deformed lattice planes. The motion, which brings a spatially fixed reference crystal into coincidence with the "tangential" crystal, is the degeneracy parameter at p (modulo H).

The fundamental groups of V=E(d)/H yield all the well-known line singularities in crystals: dislocations, disclinations, and dispirations. The group operations also confirm the results of defect processes, for example the equivalence of a separated disclination - antidisclination pair with a dislocation. Furthermore, singularities in curved space tessellations, which are used for models of amorphous materials, are described correctly by the homotopy groups (for polytopes on S^3 one takes V=SO(4)/H, where H designates the polytope symmetry group).

Fixing the degeneracy parameter, however, requires knowledge of the crystal structure not only at the point p, but in an entire neighborhood of p. Therefore nonlocality conditions are present - not yet completely specified -, which might invalidate the homotopy analysis. Mermin[2)] first argued, that defect classes, which have been predicted by homotopy theory, might not be realizable, or that singularities of the same class might not be continuously deformable one into the other, if the nonlocality conditions have to be taken into account. For line singularities in crystals Mermin's conjectures have not found a proof yet. However, for pointlike topological solitions, for which the results of homotopy theory have first been questioned by Gunn and Ma[10)], a modified classification has been presented in ref. 11. It materializes all the aspects of Mermin's critique and therefore is outlined in section 3.2. Further attempts to find a consistent defect classification for systems of broken translational symmetry have led to a labeling scheme for point singularities in layered systems, that is introduced in the final section 3.3.

3.2 Configurations in crystals

Pointlike topological solitons, also termed "configurations", are nonsingular fields, valued in the space of degeneracy, which are constant far away from a point. According to section 1.2.2, the labeling homotopy groups in three-space are $\pi_3(V)$. For crystals one obtains:

$$\pi_3(E(3)/H) = \pi_3(E(3)) = \pi_3(SO(3)) = \mathbf{Z} .$$

Only the rotational part SO(3) of E(3) is responsible for the stability of configurations. Therefore we can retreat to a continuum model, which registers only the primitive lattice vectors of the deformed crystal. The lattice constant approaches zero. The space of degeneracy is replaced by $GL_+(3,\mathbb{R})$, but since SO(3) is a subgroup of $GL_+(3,\mathbb{R})$, the equality $\pi_3(GL_+(3,\mathbb{R}))=\mathbb{Z}$ holds further, and the statements about stability and types of configurations remain unchanged. Accordingly, the abstract, infinite-dimensional space $\mathcal{K}$ of mappings $(\mathbb{R}^3,\infty)\to(GL_+(3,\mathbb{R}),1)$ divides into islands (connected components), each corresponding to an element of $\pi_0(\mathcal{K})=\pi_3(GL_+(3,\mathbb{R}))$, and hence labeled by an integer. Mappings on one and the same island are continuous deformable one into the other (i.e. connected by a continuous path of nonsingular mappings obeying the boundary condition). Configurations on the island numbered zero can decay into the uniform ground state (fig. 9a).

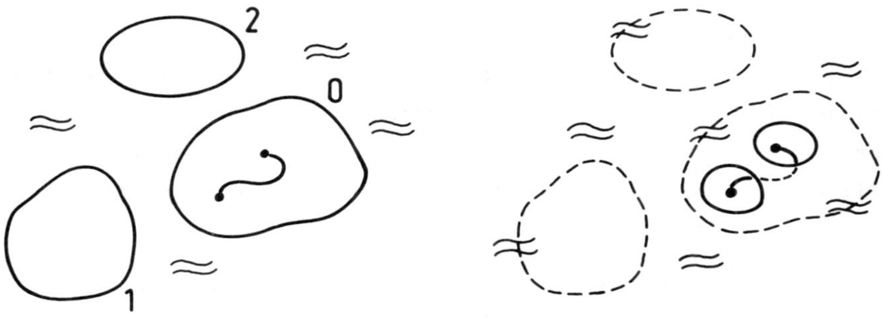

Fig. 9 Configuration space of pointlike topological solitons in crystals:
a) without integrability conditions: $\mathcal{K}$;
b) subject to integrability conditions: $\mathcal{H}$.

By definition, configurations are nonsingular, hence also free of dislocations. From elementary arguments of the continuum theory of dislocations[12] follows, that for each tripod of deformed lattice vectors

$[\underline{a}_1,\underline{a}_2,\underline{a}_3]$, the reciprocal tripod $[\underline{b}_1,\underline{b}_2,\underline{b}_3]$, $\underline{b}_i\cdot\underline{a}_j=\delta_{ij}$, must be integrable, i.e. derivable from three scalar functions f_i, $i=1,2,3$, such that $\underline{b}_i=\nabla f_i$. The integrability conditions (which are the nonlocality conditions) thus reduce the space of allowed configurations from $\mathcal{K}$ to a subspace $\mathcal{H}\subset\mathcal{K}$. They work like raising the sea level (fig. 9b): all the nonintegrable tripod fields are flooded. Only the integrable ones, sitting on the highest mountains, survive and form new islands, labeled by elements of $\pi_o(\mathcal{H})$. Some islands completely disappear, other split into subislands. Continuous paths between two tripod fields are swamped, so that configurations, which have been homotopic previously, become inequivalent under the integrability conditions.

The integrals f_i of the fields $\underline{b}_i$ of reciprocal lattice vectors form a diffeomorphism $\underline{f}:\mathbb{R}^3\to\mathbb{R}^3$ of three-space. Due to the boundary condition $\underline{f}(x\to\infty)\equiv x$, it can be transformed into a diffeomorphism of the three-sphere by sterografic projection. In general, for d-dimensional crystals, $\mathcal{H}^d$ equals the group of diffeomorphisms of S^d. The groups $\pi_o(\mathcal{H}^d)$, whose elements label the subislands of fig. 9b, have been extracted from the mathematical literature in ref. 11. Three-dimensional configurations are unstable, in contrast to the results of homotopy theory, since $\pi_o(\mathcal{H}^3)=0$. In six dimensions, however, there are 27 stable configurations, because $\pi_o(\mathcal{H}^6)=\mathbb{Z}_{28}$, although homotopy theory predicts none: $\pi_6(GL_+(6,\mathbb{R}))=0$.

The question, which island of $\mathcal{H}$ is sitting on which island of $\mathcal{K}$, is answered by the homomorphism

$$\varphi:\quad \pi_o(\mathcal{H}^d)\longrightarrow\pi_o(\mathcal{K}^d)=\pi_d(GL_+(d,\mathbb{R}))\ .$$

According to Dusa McDuff (private communication), $\varphi(\alpha)=0$ for all elements $\alpha\in\pi_o(\mathcal{H}^d)$. Hence no nontrivial homotopy class is realizable by an integrable d-pod field (all but the zero island perish). Stability of configurations is achieved only, if the trivial homotopy class splits into subclasses, which is the case for example in six dimensions.

3.3 Refined classification of point singularities in layered systems

It has been demonstrated in the preceding section, that for systems of broken translational symmetry the connectivity properties of the configuration space are not provided simply by the connectivity properties of the space of degeneracy. Therefore it is probably not possible to establish a general recipe for the classification of singularities, and each system requires its own methods. In the following it is sketched how one has to proceed to classify point singularities in layered systems of variable layer separation. The proofs have been worked out by Christoph Maier and shall be presented in a forthcoming publication.

For simplicity we assume the layers to be oriented, i.e. up and down side can be distinguished. The space of degeneracy is $V=E(3)/\{(T(2)\times\mathbb{Z})\wedge C_{\infty v}\}$. The second homotopy group $\pi_2(V)$ is isomorphic to the group $\mathbb{Z}$ of integers. Therefore homotopy theory predicts the existence of stable point singularities, labeled by integer homotopy indices.

Now a model description for defected layers is introduced, a defect equivalence is defined and the corresponding classes are established. The result makes apparent, that each homotopy class splits into an infinity of subclasses.

3.3.1 <u>Model.</u> Layers are "equipotential surfaces" f^{-1}(const), where the "potential"

$$f:\ U\to\mathbb{R}$$

is a differentiable function defined in a neighborhood U of $0\in\mathbb{R}^3$ such that

i) $|\nabla f|<\infty$ (layers do not approach to arbitrarily small distances in U);

ii) $\nabla f(x)=0$ only for $x=0$ (there is only one isolated singularity, placed at the origin);

iii) f is radially separable, which means, that in local coordinates in a neighborhood of the origin f can be separated into a radial and an angular part:

$$f(r,\Omega) = \psi(r)g(\Omega),$$

$$\psi\colon [0,\varepsilon]\to \mathbb{R}_+, \quad g\colon S^2\to \mathbb{R} .$$

The description of a layered system by a scalar function is necessary and sufficient to satisfy the integrability conditions. These forbid layers to terminate in U thus forming dislocation lines.

3.3.2 Equivalence. We say that a layer system f is replaceable by a layer system $\tilde{f}$, if a differentiable function F exists having the following properties:

i) $F = \tilde{f}$ for $r < \varepsilon$,

ii) $F = f$ for $r > \delta > \varepsilon$,

iii) F is nonsingular ($\nabla F \neq 0$) in the shell $\varepsilon < r < \delta$.

ε and δ are radii of spheres contained in U. System f is denoted equivalent to $\tilde{f}$, if f is replaceable by $\tilde{f}$, and $\tilde{f}$ is replaceable by f. Defects are in the same class, if the core of one can be replaced by the core of the other due to a local fluctuation.

3.3.3 Label for a point singularity. It is assumed that the potentials are gauged such that f(x=0)=0. Consider, then, the set of points $f^{-1}(0)\cap S^2$. The layer, which contains the origin, cuts these points out of a small sphere $S^2\subset U$ centered at the singularity. The points form several nonintersecting, disjoint circles $K_1,\ldots,K_m$. Circle K_i is oriented such that f is positive to the right, and negative to the left of K_i (fig. 10).
The classification is then performed by the following

3.3.4 Theorem. Given a layer system f, characterized by a set $K_1,\ldots,K_m$ of oriented circles and designated $f(K_1,\ldots,K_m)$, and a layer system $\tilde{f}(\tilde{K}_1,\ldots,\tilde{K}_{\tilde{m}})$. Then f is equivalent to $\tilde{f}$ if and only if $m=\tilde{m}$ and

i) for $m = \tilde{m} = 0$:

f and $\tilde{f}$ are either both positive or both negative on S^2;

ii) for $m = \tilde{m} > 0$ (after suitable labeling of the circles): K_j is to the right of K_i if and only if $\tilde{K}_j$ is to the right of $\tilde{K}_i$.

Thus the class of the singularity is specified by the topology of the intersection, which the defected layer forms with a small sphere.

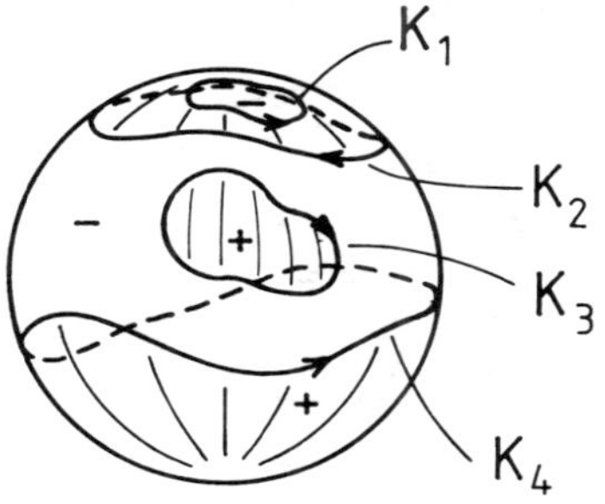

Fig. 10 System of oriented circles $K_1,\ldots,K_m$ characterizing a layer system $f(K_1,\ldots,K_m)$.

3.3.5 Comparison with the homotopic defect classification. The homotopy index n of the layer system can be read from the circle system in the following way: denote the number of sections on the sphere where f is positive by m_+, the number of sections where f is negative by m_-. It is not proven here, but easy to check for simple systems, that $n=m_+-m_-$. Fig. 11a displays the circle system for a sink: $m_+=0, m_-=1$, $n=-1$. The circle system of an inequivalent singularity having the same homotopy index is constructed by adding in a negative section a positive component, which encloses a negative one. Thus m_+ and m_- are increased by 1, but n is not changed. In this way one obtains from fig. 11a the circle system of fig. 11b, and from fig. 11b that of fig. 11c. Each homotopy class contains an infinity of subclasses, and they might become still more, if by lifting restrictions i) and iii) of section 3.3.1 the number of allowed layer systems is increased.

Smectic-A layers cut the circle system of fig. 11a out of a large sphere centered about the core of a focal conic domain. For a sphere close to the core, in the toric region, one obtains the circles of fig. 11b. In the sense of the refined classification, the two patterns label inequivalent singularities. Therefore the layers in between the two spheres have to be defected. Nature develops the inevitable singularity in the form of an ellipsoid, which sticks to the covering glass. These ellipsoids compose the focal conic texture, which is the typical appearence of smectic-A liquid crystals.

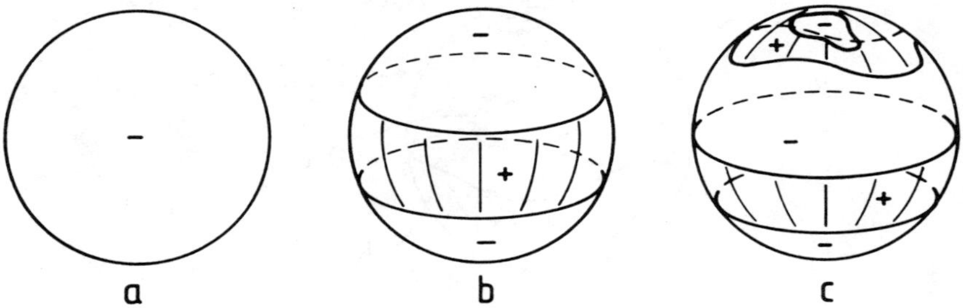

Fig. 11 Circle systems for three different point singularities of same homotopy index n=-1.

References

1. De Gennes, P.G., "The Physics of Liquid Crystals", Clarendon Press, Oxford (1975).
2. Mermin, N.D., Rev. Mod. Phys. 51, 591 (1979).
3. Michel, L., Rev. Mod. Phys. 52, 617 (1980).
4. Trebin, H.-R., Adv. Phys. 31, 195 (1982).
5. Volovik, G.E., Pis'ma Zh. Eksp. Teor. Fiz. 28, 65 (1978) [JETP Lett. 28, 59 (1978)].
6. Mermin, N.D., Mineyev, V.P., and Volovik, G.E., J. Low Temp. Phys. 33, 117 (1978).
7. Kutka, R. and Trebin, H.-R., J. Physique Lett. 45, 1119 (1984). An extensive article is in preparation.
8. Yu, L.J. and Saupe, A., Phys. Rev. Lett. 45, 1000 (1980).
9. Steenrod, N., "The Topology of Fiber Bundles", Princeton University Press (1951).

10. Gunn, J.M.F. and Ma, K.B., J. Phys. C 13, 963 (1980).
11. Trebin, H.-R., Phys. Rev. Lett. 50, 1381 (1983).
12. Kröner, E., Arch. Ration. Mech. Anal. 4, 18 (1960).

* Present address: Institut für Theoretische und Angewandte Physik der Universität Stuttgart, Pfaffenwaldring 57, D-7000 Stuttgart 80, F.R.G.

REPRESENTATIONS OF SYMMETRIC AND OF GENERAL LINEAR GROUPS: TABLES, PROBLEMS, TRENDS AND APPLICATIONS*

Adalbert Kerber

Universität Bayreuth, Lehrstuhl II für Mathematik,
Postfach 3008, D-8580 Bayreuth
GERMANY

ABSTRACT

Recent and forthcomming numerical results on representations of symmetric and general linear groups are described, related open problems and a few applications are mentioned. A result on the enumeration of strata is given.

It is the intention of my talk to describe what my co-workers and I have recently done, are doing and will do in the near future concerning numerical information on ordinary representations of finite symmetric groups and on the ordinary irreducible polynomial representations of the general linear groups over the complex field. Some of the open problems in this field will be mentioned. It will be pointed to a new trend in the representation theory of symmetric groups where tableaux are replaced by polynomials (as it is already known to physicists in the unitary group case, where the boson polynomials do the job). A few hints to recent applications will be given, and finally I shall mention a method which allows to evaluate the length of strata which have shown up in several of the foregoing talks, in particular in the talks of L. Michel.

* Partially supported by the Deutsche Forschungsgemeinschaft (Ke 201/9-1)

1. NOTATION

Let S_n denote the symmetric group on the set $\underline{n} := \{1,\ldots,n\}$. A *partition* of n is a sequence $\lambda = (\lambda_1,\lambda_2,\ldots)$ of natural numbers λ_i which satisfy $\Sigma\,\lambda_i = n$ and $\lambda_i \geqslant \lambda_{i+1}$, for all i. We shall indicate this by writing

$$\lambda \vdash n.$$

Each such sequence λ gives rise to a *Young subgroup*

$$S_\lambda := S_{\{1,\ldots,\lambda_1\}} \oplus S_{\{\lambda_1+1,\ldots,\lambda_1+\lambda_2\}} \oplus \ldots$$

consisting of the elements $\pi \in S_n$ which keep all the subsets $\{\sum_1^{j-1} \lambda_i+1,\ldots, \sum_1^{j} \lambda_i\}$ fixed.

S_λ has two trivial representations, the *identity representation* IS_λ and the *alternating representation* AS_λ, which maps π onto its sign:

$$IS_\lambda : \pi \to 1,\ AS_\lambda : \pi \to \mathrm{sgn}\,\pi.$$

IS_λ induces the *Young representation* $IS_\lambda{\uparrow}S_n$ of S_n, its character is called the *Young character* corresponding to λ and it is denoted by

$$\xi^\lambda .$$

If λ is a partition of n, it can be visualized by its *Young diagram*, for example the partition $(3,2,1^2) := (3,2,1,1,0,\ldots)$ corresponds to the diagram

```
x x x
x x
x
x
```

The lengths of the columns of the diagram corresponding to λ form the *associated partition* λ', e.g. $(3,2,1^2)' = (4,2,1)$. Using this notation we can now formulate the basic theorem on the representation theory of symmetric groups which makes use of the fact that $IS_\lambda{\uparrow}S_n$ and $AS_{\lambda'}{\uparrow}S_n$ have exactly one irreducible constituent in common and they both contain it with multiplicity 1. Denoting this common constituent by $IS_{\lambda}\uparrow S_n \cap AS_{\lambda'}\uparrow S_n$, we have the well known

1.1 Fundamental Theorem:

i) $\{[\lambda] := IS_{\lambda}\uparrow S_n \cap AS_{\lambda'}\uparrow S_n \mid \lambda\vdash n\}$ *is a complete system of ordinary irreducible representations of* S_n.

ii) $\{\xi^{\lambda} \mid \lambda\vdash n\}$ *is a* $\mathbb{Z}$ *-basis of the ring of characters of* S_n.

The character of $[\lambda]$ will be denoted by

$$\zeta^{\lambda} .$$

2. TABLES AND PROBLEMS

Let me first shortly mention the tables published in my joint book with G.D. James[1].

2.1 Character tables:

Tables containing the values of the ordinary irreducible characters of S_n can be found for $n \leqslant 10$. Such tables (and further ones) can also be found elsewhere. As there are many recursive and direct methods available for their evaluation, there is no problem. The corresponding tables for the alternating groups follow directly. But there are also tables of characters for the wreath products $S_m \wr S_n$, $mn \leqslant 10$. Here are still open problems as there are no such nice formulae available as in the symmetric group case.

2.2 Representing matrices:

The matrices representing the transposition (12) and the cycles (1...n) and (n...1) are given for $n \leqslant 7$ and each $[\lambda]$, so that in principle the matrix for each $\pi \in S_n$ can be evaluated. The chosen form is the rational integral one and in chapter 3 of the book there can be found transformation matrices which yield the seminormal or the orthogonal form which are the ones a physicist may prefer. The algorithm is not very complicated[2,3].

2.3 Inner tensor products:

The decompositions of the representations $[\lambda] \otimes [\mu]$, $\lambda \vdash n$, $\mu \vdash n$, of S_n are given, for $n \leqslant 8$. Such tables can be also found elswhere, they are easy to compute, but explicit formulae are still missing. *This is one of the main problems in ordinary representation theory of* S_n *which is open.*

2.4 Induction of outer tensor products:

As there is a natural embedding of the direct product $S_m \times S_n$ into S_{m+n} it is natural to ask for the decomposition of the corresponding induced representation

$$[\lambda][\mu] := [\lambda]\#[\mu]\uparrow S_{m+n},$$

for $\lambda \vdash m$, $\mu \vdash n$. Since this is covered by the famous Littlewood-Richardson-rule, there is no problem.
The tables given are for $m+n \leqslant 9$.

2.5 Symmetrization and permutrization:

If V denotes a vector space which affords $[\lambda]$, $\lambda \vdash m$, then the tensor power $\overset{n}{\otimes} V_\lambda$ is in a natural way an (S_m, S_n)-bimodule, i.e. S_m acts from the left, S_n from the right, and these actions commute. This implies (by Schur's lemma):

(i) If the representation of S_n which is afforded by $\overset{n}{\otimes} V_\lambda$ contains the irreducible constituent $[\mu]$ (of dimension f^μ), $\mu \vdash n$, n_μ-times, then the corresponding subspace of dimension $n_\mu f^\mu$ affords f^μ-times a representation of S_m, which is denoted by

$$[\lambda] \boxdot [\mu]$$

and called the *symmetrization of* $[\lambda]$ by $[\mu]$. Its dimension is n_μ.

(ii) If the representation $\otimes[\lambda]$ of S_m which is afforded by $\otimes V_\lambda$ contains the irreducible constituent $[\nu]$, $\nu \vdash m$, with multiplicity d_ν, then the corresponding subspace of dimension $d_\nu f^\nu$ affords f^ν-times a representation of S_n, which is denoted by

$$[\lambda] \mathbin{\triangle}_n [\nu]$$

and called the n-fold *permutrization* of $[\lambda]$ by $[\nu]$. Its dimension is d_μ.

(iii) The multiplicities of the irreducible constituents of symmetrizations and permutrizations obey the following important duality theorem[4]:

$$([\lambda] \boxdot [\mu],[\nu]) = ([\lambda] \mathbin{\triangle}_n [\nu],[\mu]),$$

for each $\lambda \vdash m$, $\mu \vdash n$, $\nu \vdash m$. Hence these multiplicities can be put into matrices such that the rows belong to symmetrizations, the columns to permutrizations. The tables show the cases $m \leqslant 8$, $n \leqslant 5$. But there are no explicit formulae for such multiplicities known and therefore *this is another case of an important and open problem.*

The fact that symmetrization can be carried out is the main reason for the great success of representation theory of symmetric groups in physics. The less well known procedure of permutrization of representations can be used to elucidate the so-called 3j-symbols.

2.6 Missing tables:

There exist natural embeddings of $S_m \wr S_n$ into both S_{mn} and S_{m^n}. To $\lambda \vdash m$ and $\mu \vdash n$ there corresponds an irreducible representation $(\lambda;\mu)$ of $S_m \wr S_n$ (see chapter 4 of the book) and hence there arises the question for the decompositions of

(i) the *plethysm* $[\lambda] \odot [\mu] := (\lambda;\mu)\uparrow S_{mn}$, and of
(ii) the *exponentiation* $[\lambda] \triangledown [\mu] := (\lambda;\mu)\uparrow S_{m^n}$.
There is very little known about these[5].

After the publication of these tables further ones were computed and published in a joint paper with K.-J. Thürlings[6]:

2.7 Character polynomials:

For each $\lambda \vdash n$ there exists a polynomial[7] $p(x_1,\ldots,x_n)$ such that, if a_i, $1 \leqslant i \leqslant n$, denotes the number of cyclic factors of length i of π, we have for the character ζ^λ of $[\lambda]$:

$$\zeta^\lambda(\pi) = p(a_1,\ldots,a_n).$$

But the main point is that this polynomial is independent of λ_1, the length of the first row of $[\lambda]$, i.e. p yields also the character of $[\lambda_1+k,\lambda_2,\ldots]$ and hence an *infinite* number of irreducible characters. We published 97 of these polynomials which cover the character tables of S_n, $n \leqslant 13$.

2.8 Young polynomials:

These are the corresponding polynomials for Young characters ξ^λ.

2.9 Young characters and their decompositions:

They are also shown for S_n, $n \leqslant 10$, and helpful for combinatorial purposes. For example the inner product $\langle\xi^\lambda,\xi^\mu\rangle$ is equal to the number of matrices with nonnegative integral entries and row sums λ_i and column sums μ_j.

2.10 Foulkes characters:

H.O. Foulkes discovered[8] that there exist certain characters

$$\chi^{n,i},\ 0 \leqslant i \leqslant n-1$$

of S_n which have the following remarkable properties:

(i) $\sum_i \chi^{n,i}$ is the regular character, i.e. for each $\pi \in S_n$ we have

$$\sum_i \chi^{n,i}(\pi) = \begin{cases} n!, & \text{if } \pi = 1 \\ 0, & \text{otherwise.} \end{cases}$$

(ii) Each character χ of S_n, the value of which on $\pi \in S_n$ does only depend on the number of cyclic factors of π, is a unique $\mathbb{Q}$-linear combination of the $\chi^{n,i}$.

These characters are tabulated for $n \leqslant 10$. They obey easy recursion relations. Recent applications to the invariant quantization of the relativistic free string (K.-H. Rehren) have shown that these characters are of interest in physics, too.

3. THE POLYNOMIAL REPRESENTATIONS OF THE GENERAL LINEAR GROUPS

Having described the tables published in 1981 and 1983 on the ordinary representations of S_n (I left out the tables on the modular representations since I do not know of applications to physics yet) let me describe what we are doing at the moment in this field.

Our next aim is the evaluation of the polynomials for the irreducible polynomial representations of the general linear groups.

The famous discovery of I. Schur[9,10] was that to each partition $\mu \vdash n$ there corresponds exactly one equivalence class $\langle\mu\rangle$ of ordinary irreducible polynomial representations of the general linear group $GL(m,\mathbb{C})$. That

$\langle\mu\rangle$ is polynomial means that the elements of the matrix $\langle\mu\rangle(x_{ik})$ representing the element $(x_{ik}) \in GL(m,\mathbb{C})$ are polynomials in the x_{ik}, in fact they are homogeneous of degree n.

These representations $\langle\mu\rangle$ are in fact symmetrizations as described in 2.5: Replacing $[\lambda]$ of S_m by the identity map $id:(x_{ik}) \to (x_{ik})$ which obviously is a representation of $GL(m,\mathbb{C})$, one obtains

3.1 $$\langle\mu\rangle = id \boxdot [\mu].$$

D.E. Littlewood described a method which yields the representing matrices[11,12], a few examples may illustrate it, further ones will be published in due course.

Recall that the dimension of $\langle\mu\rangle$ of $GL(m,\mathbb{C})$ is equal to[1]:

$$\frac{f^\mu}{n!} \prod_{1 \leqslant i \leqslant \mu_1'} \prod_{1 \leqslant j \leqslant \mu_i} (m-i+j).$$

Thus

(i) The irreducible polynomial representations of $GL(1,\mathbb{C})$ and of the form $\langle n\rangle$, $n \in \mathbb{N}$, and the matrix representing $(x_{11}) \in GL(1,\mathbb{C})$ is

$$\langle n\rangle(x_{11}) = (x_{11}^n),$$

for each n.

(ii) The irreducible polynomial representations of $GL(2,\mathbb{C})$ correspond to the partitions λ, which have at most two summands $\lambda_i > 0$. The first ones are

a) $$\langle 0\rangle\begin{pmatrix} x_{11} & x_{12} \\ x_{21} & x_{22} \end{pmatrix} = (1).$$

b) $$\langle 1\rangle\begin{pmatrix} x_{11} & x_{12} \\ x_{21} & x_{22} \end{pmatrix} = \begin{pmatrix} x_{11} & x_{12} \\ x_{21} & x_{22} \end{pmatrix}.$$

c) $$\langle 2\rangle(x_{ik}) = \begin{bmatrix} x_{11}^2 & x_{11}x_{12} & x_{12}^2 \\ 2x_{11}x_{21} & x_{12}x_{21}+x_{11}x_{22} & 2x_{12}x_{22} \\ x_{21}^2 & x_{21}x_{12} & x_{22}^2 \end{bmatrix}$$

d) $$\langle 1^2\rangle(x_{ik}) = (x_{11}x_{22}-x_{12}x_{21}) = \det(x_{ik}).$$

e) $$\langle 3\rangle(x_{ik}) = \begin{bmatrix} x_{11}^3 & x_{11}^2x_{12} & x_{11}x_{12}^2 & x_{12}^3 \\ 3x_{11}^2x_{21} & 2x_{11}x_{12}x_{21}+x_{11}^2x_{22} & 2x_{11}x_{12}x_{22}+x_{12}^2x_{21} & 3x_{12}^2x_{22} \\ 3x_{11}x_{21}^2 & 2x_{11}x_{21}x_{22}+x_{12}x_{21}^2 & x_{11}x_{22}^2+2x_{12}x_{21}x_{22} & 3x_{12}x_{22}^2 \\ x_{21}^3 & x_{2}^2x_{22} & x_{21}x_{22}^2 & x_{22}^3 \end{bmatrix}$$

and so on.

(iii) The irreducible polynomial representations of $GL(3,\mathbb{C})$ correspond to the partitions λ, where $\lambda_1' \leqslant 3$. The first ones are

a) $$\langle 0\rangle(x_{ik}) = (1)$$

b) $$\langle 1\rangle(x_{ik}) = (x_{ik}).$$

c) $$\langle 1^2\rangle(x_{ik}) = \begin{bmatrix} -x_{12}x_{21}+x_{11}x_{12} & -x_{13}x_{21}+x_{11}x_{13} & -x_{13}x_{22}+x_{12}x_{23} \\ -x_{12}x_{31}+x_{11}x_{32} & -x_{13}x_{31}+x_{11}x_{33} & -x_{13}x_{32}+x_{12}x_{33} \\ -x_{22}x_{31}+x_{21}x_{32} & -x_{23}x_{31}+x_{21}x_{33} & -x_{23}x_{32}+x_{22}x_{33} \end{bmatrix}$$

and so on.

4. LETTER-PLACE ALGEBRAS

The numerical results mentioned above were obtained with the aid of computer programs using classical methods from ordinary representation theory of symmetric groups. These methods although being quite explicit in many cases have a certain disadvantage which has recently been overcome by a new approach which I would like to recommend also for applications to sciences. The classical approach to representation theory uses bases of the irreducible representation spaces consisting of standard Young tableaux as you all know. These tableaux are not difficult to handle but if one really wants to know what they "are", one finds out that they come in along a recursive counting argument concerning the dimension of ordinary irreducible representations of S_n. Thus one either has to agree that they are of only a formal nature or one has to consider them as chains in *Young's lattice* of diagrams. E.g. the five standard Young tableaux

```
123    124    125    134    135
45  ,  35  ,  34  ,  25  ,  24
```

forming a basis of [3,2] can be identified with the five chains leading from the empty diagram Ø to the diagram xxx/xx as it is indicated below using double bars:

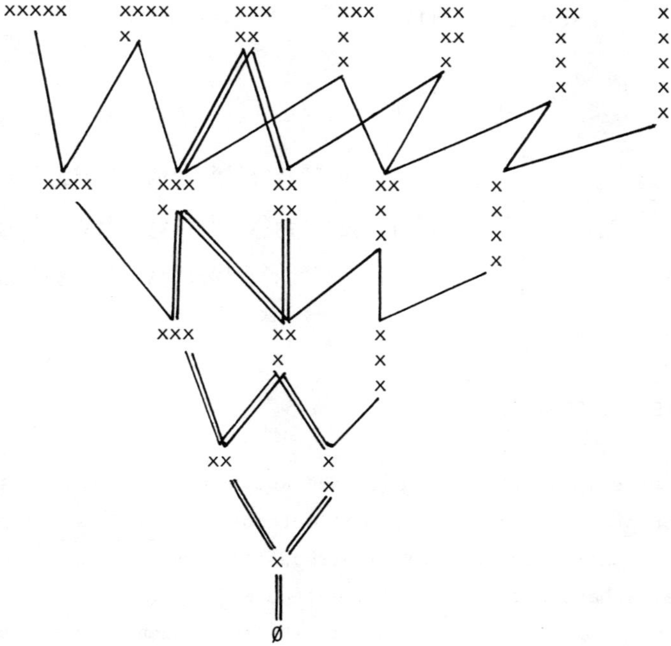

The new approach mentioned above is due to M. Clausen[13,14] and it is based on a very important theorem due to Rota et al.[15], derived for invariant theoretic purposes. It associates with each pair of tableaux of same shape λ (which may contain repetitions and which also may be nonstandard) a certain polynomial a so-called *bideterminant*. And Rota's theorem says that and how it can be written as a linear combination of *standard bideterminants*, i.e. bideterminants belonging to pairs of standard tableaux (i.e. strictly increasing along the rows and weakly along the columns). *Thus we have a basis consisting of polynomials associated with standard tableaux*. These polynomials are polynomials in doubly indexed indeterminates y_{ik}, the groups act on the indices. Let me briefly describe this, for details see the cited references.

Take a pair (S,T) of λ-tableaux with elements $s_{ij} \in \underline{m}$, $t_{ij} \in \underline{n}$, say

$$(S,T) = \begin{pmatrix} s_{11} \ldots\ldots s_{1\lambda_1} & t_{11} \ldots\ldots t_{1\lambda_1} \\ s_{21} \ldots s_{2\lambda_2} & , \ t_{21} \ldots t_{2\lambda_2} \\ \ldots & \ldots \end{pmatrix}$$

and form the corresponding *bideterminant* (in the indeterminates $y_{u,v}$, $1 \leqslant u \leqslant m$, $1 \leqslant v \leqslant n$):

$$(S|T) := \prod_i (s_{i1} \ldots s_{i\lambda_i} \mid t_{i1} \ldots t_{i\lambda_i}),$$

where

$$(s_{i1} \ldots s_{i\lambda_i} \mid t_{i1} \ldots t_{i\lambda_i}) := \begin{cases} 1, \text{ if } \lambda_i = 0 \\ \det(y_{s_{ij},t_{ik}}), \text{ if } \lambda_i > 0. \end{cases}$$

For example

$$\left(\begin{array}{c|c} 34 & 56 \\ 11 & 78 \\ 2 & 9 \end{array}\right) = \det\begin{bmatrix} y_{35} & y_{36} \\ y_{45} & y_{46} \end{bmatrix} \det\begin{bmatrix} y_{17} & y_{18} \\ y_{17} & y_{18} \end{bmatrix} \det(y_{29})$$

$$= (y_{35}y_{46}-y_{45}y_{36})(y_{17}y_{18}-y_{17}y_{18})y_{29}$$

$$= 0.$$

I cannot go into details since this would need very many pages. Let me just mention that bases arising from standard bideterminants for the irreducible representation spaces of both S_n and $GL(m,\mathbb{C})$ are known. Furthermore it is astonishing to see how easy very important theorems like Young's rule and the Littlewood-Richardson rule can be derived along these lines. Moreover many results can be derived without assumptions on the ground field. Standard bideterminants were already used by J. Louck[16] et. al. for the irreducible representations of the unitary groups, and the letter-place-algebra approach can very well serve as a unification of his approach and the one by P. Kramer[17].

5. THE ENUMERATION OF STRATA

In several of the foregoing talks the notion of *stratum* was mentioned which means the union of all the orbits of a finite group Γ on a set S the elements of which have conjugate stabilizer subgroups. Let me briefly describe a certain situation which we have studied in detail and where we know how to enumerate strata. (A physical example of this situation occurs in the talk by T. Lulek.)

5.1 The general Ansatz:

Choose suitabe finite sets X,Y and form

$$Y^X := \{f:X \to Y\}.$$

A group G which already acts on X then operates in a natural way also on Y^X:

$$g:f \to f\circ g^{-1} \text{ (composition)}$$

Many structures in mathematics and physics can be defined as orbits of such actions (see the talk by T. Lulek).

5.2 Example: graphs on n points

Take $X := \{1,\dots,\binom{n}{2}\}$, the set of *pairs of points*, and put $Y := \{0,1\}$. Then $f \in Y^X$ is a *labelled graph*. The symmetric group S_n acts on the points, hence also on the pairs of points and therefore on Y^X as it is described in 5.1. A *graph* on n points is an orbit of S_n on Y^X.

5.3 Problems

Questions which arise in connection with such an action of G on Y^X are among many others:

(i) What is the total number of orbits?

(ii) What is the number of orbits with given weight (e.g. graphs with given number of edges)?

(iii) What is the number of orbits in a given *stratum* (i.e. the stabilizers of elements lie in a certain class of conjugate subgroups of G)?

(iv) Construct a transversal of the orbits.

The original problem which gave rise to this part of combinatorial theory of enumeration stems from chemistry: how many isomers exist for a given gross formula? There is still no satisfactory solution know! Let me just describe the solution of 5.3 (iii):

5,4 The length of strata

Theorem: *Let Γ denote a finite group acting on a finite set S, let $\tilde{U}_1,\dots,\tilde{U}_d$ denote the conjugacy classes of subgroups of Γ, choose $U_i \in \tilde{U}_i$, and let $w: S \to R$ be a mapping from S into a commutative ring containing $\mathbb{Q}$, that is constant on the orbits ω_ν, w_ν the value there. If $s_\nu \in \omega_\nu$ is a representative of the orbit ω_ν while Γ_s denotes the stabilizer of s, $N_\Gamma(\Delta)$ the normalizer of the subgroup Δ and μ the Möbius function on the subgroup lattice, then we have*

$$\sum_{\substack{\nu \\ \Gamma_{s_\nu} \in \tilde{U}_i}} w_\nu = \frac{|U_i|}{|N_\Gamma(U_i)|} \sum_{V \leqslant \Gamma} \mu(U_i, V) \sum_{\substack{s \\ V \leqslant \Gamma_s}} w(s).$$

In case $\Gamma := G$ acts on $S := Y^X$ we obtain (by putting $w: s \to 1$):

Corollary:

The numbers n_i *of orbits of* G *on* Y^X *which are of type* $\tilde{U}_i$, *i.e. the lengths of the strata of types* $\tilde{U}_i$, *satisfy the equation*

$$\begin{bmatrix} \vdots \\ n_i \\ \vdots \end{bmatrix} = {}^t\Omega(G)^{-1} \begin{bmatrix} \vdots \\ |Y|^{|X/U_i|} \\ \vdots \end{bmatrix}$$

if $|X/U_i|$ *denotes the number of orbits of* U_i *on* X, *and* $\Omega(G) = (\omega_{ik})$ *is the table of marks of* G, *i.e.*

$$\omega_{ik} := \frac{|N_G(U_k)|}{|U_i|} \sum_{\tilde{U}_k \ni V \leqslant U_i} 1.$$

The table of marks of S_4 looks as follows[6]:

$$\begin{bmatrix}
24 \\
12 & 2 \\
12 & 0 & 4 \\
8 & 0 & 0 & 2 & & & & 0 \\
6 & 0 & 2 & 0 & 2 \\
6 & 0 & 6 & 0 & 0 & 6 \\
6 & 2 & 2 & 0 & 0 & 0 & 2 \\
4 & 2 & 0 & 1 & 0 & 0 & 0 & 1 \\
3 & 1 & 3 & 0 & 1 & 3 & 1 & 0 & 1 \\
2 & 0 & 2 & 2 & 0 & 2 & 0 & 0 & 0 & 2 \\
1 & 1 & 1 & 1 & 1 & 1 & 1 & 1 & 1 & 1 & 1
\end{bmatrix}$$

Further tables will be published soon[18]

REFERENCES

1. G.D. James and A. Kerber: The representation theory of the symmetric group. Encyclopedia of Mathematics and its Applications, Vol. 16, Addison-Wesley 1981.
2. H. Boerner: Darstellungen von Gruppen mit Berücksichtigung der Bedürfnisse der modernen Physik. Springer-Verlag 1955.
3. J.M. Clifton: A simplification of the computation of the natural representation of the symmetric group S_n. Proc. Amer. Math. Soc. 83 (1981), no. 2, 248-250.
4. N. Esper: Die Symmetrisierungsmatrix. Dissertation, Aachen 1976.
5. F. Sänger: Plethysmen von irreduziblen Darstellungen symmetrischer Gruppen. Dissertation, Aachen 1980.
6. A. Kerber and K.-J. Thürlings: Symmetrieklassen von Funktionen und ihre Abzählungstheorie. Bayreuther Mathematische Schriften 12 (1983), 235 pp., 15 (1983), 338 pp.
7. W. Specht: Die Charaktere der symmetrischen Gruppe. Math. Z. 73 (1960), 312-329.
8. H.O. Foulkes: Eulerian numbers, Newcomb's problem and representations of symmetric groups. Discrete Mathematics 30 (1980), 3-49.
9. I. Schur: Über eine Klasse von Matrizen, die sich einer gegebenen Matrix zuordnen lassen. Dissertation, Berlin 1901.
10. I. Schur: Über die rationalen Darstellungen der allgemeinen linearen Gruppe. Berl. Ber. 1927, 58-75.
11. D.E. Littlewood: The construction of invariant matrices. Proc. London Math. Soc. (2) 43 (1937), 226.
12. D.E. Littlewood: The theory of group characters. At the Clarendon Press 1940.
13. M. Clausen: Letter-Place-Algebren und ein charakteristik-freier Zugang zur Darstellungstheorie symmetrischer und voller linearer Gruppen. Bayreuther Mathematische Schriften 4 (1980), 133 pp.
14. M. Clausen: Letter place algebras and a charakteristic-free approach to the representation theory of the general linear and symmetric groups. Advances in Math. 33 (1979), 161-191, 38 (1980), 152-177.

15. J. Désarménien and J.P.S. Kung and G.-C. Rota: Invariant theory, Young bitableaux and combinatorics. Advances in Mathematics 27 (1978), 63-92.

16. J.S. Louck: Recent Progress Toward a Theory of Tensor Operators in the Unitary Groups. Amer. J. Physics, Vol. 38 (1970), 3-42.

17. P. Kramer and G. John and D. Schenzle: Group Theory and the Interaction of Composite Nucleon Systems. Viehweg 1981, Braunschweig, viii + 224.

18. A. Kerber and K.-J. Thürlings: Symmetrieklassen von Funktionen und ihre Abzählungstheorie III. Bayreuther Mathematische Schriften (in preparation).

STUDIES IN PHASE TRANSITIONS AND SOLITONS

R. Chatterjee
Department of Physics
The University of Calgary
Calgary, Alberta, Canada
T2N 1N4

1. INTRODUCTION

Landau theory of phase transitions has been extensively applied to interpret various phenomena in all different branches of physics. In solid state physics, Landau theory is used by Cowley (1980), Blinc (1981) and many others, to provide the physical interpretation of order-disorder phase transitions due to the changes of the crystallographic structure of the material. These structural phase transitions are associated with an ordering of some variable which is disordered in the high temperature phase. The lower temperature phase which is the ordered phase belongs to a symmetry group G_o which is a subgroup of the high temperature symmetry group G. The ordered parameter of the structural phase transition may be determined experimentally in case of a single crystal by EPR, NMR technique as shown by Kabayashi (1973), Blinc et al. (1983).

Tuszynski, Paul and Chatterjee (1984) extended the Landau-Ginzburg model of phase transitions by assuming the order parameter as a complex time and space dependent function. The resulting evolution equation is one-dimensional nonlinear partial differential equation similar to nonlinear Schrodinger equation of third power of potential. They solved it exactly by assuming the simplest possible orbit following the method of Boyer, Sharp and Winternitz (1976). This type of nonlinear Schrodinger equation also appears in ease of biological systems such as membrane where phase transition takes place due to Bose condensation and the solution of which depends on different types of parameters as shown by Tuszynski, Paul, Chatterjee and Sreenivasan (1984). This nonlinear Schrodinger equation belongs to the class of nonlinear partial differential equations which are studied by numerous authors (cf. Ablowitz, Kaup, Newell and Segur 1974). Ablowitz, Segur (1977) showed further that this class of nonlinear evolution equations are related to Painlevé ordinary differential equations whose solutions are already known. They further (1978) provided a method of linearization of these nonlinear equations by the Painlevé transcendents.

In this paper a systematic technique of solving nonlinear Schrodinger equation of third power of potential will be provided following the similarity reduction method of Bluman and Cole (1974), Boyer, Sharp and Winternitz (1976) etc. by one parameter Lie group of transformations. It will also be shown that this equation can be reduced to Painlevé trancendental equations. Attention will be drawn to the significance of the parameters to the solutions in the form of soliton.

2. THEORY

Tuszynski <u>et al</u>. (1984, 1984) obtained a nonlinear partial differential equation of the ordered parameter η similar to nonlinear Schrodinger equation from two different models of phase transitions which is given as

$$i\frac{\partial\eta}{\partial t} + p\frac{\partial^2\eta}{\partial x^2} + r|\eta|^2\eta = 0 \qquad (1)$$

where the parameters p and r depends on the particular model chosen. In case of Landau-Ginzburg model of phase transitions, the parameter p is related to the dissipation constant and the parameter r can be expressed in terms of the coefficients of the order parameters of Landau's expansion. Similarly for the biological systems the model put forward by Frohlich and Davydov to interpret the phenomenon of Bose Condensation, these parameters p and r are related to the coupling coefficients of the phonons and the heat bath (Tuszynski <u>et al</u>. 1984). Equation (1) is similar to one dimensional Schrodinger equation (for cubic potential)

$$i\,\Psi_t + p\,\Psi_{xx} + r|\Psi|^2\Psi = 0 \tag{2}$$

where p and r are constants. Equation 2 can be reduced to ordinary differential equation by similarity transformation (Bluman and Cole 1974). In this method one applies one parameter (ε) Lie group of infinitesimal transformation in (x, t, Ψ) space

$$\begin{aligned} x' &= x + \varepsilon\,\xi(x, t, \Psi) + O(\varepsilon^2) \\ t' &= t + \varepsilon\,\tau(x, t, \Psi) + O(\varepsilon^2) \\ \Psi' &= \Psi + \varepsilon\,\eta(x, t, \Psi) + O(\varepsilon^2) \end{aligned} \tag{3}$$

using the invariant criteria of equation (2) one obtains the infinitesimal elements as

$$\begin{aligned} \xi &= \alpha x + \beta t + \gamma \\ \tau &= 2\alpha t + \delta \\ \eta &= -\alpha\Psi + 1/2\ i\beta\Psi x + i\lambda\Psi \end{aligned} \tag{4}$$

where α, β, δ and λ are five arbitrary parameters. Similarity variables are obtained from Lagrange characteristic equations

$$\frac{dx}{\xi} = \frac{dt}{\tau} = \frac{d\Psi}{\eta} \tag{5}$$

If we choose a particular combination of these five arbitrary parameters, we get various similarity variables (or orbits) depending on these parameters and these orbits which lead to a particular type of Painlevé equations which are shown as follows.

Orbit 1: If $\gamma \neq 0$, $\lambda \neq 0$, $\delta = 1$ and $\alpha = \beta = 0$ in Eq. (4) then we get

$$\xi = \gamma,\ \tau = 1,\ \eta = i\lambda\Psi \tag{6}$$

If we solve Eq. (5) by using Eq. (6), we get similarity variable $\zeta(x,t) = x-\gamma t$ and the forms obtained by integrating Eq. (5) as

$$\begin{aligned} \Psi &= f(x - \gamma t)\, e^{i\left[\frac{\gamma}{2p}(x-\gamma t) - \lambda t\right]} \\ &= f(\zeta)\, e^{i\left[\frac{\gamma}{2p}\zeta - \lambda t\right]} \end{aligned} \tag{7}$$

If we apply this value of Ψ from Eq. (7) in NSE of Eq. (2), we get

$$p\, f''(\zeta) \;+\; f(\zeta)\left[r|f(\zeta)|^2 \;+\; \lambda \;+\; \frac{\gamma^2}{4p}\right] = 0 \tag{8}$$

This Eq. (8) is exactly the form obtained for this orbit by Tuszynski et al., (1984). This Eq. (8) can be solved exactly by elliptic integral and the solution f(ζ) can be expressed in terms of solitons as shown by Tuszynski et al. (1984).

Orbit 2: If $\beta \neq 0$, $\delta = 1$ and $\alpha = \gamma = \lambda = 0$ then $\xi = \beta pt$, $\tau = 1$, $\eta = \frac{i}{2}\ \Psi\beta x$ and the similarity variable is

$$\zeta(x,t) = x - \frac{1}{2}\beta' t^2$$

and the form is $\Psi = f(\zeta)\, e^{i\left(\frac{\beta' xt}{2} - \frac{1}{6}\beta'^2 t^3\right)}$.

The NSE of Eq. (2) becomes

$$pf'' - \left(\frac{\beta x}{2} - \frac{1}{4}\beta^2 p^2 t^2\right) f + r|f|^2 f = 0$$

$$pf'' - \frac{1}{2}\beta\,\zeta\, f \;+\; r|f|^2 f = 0 \tag{9}$$

Use the following transformation

$$\zeta = \left(\frac{2P}{\beta'}\right)^{1/3} \qquad \zeta = \chi\rho$$

$$f(\zeta) = -i\,(2p)^{1/6}\left(\frac{\beta'}{r}\right)^{1/2} g(\zeta)$$

$$= \frac{g}{\chi K}$$

where χ and K are two constants.

Eq. (9) reduces to Painlevé (P_{II}) second equation.

$$\frac{d^2 g(\zeta)}{d\rho^2} = \rho g + 2g^3 \tag{10}$$

Orbit 3: Boiti and Pempinelli (1980) showed that this nonlinear Schrodinger Eq. (2) can be reduced to Painlevé fourth equation. If in our choice of parameters, we use the similarity variable $\zeta = \frac{x}{\sqrt{2}}$ and the form

$$\Psi = f(\zeta)\ e^{i \frac{\lambda}{2} \log t}$$

then the ordinary differential equation thus obtained can be reduced to Painlevé fourth equation by another suitable transformation. The characteristics of the fourth Painlevé equation has been discussed in details by Lukashevich (1967) where from the singularity analysis he has shown the method of obtaining a series of solutions of this equation by the suitable choice of parameters, so one could obtain a class of solutions from a nonlinear partial differential equation by the choice of orbits which results as a consequence of the choice of the parameters of the equation.

3. CONCLUSION

From the above analysis it is obvious that for the different values of the parameters in the similarity variable we get different ordinary differential equations which can be reduced to one of the six Painlevé equations, so N-soliton solutions can be obtained which depends on the values of the parameters. Taijiri et al. (1983) showed that by following this technique, it is possible to reduce three dimensional Schrodinger equation to one dimensional Schrodinger equation and finally to Painlevé equation. But there are some differences in the values of the parameters in the similarity variables in this process of reduction. One particular parameter may depend on the three dimensional equation which is zero in the one dimensional equation. The significance of this effect of parameter are not known yet.

4. REFERENCES

Cowley, R.A. (1980). Advances in Physics 29, 1-110.

Blinc, R. et al. (1981). Phys. Rev. Letts. 46, 1406.

Kobayashi, T. (1973). Jour. Phys. Soc. Japan 35, 558.

Blinc, R., Prelovsek, P. (1983). Phys. Rev. B 27, 5404.

Tuszynski, J.A., Paul, R. and Chatterjee, R. (1984). Phys. Rev. B 29, 380.

Boyer, C.P., Sharp, R.T. and Winternitz, P. (1976). Jour. Math. Phys.

17. 1439.

Tuszynski, J.A., Paul, R., Chatterjee, R. and Sreenivasan, S.R. (1984). Phys. Rev. 30A, 2666.

Ablowitz, M.J., Kaup, D.J., Newell, A.C. and Segur, H. (1974). Stud. Appl. Math. 53, 249.

Ablowitz, M.J., Segur, H. (1977). Phys. Rev. Letts. 38, 1103.

Ablowitz, M.J., Ramani, A. and Segur, H. (1978). Lett. Nuov. Cim. 23, 333.

Blumen, G.W. and Cole, J.D. (1974). "Similarity Methods for Differential Equations", Springer-Verlag, N.Y.

Tajiri, M. (1983). Jour. Phys. Soc. Japan 52, 1908.

Lukashevich, N.A. (1967). Differential Equations 3, No. 5, 771-780.

SYMMETRY PROPERTIES OF MODULATED CRYSTALS

A. Janner and T. Janssen
Institute for Theoretical Physics
Toernooiveld, 6525 ED Nijmegen, The Netherlands.

1. EUCLIDEAN CRYSTALLOGRAPHY

Lattice periodicity is the basic concept occurring in crystallography. Accordingly a crystal can be characterized by a charge distribution $\rho(\vec{r})$ whose Fourier expansion involves wave vectors $\vec{k}$ which are elements of a reciprocal lattice Λ^*:

$$\rho(\vec{r}) = \sum_{\vec{k}\in\Lambda^*} \hat{\rho}(\vec{k})\, e^{i\vec{k}\vec{r}} \tag{1.1}$$

where

$$\vec{k} = \sum_{i=1}^{3} z_i \vec{a}_i{}^* \equiv (z_1, z_2, z_3) \in \Lambda^* \tag{1.2}$$

is an integral linear combination of three basic periodicities with wave vectors $\vec{a}_i{}^*$ for i=1,2,3. It is also required that these wave vectors span the 3-dimensional Euclidean space. Thus one has:

$$\Lambda^* = Z^3 \quad \text{and} \quad \text{span}\,\{\Lambda^*\} = R^3 \tag{1.3}$$

i.e. Λ^* is a free Z-module of rank and of dimension three. The rank denotes the number of free generators, whereas the dimension is that of the vector space spanned by those generators. Crystal symmetry is defined as the group G of those Euclidean transformations

$$g = \{R\,|\,\vec{v}\} \in E(3) \quad \text{with} \quad g\vec{r} = R\vec{r} + \vec{v} \tag{1.4}$$

leaving the charge density ρ invariant:

$$\rho(g^{-1}\vec{r}) = \rho(\vec{r}) \qquad g \in G. \tag{1.5}$$

The symmetry condition (1.5) can equally well be formulated in terms of the Fourier components according to:

$$\hat{\rho}(\vec{k}) = \hat{\rho}(R\vec{k})\, e^{iR\vec{k}\vec{v}} \tag{1.6}$$

for g as in eq. (1.4). It follows that G is a 3-dimensional space group, whose subgroup of translations generates a lattice Λ, which also is a free Z-module of rank and dimension three. The point group K of G consists of all homogeneous parts R of g and is a subgroup of the orthogonal group O(3). The non-homogeneous parts $\vec{v}(R)$ are called non-primitive translations if they are not lattice translations, which are the primitive translations. A general element of G can thus be written as:

$$g = \{\, R \mid \vec{n} + \vec{v}(R) \,\} \tag{1.7}$$

and G can be specified in terms of K, of Λ and of a system of non-primitive translations $\vec{v}(K)$.

2. INCOMMENSURATE CRYSTAL PHASES

Within a same phase and because of thermal expansion the symmetry group of a crystal changes as a function of temperature, but it remains in the same isomorphism class:

$$\Lambda = \Lambda(T) \quad \text{thus} \quad G = G(T) \quad \text{but} \quad G(T) \approx G(T') \tag{2.1}$$

within the same phase. It is only at a phase transition that the isomorphism class of the space group possibly changes.

2.1 The modulated crystal structures

Fairly often structural phase transitions are characterized by the appearance of a periodic deformation called modulation. In Fig. 1 two very common types of modulation are represented in a schematic way. In real crystals both may occur at the same time, but usually one is dominant.

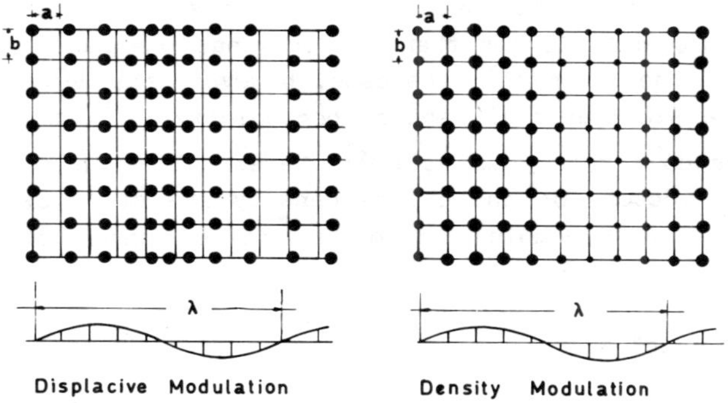

Fig. 1 Two forms of modulated crystals
(a) Displacive modulation. (b) Density modulation.

The periodic deformation due to the modulation gives rise to additional wave vectors in the Fourier expansion, and accordingly to additional Bragg spots in the crystal diffraction pattern. Normally (but not always) these have weaker intensities than the Bragg spots associated with the average structure. The diffraction spots of the average structure are therefore called main reflections , whereas the spots due to the modulation are called satellite reflections [1].

The components of the Fourier wave vectors, and thus of the position vectors in reciprocal space of the Bragg reflections, when expressed with respect to a basis $\vec{a}_i^*$, for i=1,2 and 3 of the reciprocal lattice Λ^* are called indices. Main reflections have integral indices (traditionally denoted by h, k, l), whereas that is in general not the case for satellite reflections. If the indices of the latter are rational numbers , this means that the modulation periodicity (or periodicities) fits with an element of the lattice Λ of the average structure, and is thus called commensurate modulation. The corresponding modulated crystal forms then a superstructure.

In 1964 P.M. de Wolff and collaborators found in γ-Na_2CO_3 the existence of an incommensurate modulation as well, associated with irrational indices, the non-rationality being imposed by a continuous dependency of the indices of satellite reflections on the temperature [2]. In the case of a superstructure, the lattice periodicity (and thus the space group symmetry) is still present, the unit cell being only larger than the original one, and that explains the name. In the incommensurate case, however, the unit cell becomes infinite, the 3-dimensional lattice periodicity is lost and therefore the space group symmetry as well [3]. Nevertheless in both cases the Fourier wave vectors of the charge density are still expressible as an integral linear combination of a finite number of basic wave vectors:

$$\vec{k} = h\vec{a}^* + k\vec{b}^* + l\vec{c}^* + m_1\vec{q}_1 + m_2\vec{q}_2 + \dots + m_d\vec{q}_d \qquad (2.2)$$

and are thus elements of a free Z-module M*, which is 3-dimensional (as it generates a 3-dimensional vector space) but is of rank 3+d, where d is the number of independent modulations.

2.2 Problems associated with incommensurate crystals

The existence in nature of incommensurate crystal phases leads to the following considerations:

1) The crystal symmetry is not necessarily described by one of the 230 space groups.

2) The Fourier wave vectors of the charge density still form a free Z-module but not necessarily a reciprocal lattice.

3) The Euclidean symmetry of an incommensurate crystal is fairly low and does not explain the regularities of the diffraction pattern (main reflections, satellite reflections and possibly systematic extinctions).

Nevertheless the physical properties of incommensurate crystals are essentially the same as the well known ones of commensurate crystal phases. Therefore the questions arise:

- What is a crystal ?
- Which symmetry is a "good" one ?

An attempt to answer these questions has brought to the so called superspace approach [4], which will be outlined in the following section.

3. THE SUPERSPACE APPROACH

The basic idea of the superspace approach is to treat the basic periodicities generating the Fourier wave vectors as independent ones, embedding the corresponding vectors as a basis of a (3+d)-dimensional reciprocal lattice in a Euclidean space of same dimension. The structural information contained in the corresponding Fourier coefficient is conserved during this embedding [5].

3.1 Crystal embedding in superspace

Considered is a crystal described in terms of a charge density of the form:

$$\rho(\vec{r}) = \sum_{\vec{k}\in M^*} \hat{\rho}(\vec{k})\, e^{i\vec{k}\vec{r}} \qquad (3.1)$$

where

$$\vec{k} = \sum_{\nu=1}^{3+d} z_\nu \vec{a}_\nu^{\,*} = (z_1, z_2, \ldots, z_{3+d}) \quad \text{element of} \quad M^* \qquad (3.2)$$

so that M* is a 3-dimensional free module of rank 3+d. Denoting by FT_3 and by FT_{3+d} the 3-dimensional Fourier transform and the (3+d)-dimensional one, respectively, we can write for the charge density of the crystal:

$$\rho(\vec{r}) = FT_3\, \hat{\rho}(z_1,\ldots,z_{3+d}) \quad . \qquad (3.3)$$

The embedding in the (3+d)-dimensional Euclidean space (called the superspace) defines by a (3+d)-dimensional Fourier transform a density function $\rho_s(\vec{r}_s)$ that we call the supercrystal. The supercrystal density of a crystal as in eq. (3.3) is then given by:

$$\rho_s(\vec{r}_s) = FT_{3+d}\, \hat{\rho}\, (z_1,\ldots\ldots,z_{3+d}) \qquad (3.4)$$

with correspondingly the same Fourier components $\hat{\rho}$ as in eq. (3.3). Note however that the same set of 3+d integral indices $(z_1,\ldots\ldots,z_{3+d})$ in eq. (3.3) refers to a 3-dimensional wave vector element of M* given as in eq. (3.2), whereas in eq. (3.4) it refers to a vector of a reciprocal lattice Σ* in 3+d dimensions expressible as:

$$(z_1,\ldots\ldots,z_{3+d}) = z_1\vec{a}^*_{s1} + \ldots\ldots + z_{3+d}\vec{a}^*_{s(3+d)} \qquad (3.5)$$

where the lattice vectors $\vec{a}^*_{s\nu}$, for $\nu = 1,\ldots,3+d$ have as projection in the 3-dimensional subspace of the crystal the Z-module basis vectors $\vec{a}^*_{\nu}$, respectively. As intersection is a dual operation with respect to projection, in the direct space the crystal density $\rho(\vec{r})$ is obtained from that of the corresponding supercrystal by taking the 3-dimensional intersection. That relation is illustrated in Fig. 2.

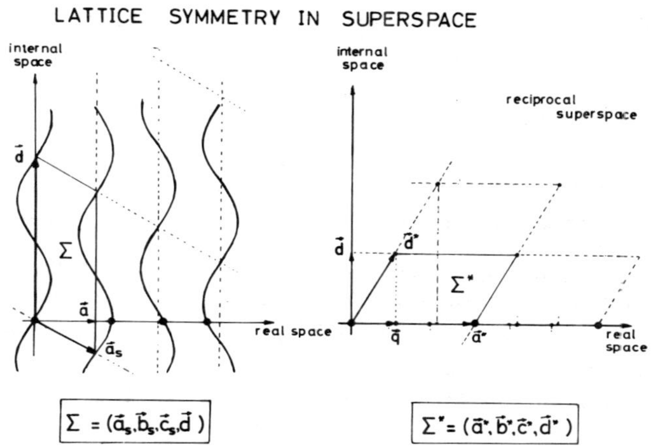

Fig. 2 Lattice symmetry of the supercrystal pattern describing a modulated crystal.

The (3+d)-dimensional superspace V_s can naturally be decomposed in a direct sum of the original 3-dimensional Euclidean space V_o now denoted (as subspace) as "external" or "positional" space V_E and an additional d-dimensional Euclidean space denoted by V_I and called "internal":

$$V_o \quad \rightarrow \quad V_s = V_E \oplus V_I \quad . \tag{3.6}$$

Corresponding to that splitting we write for vectors in the superspace

$$\vec{r}_s = (\vec{r}, \vec{r}_I) \quad \text{and} \quad \vec{k}_s = (\vec{k}, \vec{k}_I) \quad . \tag{3.7}$$

The supercrystal charge density takes the form:

$$\rho_s(\vec{r}_s) = \sum_{\vec{k}_s \in \Sigma^*} \hat{\rho}(\vec{k}_s)\, e^{i\vec{k}_s \vec{r}_s} \quad . \tag{3.8}$$

The crystal symmetry is then defined as the group of Euclidean

(3+d)-dimensional transformations leaving ρ_S invariant:

$$\rho_S(g_S^{-1}\vec{r}_S) = \rho_S(\vec{r}_S) \tag{3.9}$$

for g_S an element of E(3+d). It then directly follows that all such symmetry transformations g_S form a (3+d)-dimensional space group G_S, called superspace group because of some additional properties 6]. In this way crystallographic symmetry is recovered, and can be used to relate e.g. systematic extinctions 7] and crystal structure 8].

Indeed, in correspondence to the splitting of the superspace into an external and an internal one we also can write:

$$\rho_S(g\vec{r}, g_I\vec{r}_I) = \rho_S(\vec{r}, \vec{r}_I) \tag{3.10}$$

with g E(3) and g_I E(d). Expressing the symmetry condition in terms of the Fourier components $\hat{\rho}$ clearly shows the difference between a space and a superspace Euclidean invariance requirement. Instead of eq. (1.6) one has:

$$\hat{\rho}(\vec{k}) = \hat{\rho}(R\vec{k})\, e^{iR\vec{k}\vec{v} + iR_I\vec{k}_I\vec{v}_I} \tag{3.11}$$

for $g_S = (\{R \mid \vec{v}\}, \{R_I \mid \vec{v}_I\})$, where use has been made of the fundamental relation among Fourier coefficients:

$$\hat{\rho}_S(\vec{k}, \vec{k}_I) = \hat{\rho}(\vec{k}) \quad . \tag{3.12}$$

Comparing eqs. (1.6) and (3.11) one sees that the extension of the condition for symmetry consists of admitting additional phase relations among the (3-dimensional) Fourier coefficients of the crystal structure. Therefore, extinction conditions due to superspace symmetry imply extinction conditions for the 3-dimensional diffraction pattern as well.

For preparing the discussion of the example which will be given as illustration in the next section, let us restrict to the one-dimensional modulation case (d = 1) and choose convenient basis sets in the space and in the superspace.

It is natural to consider the Z-module M*, which lies in V_E^*, as orthogonal projection of a (3+d)-dimensional lattice Σ^* in V_s^*. One way of doing so is to extend the Z-module basis to a lattice basis of Σ^*. In particular for the one-dimensional modulated crystal case this can be done by defining :

$$\vec{a}_s^* = (\vec{a}^*,0), \quad \vec{b}_s^* = (\vec{b}^*,0) \quad \vec{c}_s^* = (\vec{c}^*,0), \quad \vec{d}_s^* = (\vec{q},\vec{d}^*). \qquad (3.13)$$

Note that the corresponding dual basis is given by:

$$\vec{a}_s = (\vec{a},-\alpha\vec{d}), \quad \vec{b}_s = (\vec{b},-\beta\vec{d}) \quad \vec{c}_s = (\vec{c},-\gamma\vec{d}), \quad \vec{d}_s = (0,\vec{d}), \qquad (3.14)$$

where $\vec{q} = \alpha\vec{a}^* + \beta\vec{b}^* + \gamma\vec{c}^*$.

The Fourier phase in eq. (3.8) can then be interpreted as scalar product in V_s^* according to :

$$\vec{k}_s.\vec{r}_s = 2\pi(hx + ky + lz + mt) \qquad (3.15)$$

where (h,k,l,m) are the components of $\vec{k}_s = (\vec{k},\vec{k}_I)$ and (x,y,z,t) those of $\vec{r}_s = (\vec{r},\vec{t})$, when referred to the dual bases (3.13) and (3.14), respectively.

3.2 Superspace groups

Let us simply quote the main properties of superspace groups and refer for more details to a number of papers published elsewhere [6] [9].

i) The lattice Σ of the superspace group G_s is completely determined by the basis vectors $\vec{a}_1^*,\vec{a}_2^*, \vec{a}_3^*,\vec{q}_1,\ldots,\vec{q}_d$ of M*. This basis is supposed to be linearly independent on the rationals, whereas $\vec{a}_1^*,\vec{a}_2^*$ and $\vec{a}_3^*$ span the positional (reciprocal) 3-dimensional space.

ii) The elements R_s of the point group K_s of G_s are orthogonal transformations in V_s leaving both subspaces V_E and V_I invariant.

It then follows that :

1) The point group K_s has elements $R_s = (R, R_I)$, where the groups K and K_I formed by all elements R and R_I, respectively, are crystallographic point groups of V_E and V_I, respectively. Furthermore K and K_s are isomorphic.

2) The superspace group G_s has elements $g_s = (g, g_I)$, where the components g form a space group G of V_E. Note, however, that the components g_I, which are Euclidean in V_I, do not form a space group.

3) There are bases (similar to that indicated in eq. (3.14) and which are called standard) for the lattice Σ of G_s such that the point group elements R of K take the form (with integral entries) :

$$\Gamma(R) = \begin{matrix} \Gamma_E(R) & 0 \\ \Gamma_M(R) & \Gamma_I(R) \end{matrix} \tag{3.16}$$

4) The Bravais class of a lattice Σ is given by the arithmetic crystal class of Γ(K), these classes being induced by all basis transformations on standard bases.

A full list of the (3+1)-dimensional superspace groups has been published elsewhere [7].

3.3 Application to γ-Na_2CO_3

Anhydrous sodium carbonate crystallizes at 850°C in an hexagonal α-phase, it becomes monoclinic at 489°C (β-phase) and undergoes a further transition at 360°C to an incommensurate γ phase which is monoclinic modulated, with wave vector lying in the mirror plane : $\vec{q} = (\alpha, 0, \gamma)$. At -138°C it seems that there is a new phase transition (δ-phase) where the $\vec{q}$ becomes commensurate :$\vec{q} = (1/6, 0, 1/3)$.

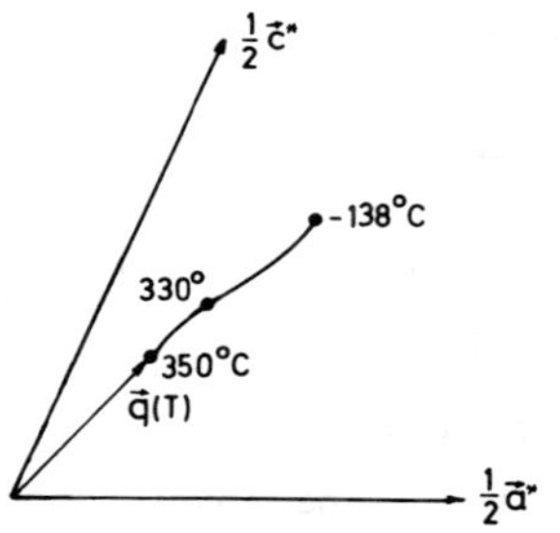

Fig. 3 Temperature dependence of the modulation in Na_2CO_3.

The γ-phase is characterized by $\vec{b}^*$ orthogonal to $\vec{a}^*$ and $\vec{c}^*$. The modulation wave vector has components α and γ which are temperature dependent; at room temperature their value is 0.182 and 0.318, respectively [10]. The following systematic extinctions are observed :

(1) h + k odd for all reflections h,k,l,m.

(2) m odd for all k = 0 reflections (i.e. h,0,l,m).

The first set of extinctions being observed for <u>all</u> reflections indicates the presence of a <u>centering translation</u> (½,½,0,0). Indeed the symmetry condition as expressed by eq. (3.11) becomes for

$\{ 1 \mid \frac{1}{2},\frac{1}{2},0,0\}$

$$\hat{\rho}(h,k,l,m) = \hat{\rho}(h,k,l,m) \exp[2\pi i(h+k)/2] \qquad (3.17)$$

implying $\hat{\rho}(\vec{K}) = 0$ for h+k odd.

Accordingly the four-dimensional lattice belongs to the Bravais class denoted by $C2/m(\alpha,0,\gamma) = P^{C2/m}_{\bar{1}\ 1}$ which is centered monoclinic [11]. The second set of extinctions is due to a glide symmetry given by $\{m_y , 1 \mid 0,0,0,\frac{1}{2}\}$ implying the condition :

$$\hat{\rho}(h,k,l,m) = \hat{\rho}(h,-k,l,m) \exp[\pi im]. \qquad (3.18)$$

Indeed then for k=0 and m odd one has :

$$\hat{\rho}(h,0,l,m) = -\hat{\rho}(h,0,l,m) = 0. \qquad (3.18)$$

Note that the four-dimensional mirror is $(R,R_I) = (m_y,1)$ because $m_y\vec{q} = \vec{q}$, whereas 2_y is associated with $\bar{1}$ because $2_y\vec{q} = -\vec{q}$, and so acts also the total inversion.

It follows that the superspace group of γ-Na_2CO_3 is $P^{C2/m}_{\bar{1}\ s}$, which is generated by the translations $(\frac{1}{2},\frac{1}{2},0,0)$, $(\frac{1}{2},-\frac{1}{2},0,0)$, $(0,0,1,0)$ and $(0,0,0,1)$ and

by the total inversion $\{\bar{1} , \bar{1} \mid 0,0,0,0\}$ and the glide $\{m_y , 1 \mid 0,0,0,\frac{1}{2}\}$. The corresponding general positions are :
In the four dimensional basis as in eq. (3.14) and with $\beta = 0$:

TABLE I

(0000), (½½00) +			
x	y	z	t
-x	-y	-z	-t
x	-y	z	½ + t
-x	y	-z	½ - t

Thus there are eight equivalent positions in the unit cell in superspace. The corresponding positions in three dimensional space, but now with respect to the basis $\vec{a}, \vec{b}$ and $\vec{c}$ and parametrically dependent on the phase variable, which is (we recall) depending on the unit cell in which the atom is situated, are :

TABLE II

$(000)(t_o)$, $(\frac{1}{2}\frac{1}{2}0)(t_o + \frac{1}{2}\alpha)$ +		
$x_o+u(t_o)$	$y_o+v(t_o)$	$z_o+w(t_o)$
$-x_o-u(-t_o)$	$-y_o-v(-t_o)$	$-z_o-w(-t_o)$
$x_o+u(t_o+\frac{1}{2})$	$-y_o-v(t_o+\frac{1}{2})$	$z_o+w(t_o+\frac{1}{2})$
$-x_o-u(-t_o+\frac{1}{2})$	$-y_o-v(-t_o+\frac{1}{2})$	$-z_o-w(-t_o+\frac{1}{2})$

Note that the centered positions are of the type :

$$x_o + \tfrac{1}{2} + u(t_o + \tfrac{1}{2}\alpha),\ y_o + \tfrac{1}{2} + v(t_o + \tfrac{1}{2}\alpha),\ z_o + w(t_o + \tfrac{1}{2}\alpha)$$

as one has to take care of the fact that the phase coordinate t_o is not an independent one, obeys the relation:

$$t_o = (\vec{r}_o + \vec{n})\,\vec{q} \tag{3.20}$$

and has thus a different value for different positions; the basic idea of superspace is precisely that of making the modulation phase coordinate an independent one.

Not all the atoms in Na_2CO_3 occupy the general eightfold position, and this imposes restrictions on the modulation wave as well. Let us consider first the monoclinic undistorted β-phase. There are four Na_2CO_3 molecules per unit cell whose atoms occupy the following Wyckoff positions of the space group C2/m (unique axis b):

TABLE III

Atom	Multiplicity	Wyckoff	Point symmetry	Coordinates
Na(1)	2	a	2/m	0,0,0
Na(2)	2	c	2/m	0,0,½
Na(3)	4	i	m	$x,0,z;\ \bar{x},0,\bar{z}$
C	4	i	m	
O(2)	4	i	m	
O(1)	8	j	1	general
O(3)	8	j	1	general

A same multiplicity is observed in the incommensurate phase as well, it involves now not equivalency of atomic point positions but of atomic strings. So to get a fourfold atomic string for y_o requires invariance of the modulation wave with respect to the glide $\{m_y\ ,\ 1\ |\ 0,0,0,\tfrac{1}{2}\}$ and this implies (in the harmonic approximation) :

$$u(t_o) = u(t_o+\tfrac{1}{2}) = -u(t_o) = 0\ ; \quad w(t_o) = w(t_o+\tfrac{1}{2}) = -w(t_o) = 0 \tag{3.21}$$

accordingly for Na(3), C and O(2) the modulation wave is transversal: $\vec{f} = (0,v,0)$.

The atoms Na(1) and Na(2) occupy each a twofold position, and this requires in addition to eq. (3.21) invariance with respect to the total inversion of the corresponding atomic string. After a good choice of the origin one has :

$$v(t_o) = -v(-t_o) \tag{3.22}$$

i.e. the modulation wave is an odd function. No such restrictions are present in the eightfold configuration.

4. ON THE RELATION BETWEEN SYMMETRY AND PROPERTIES

Incommensurability influences in one way or another all crystal properties: this influence is in general very subtle and non easily detectable, nevertheless far reaching. It involves both microscopic and macroscopic properties at the classical as well as at the quantum mechanical level. The physical laws determining these effects are in principle known. Due to the lack of lattice symmetry, however, a number of important simplifications occurring for normal crystals are no more allowed in the incommensurate case, except in a first and mostly inadequate approximation. This is the reason why very often one is reduced to perform computer calculations and for one dimensional crystal models.

The basic idea that eventually justifies superspace approach in crystal physics is that it allows a symmetry adapted description of incommensurate crystals such that group theoretical methods can be used. This requires, however, the extension to superspace of the physical considerations and an interpretation in the real space of the implications derived.

This way of doing will be illustrated in what follows. It has to be stressed that the theory is far from being complete and the considerations reported should be regarded as first attempt only.

4.1 Space group representations and superspace symmetry

Incommensurate crystal structures can be described in terms of representation of ordinary crystallographic groups. It is the approach normally adopted by people not making use of superspace groups and working within the frame of a Landau theory of order parameter. It is not the intention to discuss here the relation between the two approaches (see in particular the following lecture by T. Janssen and A. Janner), but only to explain when the description in terms of representations is a natural one, to indicate the connection with superspace symmetry and to use that connection in discussing physical properties in relation to symmetry [12].

For simplicity we will restrict here the considerations to the (3+1)-dimensional case; extension to the more general (3+d)-

dimensional case is fairly straightforwards.

Consider an incommensurate crystal described by a scalar function, i.e. its density $\rho(\vec{r})$:

$$\rho(\vec{r}) = \sum_{\vec{k}\in M^*} \hat{\rho}(\vec{k})\, e^{i\vec{k}\vec{r}}. \tag{4.1}$$

The incommensurability becomes apparent in the form of the Fourier wave vectors which are of the form :

$$\vec{k} = h\vec{a}^* + k\vec{b}^* + l\vec{c}^* + m\vec{q} \quad \text{for} \quad h,k,l,m \quad \text{integers.} \tag{4.2}$$

Here a choice has been made for the basis $\vec{a}^*$, $\vec{b}^*$, $\vec{c}^*$ of the reciprocal lattice Λ^*. In principle there is no unique prescription for this choice, but in practice there is often no difficulty in determining these "main reflections" describing an underlying basis structure with space group G and lattice Λ.

The corresponding supercrystal is then obtained by embedding in the four dimensional superspace as in eq. (3.4). We use here a slightly modified notation, namely :

$$\vec{k} = \vec{K} + m\vec{q} \quad \text{with} \quad \vec{K} \in \Lambda^* \tag{4.3}$$

so that one has :

$$\hat{\rho}_s(\vec{k}_s) = \hat{\rho}(h,k,l,m) \tag{4.4}$$

with $\vec{k}$ as in eq. (4.2) and $\vec{k}_s$ with same components but with respect to the standard basis indicated in eq. (3.13):

$$\vec{k}_s = h\vec{a}_s^* + k\vec{b}_s^* + l\vec{c}_s^* + m\vec{d}_s^*. \tag{4.5}$$

Invariance of ρ_s with respect to the element g_s of the superspace group G_s imposes the condition, like in eq. (3.11):

$$\hat{\rho}(\vec{k}) = \hat{\rho}(R(\vec{K}+m\vec{q}))\, \exp[iR(\vec{K}+m\vec{q})\vec{t}+i\varepsilon m\Delta] \tag{4.6}$$

where :

$$g_s=(g,g_I),\ g=\{R\ |\vec{t}\},\ g_I=\{\varepsilon\ |\vec{t}_I\},\quad \Delta=\vec{d}^*\vec{t}_I$$

and $\varepsilon = \pm 1$ according to

$$R\vec{q} \equiv \varepsilon\vec{q}(\mathrm{mod}\Lambda^*) \quad . \tag{4.7}$$

It is convenient to decompose $\rho(\vec{r})$ as follow :

$$\rho(\vec{r}) = \sum_m \rho_m(\vec{r}) \tag{4.8}$$

with

$$\rho_m(\vec{r}) = \sum_{\vec{K}\in\Lambda^*} \hat{\rho}(\vec{K}+m\vec{q})\ \exp[i(\vec{K}+m\vec{q})\vec{r}] \quad . \tag{4.9}$$

It then follows that $\rho_o(\vec{r})$ is invariant with respect to the space group G having as elements the external components $g = \{R\ |\vec{t}\}$ of the superspace group G_s. The other ρ_m are, of course, not invariant with respect to G, but transform in a simple way. Indeed defining T_g as substitutional operator on scalar functions $f(\vec{r})$:

$$T_g\ f(\vec{r}) = f(g^{-1}\vec{r}) \tag{4.10}$$

one finds for the external part g of g_s G_s as above :

$$T_g\ \rho_m(\vec{r}) = \rho_{\varepsilon m}(\vec{r})\exp[i\varepsilon m\Delta]. \tag{4.11}$$

On the other side the component $\rho_m(\vec{r})$ belongs to the irreducible representation of the translation group Λ with character $\chi_m(\vec{n})=\exp[-im\vec{q}\vec{n}]$ and can be expressed in terms of basis functions of irreducible representations of G as

$$\rho_m(\vec{r}) = \sum_\mu c_{m\mu}\Phi_{m\mu}(\vec{r}). \tag{4.12}$$

The components $\rho_{m\mu}$ transform under an element $g = \{R\ |\vec{t}\}$ of the group of $\vec{q}$ (which is also the group of $m\vec{q}$ because of incommensurability) as

$$(T_g c)_{m\mu} = \sum_\nu e^{-im\vec{q}\cdot\vec{t}} D_m(R)_{\mu\nu} c_{m\nu} \quad , \tag{4.13}$$

where D(R) is a (projective or ordinary) representation of the point group of $\vec{q}$.

Comparison with eq. (4.11), and noticing that now as $R\vec{q} \equiv \vec{q} (\mathrm{mod}\Lambda^*)$ one has $\varepsilon = 1$, gives:

$$\sum_\nu e^{-im\vec{q}\vec{t}} D_m(R)_{\nu\mu} = e^{im\Delta} \tag{4.14}$$

thus with N the dimension of $D_m(R)$ and χ_m the corresponding character one has :

$$\chi_m(R) = N\ e^{im(\vec{q}\vec{t} + \Delta)} = \chi(R)^m \quad . \tag{4.15}$$

This relation can be simplified further. The lattice of G_s is generated by the basis given in eq. (3.14), so that the translational part of g_s as above can be written :

$$\vec{t}_s = t_1\vec{a}_s + \ldots. + t_4\vec{d}_s = (\vec{t},\vec{t}_I) \quad \text{with} \tag{4.16}$$
$$\vec{t} = t_1\vec{a} + \cdot\cdot\cdot + t_3\vec{c} \quad \text{and} \quad \vec{t}_I = (-\vec{t}\vec{q} + 2\pi t_4)\vec{d} \quad \text{thus}$$

$$\Delta = \vec{t}_I\vec{d}^* = 2\pi t_4 - \vec{t}\vec{q} \tag{4.17}$$

yielding finally the simple result :

$$\chi(R) = N\ e^{2\pi i t_4} \tag{4.18}$$

This is the key relation between superspace symmetry and behaviour with respect to (basis) space group transformations. We illustrate this on the concrete example of incommensurate $ThBr_4$.

4.2 Application to $ThBr_4$

This crystal has been studied by neutron diffraction 13]. The room temperature crystal structure of $ThBr_4$, (β phase) has space group $G_o = I4_1/amd = D_{4h}^{19}$. The thorium ions are at Wyckoff position 0,0,0 and 0,½,¼ in a site with $D_{2d} = \bar{4}2m$ point symmetry; the bromine anions are at the 16h position with coordinates 0,y,z . Each thorium is surrounded by eight bromines ions (Fig. 4).

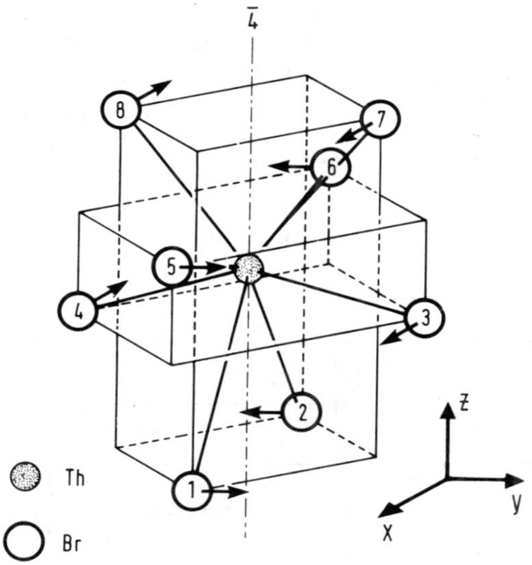

Fig. 4 The unit cell of the average structure of $ThBr_4$. The direction of the displacements in the modulated phase is indicated

The space group G_o is generated by the tetragonal body centered lattice translations with basis vectors avec, $\vec{b}$, $\vec{c}$ and in addition by:

$$g_1=\{4_z\ |0,\tfrac{1}{2},\tfrac{1}{4}\} = \{4_z\ |\vec{t}\},\ g_2=\{m_x\ |0\} \qquad (4.19)$$

$$\text{and } g_3=\{m_z\ |0,\tfrac{1}{2},\tfrac{1}{4}\}=\{m_z\ |\vec{t}\}.$$

At T_c = 95K the crystal undergoes an incommensurate phase transition due to the condensation of a soft phonon as observed by neutron scattering [13]. The low temperature phase is modulated accordingly with wave vector $\vec{q}$ given by :

$$\vec{q} = \gamma\vec{c}^* = 0.310\ \vec{c}^* \quad . \qquad (4.20)$$

The soft mode responsible for the transition transforms according to the irreducible representation τ^4, in Kovalev's notation [14], of the group of $\vec{q}$, which in the present case is $G_{\vec{q}} = I4_1md$ with point group $K_{\vec{q}}$. The corresponding character table is :

TABLE IV

Kovalev	h_{14}	h_4	h_{15}	h_{26}
Seitz	$\{C_{4z}\ \|\vec{t}\}$	$\{C_{2z}\ \|\vec{t}\}$	$\{C_{4z}^3\ \|\vec{t}\}$	$\{\sigma_{yz}\ \|0\}$
Inter.	$\{4_z\ \|\vec{t}\}$	$\{2_z\ \|0\}$	$\{4_z^3\ \|\ \vec{t}\}$	$\{m_x\ \|0\}$
τ^4	$\overline{1}$	1	$\overline{1}$	$\overline{1}$
Kovalev	h_{37}	h_{27}	h_{40}	
Seitz	$\{\sigma_d\ \|\vec{t}\}$	$\{\sigma_{xz}\ \|0\}$	$\{\sigma_{d'}\ \|\vec{t}\}$	
Inter.	$\{m_d\ \|\vec{t}\}$	$\{m_y\ \|0\}$	$\{m_{d'}\ \|\vec{t}\}$	
τ^4	1	$\overline{1}$	1	

In this table alternative conventional notations for the same group elements are indicated.

The superspace follows immediately. Admitted are elements $g_s = (g, g_I) = \{R, \varepsilon \mid \vec{t} + \vec{t}_I\}$ such that :

1) $g = \{R \mid \vec{t}\}$ is an element of $G_o = I4_1/amd$

2) There exists an internal translation $\vec{t}_I$ which "compensates" to one the character values (of the point group elements R) different from one; Eqs. (4.17) and (4.18) have thus to be satisfied.

In the present case this all simply implies an additional internal non-primitive translation $s = t_4\vec{d} = \frac{1}{2}\vec{d}$ for the elements with character -1 in the table given above. Note that geometrically speaking the translation s corresponds to a shift by π of the modulation wave.

Assuming that the incommensurable structure is that given by a "frozen-in" soft phonon we get, first of all for the translations of the superspace group those of the lattice generated by :

$$\vec{a}_s = (\vec{a}, 0),\ \vec{b}_s = (\vec{b}, 0),\ \vec{c}_s = (\vec{c}, -\gamma\vec{d}) \text{ and } \vec{d}_s = (0, \vec{d}) \tag{4.21}$$

with $\vec{a}$, $\vec{b}$, $\vec{c}$ tetragonal as above and $\vec{d}$ unit vector along the internal axis, together with the tetragonal centering translation :

$$(\tfrac{1}{2}, \tfrac{1}{2}, \tfrac{1}{2}, 0) = \tfrac{1}{2}\vec{a}_s + \tfrac{1}{2}\vec{b}_s + \tfrac{1}{2}\vec{c}_s \tag{4.22}$$

The corresponding Bravais class [11] is denoted by :

$$P^{I4/mmm}_{1\ \bar{1}11} = I4/mmm(0,0,\gamma) \tag{4.23}$$

The corresponding Bravais lattice has as generators for its (holohedral) point group the (4-dimensional) rotations :

$$R_1 = (4_z, 1),\ R_2 = (m_x, 1) \text{ and } R_3 = (m_z, \bar{1}). \tag{4.24}$$

In addition to the lattice translations the superspace group of $ThBr_4$

has (according to the characters of the representation τ^4) the generators:

$$g_{1s} = \{4_z\ ,\ 1\ |0,½,¼,½\},\quad g_{2s} = \{m_x\ ,\ 1\ |0,0,0,½\}. \tag{4.25}$$

The non-primitive translations of g_{1s} and g_{2s} depend on the choice of the origin in the external space (here it is at a point with $\bar{4}2m$ symmetry), but are independent on the choice of the origin in the internal space (i.e.the choice of the phase of the modulation wave). This is no longer the case for the generator

$$g_{3s} = \{m_z\ ,\ \bar{1}\ |0,½,¼,t_4\} \sim \{m_z\ ,\ \bar{1}\ |0,½,¼,0\} \tag{4.26}$$

whose t_4 component can always be transformed away. Accordingly the superspace group of the low temperature phase of $ThBr_4$ is

$$G_s = P_{s\ \ \bar{1}s1}^{I4_1/amd} \tag{4.27}$$

As indicated in the full list of the (3+1)-dimensional superspace groups [7], this is the only one superspace group with Bravais class as above and non-primitive internal translations. Such a superspace group implies the following conditions for reflections :

TABLE V

h,k,l,m	condition for $\hat{\rho}(h,k,l,m)\neq 0$	
general	h+k+l = 2n	centering (½,½,½,0)
h,h,l,m	2h+l = 4n	$\{m_d\ 1\ \mid 0,½,¼,0\}$
h,0,l,m	m = 2n	$\{m_y\ 1\ \mid 0,0,0,½\}$
0,k,l,m	m = 2n	$\{m_x\ 1\ \mid 0,0,0,½\}$

The corresponding systematic extinction rules are compatible with the experimental data obtained from neutron diffraction.

4.3 Perturbative approach

As already suggested by the diffraction pattern of an incommensurate crystal where the main reflections describing an average structure are, as a whole, stronger than the satellite reflections due to the deviations from that average, it is natural to treat incommensurability as a perturbation. One should be aware that this way of doing, while excellent for a number of systems and physical properties, it clearly has its limitations also.

In what follows we consider the crystal potential of an incommensurate crystal. We first show that, as expected, such a potential has an analogous Z-module Fourier decomposition as the crystal density function, and that both share the same superspace group symmetry. We then discuss how this leads to selection rules for matrix elements appearing in a quantum mechanical perturbation scheme.

Consider, as an example the Coulomb potential arising from the charge density ρ of an incommensurate crystal with superspace group G_s. We have :

$$V(\vec{r}) = \int \frac{\rho(\vec{r}\,')}{|\vec{r} - \vec{r}\,'|} d^3r' \quad . \tag{4.28}$$

In this expression it is assumed that the boundary effects for a macroscopic crystal are negligible once compensated by an appropriate background charge. The Fourier transform of such a convolution is simply:

$$\hat{V}(\vec{k}) = \frac{\hat{\rho}(\vec{k})}{|\vec{k}|^2} \qquad \text{for } \vec{k} \neq 0. \tag{4.29}$$

The potential has thus the same Z-module structure M* as the charge density. Note, however, that in eq. (4.29) and because of the incommensurability the possibility is implicitly present of a pathological behaviour. This because in M* within any arbitrary small sphere around the origin in reciprocal space there is always an infinite number of wave vectors $\vec{k}$.

In many cases this pathological behaviour does not occur. So e.g. for smooth displacive modulation the small $\vec{k}$'s do not play a role because $\hat{\rho}(\vec{k})$ goes much stronger to zero than the corresponding $|\vec{k}|$'s (typically exponentially in the case of harmonic modulation).

Disregarding thus pathological situations, the scalar potential $V(\vec{r})$ can be embedded in the superspace exactly as the charge density. It then has the same superspace group symmetry.

Therefore the considerations of the previous section can be applied to the potential $V(\vec{r})$ as well, and with respect to the basis space group G it transforms in the same simple way as $\rho(\vec{r})$.

This allows the derivation of selection rules for matrix elements occurring in a perturbation scheme. Consider the basis structure (which often has the same symmetry as the high temperature crystal phase) as representing the unperturbed system with (3-dimensional) space group G_o. The unperturbed energy eigenstates $|\psi_{\nu i}\rangle$ can then be chosen in such a way that they transform according to irreducible representations Γ_ν of G_o. One can also decompose the perturbation potential into irreducible components labeled by m as done above for the charge density.

A general perturbative matrix element has then the form :

$$\langle\psi_{\nu 1} \mid V_m \mid \psi_{\mu 2}\rangle = \int d^3r \; \psi^*_{\nu 1}(\vec{r}) V_m(\vec{r}) \psi_{\mu 2}(\vec{r}) \qquad (4.30)$$
$$= \int d^3 \psi^*_{\nu 1}(g^{-1}\vec{r}) V_m(g^{-1}\vec{r}) \psi_{\mu 2}(g^{-1}\vec{r})$$

for g G, which is the space group of all external components of the elements of the superspace group G_s, and thus according to the general theory [6] a proper or improper subgroup of G_o.

Condition for a non-zero matrix element is that the representation according to which the integrand transforms

$$\Gamma^\dagger_\nu \otimes D_m \otimes \Gamma_\mu \qquad (4.31)$$

contains the identical representation.

Along this way one can treat e.g. optical properties of transition metal ions in incommensurate crystals. In that case the unperturbed wave functions are atomic orbitals and the space group elements g to be considered are those of the site symmetry of the transition ion.

The same idea has been applied in the analysis of de Haas van Alphen measurements in metallic $Hg_{3-\delta}AsF_6$ [15]. In that case the unperturbed states are Bloch states of the commensurate part of the crystal potential and the remaining (incommensurate) part is treated as perturbing potential causing transitions among the Bloch states and modifying accordingly the shape of the Fermi surface.

All this theory, although in principle well defined, has still to be worked out in detail and applied to a number of cases of incommensurate crystals for learning more about its applicability.

References

[1] Korekawa, M., "Theorie der Satellitenreflexe", Habilitationsschrift der Ludwig-Maximilian-Univ. Muenchen, (1967).

[2] Wolff, P.M. de, Acta Cryst. A 28 S111 (1972).

[3] Wolff, P.M. de, "The pseudo-symmetry of modulated crystal structures", Acta Cryst. A 30 777-785 (1974).

[4] Janner, A. and Janssen, T., "Symmetry of periodically distorted crystals", Phys. Rev. B 15 643-658 (1977).

[5] Janner, A. and Janssen, T., "Symmetry of incommensurate crystal phases. I Commensurate basic structures", Acta Cryst. A 36 399-408 (1980).

[6] Janner, A. and Janssen, T., "Superspace groups", Physica A 99 47-76 (1979).

[7] Wolff, P.M. de, Janssen, T. and Janner, A., "The superspace groups for incommensurate crystal structures with a one-dimensional modulation", Acta Cryst. A 37 625-636 (1981).

[8] Yamamoto, A., "Structure factor of modulated crystal structures", Acta Cryst. A 38 87-92 (1982).

[9] Janner, A., "The role of superspace groups in crystal physics", in Symmetries and properties of non-rigid molecules: A comprehensive survey, Maruani J. and Serre

Josiane ed(s), 461-486, Elsevier, Amsterdam, Studies in Physical and Theoretical Chemistry, Vol. 23, (1983).

[10] Aalst, W. van, Hollander, J. den, Peterse, W.J.A.M. and Wolff, P.M. de, "The modulated structure of γ-Na_2CO_3 in a harmonic approximation", Acta Cryst. B 32 47-58 (1976).

[11] Janner, A., Janssen, T. and Wolff, P.M. de, "Bravais classes for incommensurate crystal phases", Acta Cryst. A 39 658-666 (1983).

[12] Janssen, T. and Janner, A., "Superspace groups and representations of ordinary space groups: alternative approaches to the symmetry of incommensurate crystal phases", Physica A 126 163-176 (1984).

[13] Bernard, L., Currat, R., Delamoye, P., Zeyen, C.M.E., Hubert, S. and Kouchkovsky, R. de, "Neutron scattering investigation of incommensurate $ThBr_4$", J. Phys. C. Solid State Phys. 16 433-456 (1983).

[14] Kovalev, O.V., "Irreducible Representations of the Space Groups", Gordon and Breach, New York (1964).

[15] Buiting, J.J.M., Janner, A. and Weger, M., "Superspace symmetry of $Hg_{3-d}AsF_6$", Preprint, (1985).

Acknowledgement
The stimulating collaboration with P.M. de Wolff is gratefully acknowledged.

SYMMETRY CHANGES IN CRYSTAL STRUCTURES

T.Janssen and A.Janner
Institute for Theoretical Physics
Toernooiveld, 6525 ED Nijmegen, The Netherlands.

As we have heard in other lectures in this school there is at present a large number of compounds known which show in a certain temperature interval an incommensurate phase. This interval may be small, like 1.5 degrees in $NaNO_2$, or very large, for example, several hundreds of degrees in Na_2CO_3. Usually there are at least two phase transitions: a transition at T_i from the symmetric high-temperature phase with space group symmetry to an (intermediate) incommensurate phase, and a transition at T_c from the incommensurate phase to a low-temperature superstructure. Inside the intermediate phase one may observe additional phase transitions. Moreover, the intermediate phase is not necessarily incommensurate, but may be commensurate as well.

The first problem is how to describe the intermediate phase. If it is incommensurate, it does not have space group symmetry. It can be shown, however, that there is a crystallographic group in more than three dimensions that accounts for the structure of such an incommensurate phase (See the lecture by Janner & Janssen). On the other hand one can describe the structure also using representation theory as is done in Landau theory (See Toledano's lecture). In the first chapter the relation between the two approaches will be discussed. The next question pertains to the origin of the incommensurate phase. A number of simple models has been devised for that. They will be discussed in the second chapter. Finally, in chapter three we shall talk about the relation between these two problems: the implication of symmetries of the model solutions for the symmetries of the crystal. This involves a small excursion into bifurcation theory. The concepts will be exemplified on a family of compounds with many commensurate and incommensurate phases and with a structure that can be derived from that of β-K_2SO_4.

1. IRREDUCIBLE REPRESENTATIONS OF SPACE GROUPS AND SUPERSPACE

1.1 Irreducible representations of space groups.

To fix the notation we briefly discuss the representation theory of space groups in 3 dimensions. Such a group is an extension G of the group of lattice translations Λ, isomorphic with the free abelian group of rank three Z^3, by a finite group K, isomorphic with a subgroup of O(3), called the point group. The irreducible unitary representations of Λ are given by $\{\exp(-i\vec{k}.\vec{a}) | \vec{a} \text{ in } Z^3\}$, where $\vec{k}$ is a vector in the unit cell of the dual group Λ^*. The unit cell is also called Brillouin zone. The space group G acts on the irreducible representations according to $g.\exp(-i\vec{k}.\vec{a}) = \exp(-iR\vec{k}.\vec{a})$, when $g=\{R|\vec{v}\}$ is an element of the space group G. The orbit under this action is called the star of $\vec{k}$. The isotropy group of a representation is denoted by $G_{\vec{k}}$ and is called the group of $\vec{k}$. The irreducible representations of $G_{\vec{k}}$ that, if restricted to the translation subgroup Λ, subduce a multiple of the representation labeled by $\vec{k}$ are the allowable representations. The allowable representations are then of the following form:

$$D_{\vec{k}}(\{R|\vec{v}\}) = e^{-i\vec{k}.\vec{v}}\,\Gamma(R), \tag{1.1}$$

where Γ is an irreducible projective representation of the factor group $K = G_{\vec{k}}/\Lambda$. The projective representation has a trivial factor system if the group G is symmorphic (which means that there is an origin such that all translations $\vec{v}$ in the elements $\{R|\vec{v}\}$ belong to Λ) or if $\vec{k}$ is on the boundary of the Brillouin zone (in case the unit cell is chosen as the set of points with distance to the origin smaller than or equal to any other point of Λ^*).

Finally, the irreducible representations of G are obtained from the allowable representations of the various groups $G_{\vec{k}}$ by induction. The dimension of such a representation is the product of the dimension of Γ and the number of points in the star of $\vec{k}$.

1.2 Description of modulated structures.

A modulated structure may be described by a density function $f(\vec{r})$ that can be decomposed in a Fourier series.

$$f(\vec{r}) = \sum_{\vec{k}\in M^*} \hat{f}(\vec{k})e^{i\vec{k}.\vec{r}}. \qquad (1.2)$$

For an ordinary crystal M^* is the reciprocal lattice, for an incommensurate phase it is a Z-module, which means that the vectors $\vec{k}$ are of the form

$$\vec{k} = h\vec{a}^*+k\vec{b}^*+l\vec{c}^*+\sum_{i=1}^{d} m_i\vec{q}_i, \qquad (1.3)$$

where $\vec{a}^*,\vec{b}^*,\vec{c}^*$ are the basis vectors of a lattice Λ^* in reciprocal space (See the lecture by Janner and Janssen). For simplicity we take here **d=1.** Then one can resum (1.2).

$$f(\vec{r}) = \sum_m \sum_{\vec{K}\in\Lambda^*} \hat{f}(\vec{K}+m\vec{q})\; e^{i(\vec{K}+m\vec{q}).\vec{r}} =: \sum_m f_m(\vec{r}). \qquad (1.4)$$

The component f_m transforms with a representation of G that belongs to the star of $m\vec{q}$, as one can see from the action of a translation element $\vec{a}$.

$$T_{\vec{a}}f_m(\vec{r}) = e^{-im\vec{q}.\vec{a}}\; f_m(\vec{r}). \qquad (1.5)$$

To strip the argument to its bare essentials we shall make some more, not essential, simplifications. So we suppose that the dimension of the representation Γ is one. That means that its matrices are just their characters. Consequently one can write for the action of an element of the group of $m\vec{q}$, which is also that of $\vec{q}$ for an incommensurate structure:

$$T_g f_m(\vec{r}) = e^{-im\vec{q}.\vec{v}}\chi_m(R)f_m(\vec{r}) \quad \text{with} \quad g=\{R|\vec{v}\} \text{ in } G_{m\vec{q}}. \qquad (1.6)$$

The reality of f requires that f_{-m} is equal to f^*_m and that it belongs to the same (co-)representation as f_m.

A next simplification is to assume that the inversion $I=\{-1|\vec{0}\}$ belongs to G. Then the two cosets of $G_{m\vec{q}}$ are $IG_{m\vec{q}}$ and $G_{m\vec{q}}$. The transformation of f_m and f_{-m} under all the elements of G can then be written down.

$$T_g f_m = e^{im\vec{q}.\vec{v}+i\psi}\chi_m(-R) \quad \text{for } g\in IG_{\vec{q}} , \tag{1.7}$$

$$T_g f_{-m} = e^{im\vec{q}.\vec{v}}\chi_m(R)^* f_{-m} \text{for } g\in G_{\vec{q}} ,$$

where ψ is some phase shift. In principle the characters χ_m are independent for different m. In Landau theory where the order parameter belongs to one irreducible representation, the amplitudes of the higher harmonics are determined by coupling terms. We shall see that, in order to have a one-component structure, one has to require that $\chi_m(R) = \chi_1(R)^m$. In the sequel we shall assume that and drop the subindex 1.

We have now described a modulated structure by the irreducible representation of G to which it belongs. The Landau theory does not stop here, but what comes next is not a pure symmetry reasoning. One has to introduce a free energy polynomial, up to a certain order, and to construct the possible terms in such an expansion from the invariants that are allowed by symmetry [1]. One immediate consequence of the transformation property, however, is the following. Since for g $G_{\vec{q}}$ the transformation of $f(\vec{r})$ is on one hand given by

$$T_g f(\vec{r}) = \sum_m e^{-im\vec{q}.\vec{v}}\chi(R)^m \sum_{\vec{K}} \hat{f}(\vec{K}+m\vec{q})e^{i(\vec{K}+m\vec{q}).\vec{r}}, \tag{1.8}$$

and on the other hand

$$T_g f(\vec{r}) = f(g^{-1}\vec{r}) = \sum_m \sum_{\vec{K}} \hat{f}(\vec{K}+m\vec{q}) \exp[iR(\vec{K}+m\vec{q}).(\vec{r}-\vec{v})], \tag{1.9}$$

one may draw the conclusion that for elements g for which $R(\vec{K}+m\vec{q})=\vec{K}+m\vec{q}$ one has $\hat{f}(\vec{K}+m\vec{q})=0$ unless $\chi(R)^m e^{im\vec{q}.\vec{v}}=1$. This means that there are symmetry determined systematic extinctions, although the system does not have space group symmetry.

1.3 Embedding into 3+1 dimensions.

For the same system one may use also a description in a higher-dimensional space. To that end we define a function f_s in (3+1)-dimensional space.

$$f_s(r_s) := f_s(\vec{r},\phi) := \sum_m f_m(\vec{r})e^{im\phi}. \tag{1.10}$$

The symmetry of this function is the subgroup of the direct product of G and the Euclidean group in 1 dimension that leaves f_s invariant. The argument to take the direct product instead of, for example, the Euclidean group in 4 dimensions, is that one wants to make a distinction between the main reflections (m=0 in 1.3) and the satellites (m≠0). This comes down to the requirement that main reflections are mapped on main reflections. [Actually, the argument is only valid for displacive and occupational modulation and is sometimes questionable for composite structures.]

For $g_s=(g,g_I)$ with g G and $g_I=\{\varepsilon|\Delta\}\in E(1)$, where $\varepsilon=\pm1$ one has the transformation

$$T_{g_s}f_s(r_s) = f_s(g^{-1}\vec{r},\varepsilon(\phi-\Delta)) = \sum_m \chi(R)^m e^{-im\vec{q}.\vec{v}}f_m(\vec{r})e^{im\varepsilon(\phi-\Delta)}. \tag{1.11}$$

Invariance requires that $\varepsilon=1$ and that $\chi(R)^m\exp(-im(\vec{q}.\vec{v}+\Delta))=1$. Since we have assumed that the representation Γ is one-dimensional the absolute value of $\chi(R)$ is unity.

$$\chi(R) = e^{i\xi(R)} \rightarrow \xi(R)-\vec{q}.\vec{v}-\Delta \equiv 0 \pmod{2\pi}. \tag{1.12}$$

In particular, if $g=\{1|\vec{a}\}$ is an element of the translation group one has $\Delta = -\vec{q}.\vec{a}+2\pi n$ (integer n). Hence there are four independent translations leaving f_s invariant: $(\vec{a},-\vec{q}.\vec{a})$, $(\vec{b},-\vec{q}.\vec{b}\ )$,

$(\vec{c},-\vec{q}.\vec{c})$ and $(0,2\pi)$. Therefore, the function $f_s(r_s)$ has a four-dimensional space group symmetry. It is exactly the group introduced in the lecture by Jannner and Janssen. There m was denoted by k_I and ϕ by r_I. Additional elements of the (3+1)-dimensional space group can be found from a combination of an element of $IG_{\vec{q}}$ with an element $\{\varepsilon= -1|\Delta\}$.

The extinction rules found above from the representation follow now immediately using the standard crystallographic expressions[2]. Because the Fourier transform $\hat{f}_s(k_s)$ transforms into $\hat{f}_s(R_s k_s) \exp(iR_s k_s.v_s)$ one has the consequence that for an element k_s left invariant by the 4-dimensional orthogonal transformation R_s the Fourier component $\hat{f}_s(k_s)$ vanishes unless $\exp(ik_s.v_s)=1$. This corresponds to the well known systematic extinctions which are a consequence of the existence of glide planes and screw axes in three dimensions. The importance of this observation lies in the fact that for incommensurate structures the Fourier components $\hat{f}_s$ are in one-to-one correspondence with the coefficients $\hat{f}$ in (1.2). Hence the systematic extinctions in the diffraction pattern may be interpreted in much the same way as in ordinary crystallography.

1.4 Notation of (3+1)-dimensional space groups.

Since for a symmorphic G or for $m\vec{q}$ inside the Brillouin zone the representation Γ is an ordinary representation of the factor group $G_{\vec{q}}/\Lambda$, the exponent $\xi(R)$ of the character has the value $0, \pm 2\pi/2, \pm 2\pi/3, \pm 2\pi/4$ or $\pm 2\pi/6$. In the symbol for the (3+1)-dimensional group this is indicated by putting under the symbol for an element of $G_{m\vec{q}}$ the symbol 1,s,t,q or h, respectively. If a shift Δ may be found for an element of $IG_{m\vec{q}}$ there is always an origin with respect to which $\Delta=0$. Hence under such an element we put "$\bar{1}$". If the group G is nonsymmorphic and $m\vec{q}$ lies on the zone boundary, Γ is a projective representation, but since it is one-dimensional in the case considered here, it is equivalent with an ordinary representation and the same symbols may be used.

As an example consider the (3+1)-dimensional group $P^{Pcmn}_{1s\bar{1}}$. The group G is in this case the orthorhombic group Pcmn. The prefix P (primitive) indicates that the vector $\vec{q}$ is inside the Brillouin zone. Since n is a mirror along the z-axis combined with a translation $\vec{v}=(\vec{a}+\vec{b})/2$, and the symbol in the bottom line is $\bar{1}$ (indicating $\varepsilon=-1$), the vector $\vec{q}$ points along the c-axis ($\vec{q} \doteq \gamma\vec{c}^*$). The point $(x,y,z,t):=(x\vec{a}+y\vec{b}+z\vec{c},2\pi t)$ is transformed by the generators of the group as follows: into

$$(-x,y,z+1/2,t-\gamma/2) \quad \text{by} \quad \binom{c}{1}; \tag{1.13}$$

$$(x,-y,z,t+1/2) \quad \text{by} \quad \binom{m}{s};$$

$$(x+1/2,y+1/2,-z,-t) \quad \text{by} \quad \binom{n}{\bar{1}}.$$

1.5 Commensurate phases.

Up to this point we have concentrated our attention on the incommensurate case. The (3+1)-dimensional groups, however, may also be used for the description of commensurate phases. In that case $\vec{q}$ has rational components with respect to the reciprocal lattice vectors. Then the structure has still as symmetry a three-dimensional space group, which may be derived from the (3+1)-dimensional group as the subgroup which leaves the actual crystal, i.e. the function f_s for a fixed value of the phase ϕ, invariant. We illustrate this procedure on the same group as above, $P^{Pcmn}_{1s\bar{1}}$. Suppose that the vector $\vec{q}$ is equal to $\gamma\vec{c}^*$ with rational $\gamma=r/s$, where r and s are relatively prime. The first generator combined with lattice translations along the third and fourth basis vector $n(\vec{c},-\vec{q}.\vec{c})+m(0,2\pi)$ transforms the point (x,y,z,t) into the point $(-x,y,z+m+1/2,t-m\gamma-\gamma/2+n)$. Hence it is a 3-dimensional space group element if there exist values m and n such that

$$t = t-\gamma m-\gamma/2+n \rightarrow -r-2rm-2sn = 0 . \qquad (1.14)$$

This equation has only solutions if the greatest common divisor of 2r and 2s divides r. The third generator combined with a lattice translation $m(\vec{c},-\vec{q}.\vec{c})+n(0,2\pi)$ transforms the point to $(x+1/2,y+1/2,-z+1/2+m,-t-\gamma/2-m\gamma+n)$. This is a 3-dimensional space group operation if $-r-2rm+2sn=4st=:\tau$. There is an integer solution for m and n if r/s is odd/odd and τ even or if r and s are of different parity and τ is odd. In the same way one can proceed for the other generators of the space group. The 3-dimensional space group leaving the modulated superstructure invariant is determined by r/s and by the value of τ, which is related to the phase of the modulation function. The resulting space groups are given in the following table.

TABLE I

γ	τ	3-dim. space group
$\frac{odd}{odd}$	even	$P112_1/n$
	odd	$P2_12_12_1$
	otherwise	$P112_1$
$\frac{odd}{even}$	even	$P12_1/c1$
	odd	$P2_1cn$
	otherwise	$P1c1$
$\frac{even}{odd}$	even	$P2_1/c11$
	odd	$Pc2_1n$
	otherwise	$Pc11$

All these 3-dimensional groups follow from **one** (3+1)-dimensional group. In particular, when the vector $\vec{q}$ changes, for example as a function of temperature, γ takes an infinite number of different rational values and each time the 3-dimensional symmetry changes. The symmetry, however, is also accounted for by one group in more than three dimensions. In this sense this approach allows a unifying view on modulated structures.

1.6 Landau theory.

Just as ordinary Landau theory does not stop with the symmetry description, one can go on also in the higher-dimensional space. The normal-incommensurate phase transition can only be seen as a group-subgroup transition if one enlarges also the symmetry group of the high temperature phase. In that phase the (3+1)-dimensional density function is

$$f_s(\vec{r},\phi) = \sum_{\vec{k}\in\Lambda^*} \hat{f}(\vec{k})e^{i\vec{k}.\vec{r}}, \tag{1.15}$$

because m=0 in this case. The (3+1)-dimensional symmmetry group is now the direct product GxE(1). For $T<T_i$ the density changes to $f_s=f_{os}+\delta f_s$, where f_{os} has still GxE(1) symmetry. For each value of ϕ f_s represents a possible crystal configuration. The function δf_s may be expanded into components belonging to the (nontrivial) irreducible representations of GxE(1):

$$\delta f_s = \sum_{i\alpha} c^\alpha{}_i \Phi^\alpha{}_i(\vec{r},\phi). \tag{1.16}$$

The free energy is then a functional of the function f_s, or in other words of the coefficients $c^\alpha{}_i$. For the simple case that G is the group $p\overline{1}$, the group generated by the lattice Λ and the inversion I, the irreducible representations of GxE(1) are, with the exception of those at special points, four-dimensional and labeled by the vector $\vec{q}$ and the number Q (for the representations of E(1)). Then δf_s may be written as

$$\delta f_s(\vec{r},\phi) = c_1 e^{i(\vec{q}.\vec{r}+Q\phi)}+c_2 e^{i(\vec{q}.\vec{r}-Q\phi)} + \text{c.c.} \tag{1.17}$$

Replacing the variables c_1 and c_2 by $\rho_1\exp(i\psi_1)$ and $\rho_2\exp(i\psi_2)$, respectively, the free energy expansion up to fourth order is

$$F_s = \frac{A}{2}(\rho_1{}^2+\rho_2{}^2)+\frac{B}{4}(\rho_1{}^2+\rho_2{}^2)^2 + \frac{C}{4}\rho_1{}^2\rho_2{}^2. \tag{1.18}$$

Minimizing this expression gives solutions $\rho_1 \neq 0, \rho_2 = 0$ or $\rho_1 = 0, \rho_2 \neq 0$, or $\rho_1 = \rho_2 \neq 0$. The first two have (3+1)-dimensional space group symmetry and can be seen as two equivalent embeddings of the same crystal, the third solution has no (3+1)-dimensional lattice translation symmetry.

In the usual treatment of incommensurate phases in Landau theory the order parameters $c^{\alpha}{}_{i}$ are considered as slowly varying functions of $\vec{r}$ and in the free energy gradient terms may appear (See for example the lecture by Toledano). This can also been done in higher-dimensions. Also in this case the solution with minimal energy has nonconstant order parameter if a Lifshitz invariant exists. In the present case, the existence of a Lifshitz invariant implies a transition to a phase with a higher-dimensional modulation: $d>1$ in eq.(1.3). An example of such a transition from a one- to a two-dimensional modulation occurs, for example, in biphenyl.

1.7 Concluding remarks.

The description of incommensurately and commensurately modulated phases by means of representations of 3-dimensional space groups is to a large extent equivalent to that by means of higher-dimensional space groups. Also in the latter one may proceed along the lines of the Landau theory of phase transitions. The higher-dimensional groups allow for a more unified view on the symmetry problem and are better adapted to the crystallographic usage. For example, extinction rules may be derived in just the same way as in 3 dimensions. There is a difference if one discusses the equivalence of two structures. Equivalence of representations is a too strong requirement. For example, the equivalence class of the representation changes if, due to temperature variation, the wave vector does not stay constant. To a certain extent this may be overcome using the concept of stratum of representations: representations belong to the same stratum if their isotropy groups in G are the same. This is, however, not sufficient. For example, by a change of origin the matrices of two nonequivalent representations may sometimes be transformed into each other. They belong nevertheless to different strata if the representations Γ are nonequivalent. The equivalence definition, introduced for higher-dimensional space groups, is a generalisation of the definition in use in ordinary crystallography and therefore, accounts for changes in setting or length scales.

2. MODELS FOR DISPLACIVELY MODULATED STRUCTURES.

2.1 Introduction.

The first modulated structures have been found in magnetic systems. In these the spins may form helices with a period that is incommensurate with respect to the underlying lattice. A first model for these systems was proposed by Elliott[3] in 1962. Later a number of other models have been studied. From these model studies it has become clear that the origin of modulated structures is competition between at least two forces that favour different periodicities. These models are microscopic in that they take into account the discrete structure of the crystals and interatomic forces, although the latter are usually not very realistic. However, they explain a number of features observed in experiments and not explained by phenomenological theories.

The models consist of particles on a d-dimensional lattice. The variables x_n are either the positions or the displacements from the ideal lattice sites. The potential energy is of the form

$$V = \sum_{n\in Z^d} [\Phi_o(x_n) + \sum_{s\in Z^d} \Phi_s(x_n - x_{n-s})]. \tag{2.1}$$

The first term is an interaction with a background, the second sum is over a number of neighbours s. An example is the Frenkel-Kontorova model with d=1 and interaction terms

$$\Phi_o = V_o\cos(\frac{2\pi}{a}x_n); \quad \Phi_1 = (x_n - x_{n-1} - b)^2, \tag{2.2}$$

whereas the other interactions Φ_s vanish.

A second model is a discrete version of the so-called ϕ^4 field theory[4]. In one dimension (d=1) it has $\Phi_o=0$, the variables are the displacements u_n and

$$\Phi_1 = \frac{\alpha}{2}(u_n-u_{n-1})^2+\frac{\gamma}{4}(u_n-u_{n-1})^4; \tag{2.3}$$

$$\Phi_2 = \frac{\beta}{2}(u_n-u_{n-2})^2; \quad \Phi_3 = \frac{\delta}{2}(u_n-u_{n-3})^2.$$

Introducing the difference coordinate $x_n=u_n-u_{n-1}$ in eq. (2.3) gives the potential energy

$$V = \sum_n[\frac{\alpha+2\beta+3\delta}{2}x_n^2 + \frac{\gamma}{4}x_n^4 + (\beta+2\delta)x_n x_{n-1} + \delta x_n x_{n-2}]. \tag{2.4}$$

The latter is also the expression for the potential energy of a linear chain of particles with first and second neighbour interaction, each particle sitting in an anharmonic potential well. For α sufficiently negative this well has two minima. When α goes to $-\infty$ such that the positions of the minima do not change the potential energy becomes that of a pseudo-spin system. It is the one-dimensional version of the ANNNI (axial next nearest neighbour Ising) model[5]. By choosing other units of length and energy one can choose $\gamma=1$ and either $\beta\pm1$ or $\delta=1$.

2.2 Discrete ϕ^4 model.

The ground states of the models (2.3) and (2.4) correspond, but in their dynamics they differ, because the first potential is invariant under translations and has consequently an acoustic mode whereas the other one (2.4) does not have this feature. The ground state (at T=0) satisfies $\partial V/\partial x_n=0$. Among the, possibly many, extremal points one has to look for the stable ones (the local minima) and among the latter for the one with minimal energy (the global minimum).

For the model (2.4) one has the requirement

$$(\alpha+2\beta+3\delta)x_n+x_n^3+(\beta+2\delta)(x_{n+1}+x_{n-1})+\delta(x_{n+2}+x_{n-2}) = 0. \qquad (2.5)$$

To solve this infinite set of coupled nonlinear equations one may try to find analytical solutions, which one can find if one imposes periodicity $x_{n+N}=x_n$ with N=1,2,3 or 4. Otherwise one has to rely on numerical calculations. Here there are two methods. Either one looks for periodic solutions. Then (2.5) reduces to a finite set of equations. Or one may regard (2.5) as a recurrence relation. The latter method is discussed in the following chapter.

The set of N coupled nonlinear equations may be solved by means of a Newton-Raphson method. The solutions depend still on the parameters α, and β or δ. A solution with N=20 is given for $\delta=0$ and varying α in Fig.1.

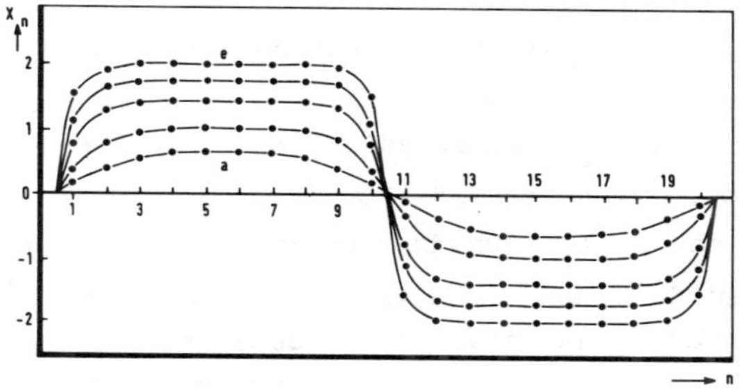

Fig.1 N=20 solution (Ref.4)

Below a certain critical value of α there is a nontrivial solution. For α in the neighbourhood of this value the amplitude is small and the form of the solution, which is the modulation function, is sinusoidal. For lower values the amplitude increases and the form becomes more rectangular. In this case the solution corresponds to a 20-fold superstructure and wave vector q=0.05. When the solution has more than 2 nodes, for instance 2s, the superstructure is still N-fold, but the modulation wave vector is s/N.

The solutions correspond only to extrema of the potential V and various solutions have to be compared in order to find the configuration with minimal energy.

When α decreases the solutions with minimal energy become less uniform (Fig.2). In fact the configuration may be described as existing of domains where the structure is nearly a periodic function with small period P (1,2,3,4 or 6) and these domains are separated by domain walls called discommensurations.

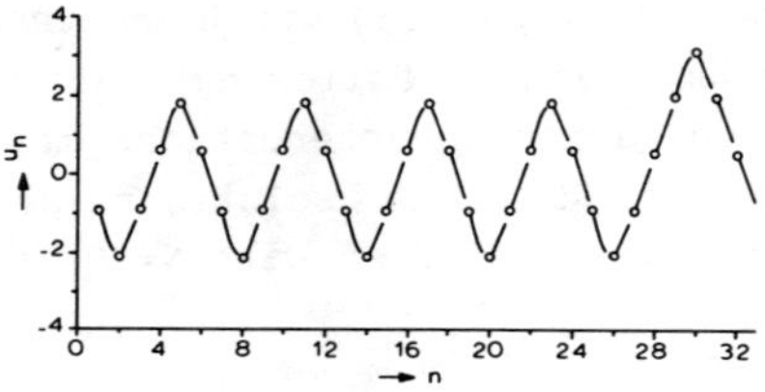

Fig.2 Discommensurations (Ref.4)

For α smaller than a critical value α_c the ground state solution is exactly periodic with period P.

When α decreases towards the value α_c, the period of the solution with minimal energy starts to shift: its modulation wave vector shifts towards the value of the superstructure (Fig.3).

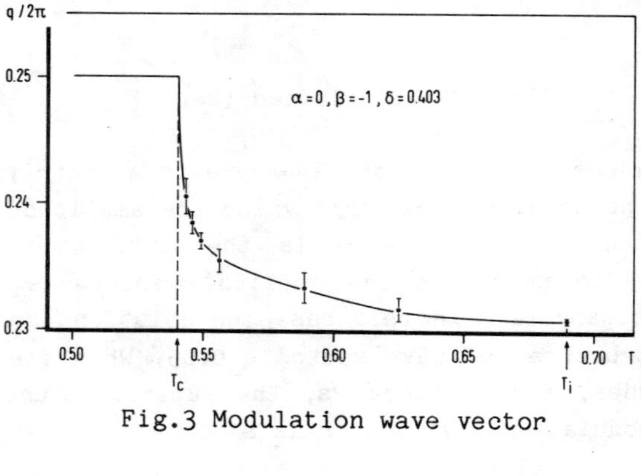

Fig.3 Modulation wave vector

By comparing the energies of the various solutions that exist simultaneously one may arrive at a diagram in parameter space indicating the structure of the ground state (Fig.4). From the figure it is clear that when one force dominates there is only a period 1 or 2 possible. In the neighbourhood, however, of the point 0,0 in the α-β-plane high order commensurate and incommensurate structures may occur (hatched regions). These hatched regions have still much internal structure (see e.g. Fig. 5 in the α-δ-plane). There is an infinite number of regions of smaller and smaller width in which a commensurate structure is the ground state. Since the parameter α may be interpreted as temperature, which can be shown using a simple mean field argument[4] this means that for changing temperature the modulation wave vector locks in at a large or even infinite number of rational values.

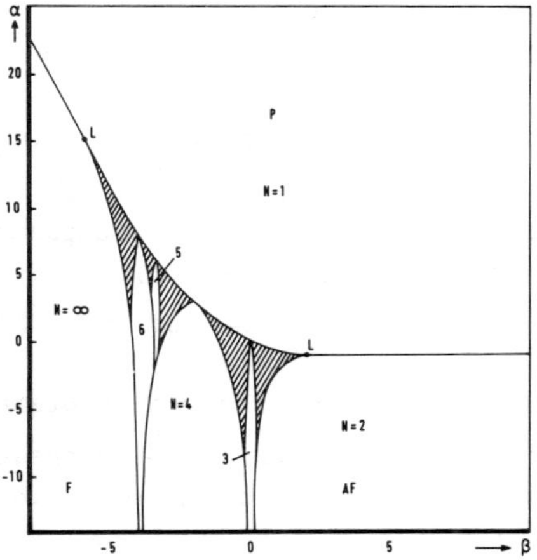

Fig.4 α-β phase diagram

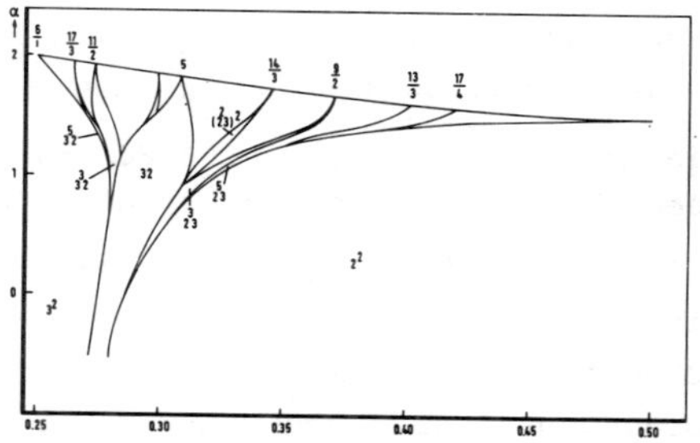

Fig.5 Detail of α-δ diagram

In a model studied by Aubry et al.[6], which includes an external field, even a lock-in at an incommensurate value is possible. One has coined for the function that has plateau´s at every rational value the name "devil´s staircase". In experiments one has indeed observed such plateau´s in the dependence of q on T. It is, however, questionable whether experiment and theory agree so well here. In the first place have devil´s stairs been found in T=0 models or molecular-field models only. In the second place the experimental situation is not so clear either. The steps are often only observed by one technique (for example with X-rays but not with neutrons) and frequently only a small number of steps can be verified with certainty. A stair with one or a small number of steps may sometimes be considered as rather devilish, but it is not exactly what is meant by that term in the literature.

2.3 Stability and excitations.

The stability of a solution to eq.(2.5) can be investigated by determining the frequencies of the small oscillations around such a solution. The squares ω^2 are the eigenvalues of the dynamical matrix $\partial^2 V/\partial x_n \partial x_m$. For the basic structure (u_n=0) of model (2.3) these eigenvalues are

$$\omega(k)^2 = \frac{1}{m} \alpha(2-2\cos(k)) + \beta(2-2\cos(2k)) + \delta(2-2\cos(3k)). \quad (2.6)$$

For decreasing values of α a dip in the dispersion curves appears which tends to zero at a wave vector k_c for $\alpha=\alpha_i$. The value of k_c is given by $\cos(k_c)=-(\beta+2\delta/4\delta)$ if the absolute value of this is smaller than 1, and oherwise $\cos(k_c)=\pm1$.

For $\alpha<\alpha_i$ the solution u_n=0 is unstable and the ground state is pseudo-periodic or periodic. The eigenfrequencies of these structures may be calculated numerically if one approximates the pseudo-periodic structure by a superstructure. For the case that k_c=s/N one has an N-fold superstructure and N branches of dispersion curves in the Brillouin zone. A large number of the excitations are just slight perturbations of the excitations in the basic structure, but there are a number of qualitatively new excitations. In Fig.6 one sees two branches for which the frequency drops to zero if the wave vector goes to zero. One is the acoustic branch.

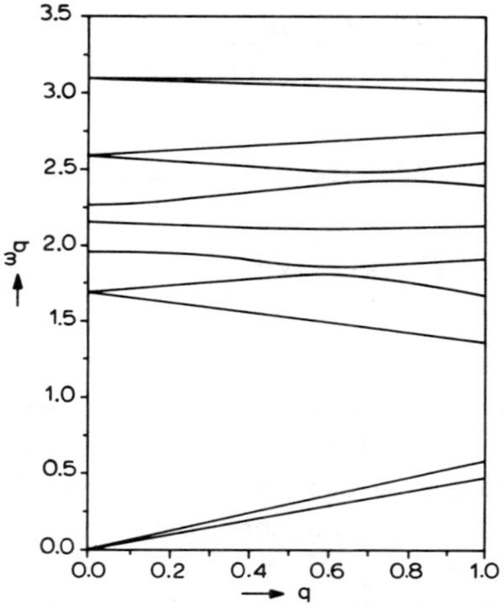

Fig.6 Dispersion curves with acoustic and phason branch (Ref.4)

For vanishing wave vector the excitation corresponds to a shift of the whole crystal. The eigenvectors of the excitations of the other linear branch give rise to motions that can be described as a shift in the phase of the modulation. In the long-wave-length limit these excitations are the phasons. Another type of excitation is an oscillation of the modulation amplitude, called amplitudon. The corresponding branch has in general a higher frequency. Both amplitudon and phason branch originate from the soft mode that drives the phase transition. At $\alpha=\alpha_i$ both branches go to zero.

2.4 The finite temperature case.

Up to this point the model considered was essentially one at T=0. For nonzero temperature one can formulate the model in a mean field approximation[7]. The particle at site n then moves in a potential

$$V_n = \Phi_o(x) + \sum_s[\Phi_s(x-\bar{x}_{n-s}) + \Phi_s(x-\bar{x}_{n+s})]. \tag{2.7}$$

The values $\bar{x}_n$ are the thermal averages of x_n obtained from

$$\bar{x}_n = \frac{\int x\exp(-V_n(x)/kT)dx}{\int \exp(-V_n(x)/kT)dx}. \qquad (2.8)$$

For the one-dimensional ϕ^4-model this leads to

$$V_n(x) = \frac{\alpha+2\beta+3\delta}{2}x^2+\frac{1}{4}x^4+(\beta+2\delta)x(\bar{x}_{n-1}+\bar{x}_{n+1})+\delta x(\bar{x}_{n-2}+\bar{x}_{n+2}). \qquad (2.9)$$

The equations (2.8) then form an infinite set of coupled nonlinear equations again which can be solved by the same means as the T=0 model above, i.e. either by looking at periodic solutions or by using the fact that (2.8) can be viewed as a recurrence relation. Among all the solutions one has to find that one that has the lowest free energy. In this way one may construct a phase diagram that is very similar to the T=0 phase diagram in parameter space.

The fluctuations around the thermodynamic ground state can be calculated as above. We shall not treat here the problem in any detail but only discuss the phason mode that exists here too. If one calculates the frequency of the phason as a function of temperature it is found that below the normal-to-incommensurate transition temperature the structure is sinusoidal and the frequency of the phason is zero. However, if one decreases the temperature the structure develops discommensurations below a temperature T_D. Below this temperature the phason frequency is finite and approaches zero for T increasing to T_D (Fig.7). The behaviour reminds of a phase transition: there is a soft mode and the structure changes. As we have discussed above one single n+1 dimensional space group may account for the symmetry of the incommensurate phase. If this is true the transition would be one without symmetry breaking. So either there is no real phase transition or one should look for additional symmetry operations that would vanish at the transition temperature.

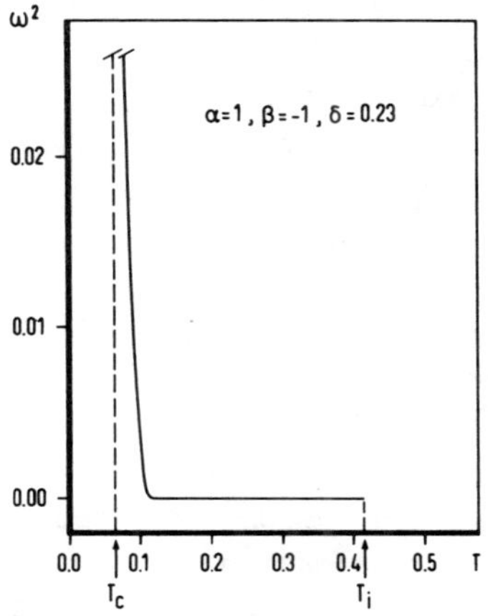

Fig.7 Frequency phason mode (Ref.12)

2.5 Excitation spectra of modulated chains.

In the previous sections the excitations of incommensurate phases have already been discussed. The spectra are in fact quite complicated. To investigate the special features occurring in these phases we have studied a very simple model: a linear chain with modulated spring constants[8].

$$V = \sum_n \alpha_n (u_n - u_{n-1})^2 \ ; \qquad \alpha_n = \alpha[1-\varepsilon\cos(2\pi\gamma n+\phi)]. \qquad (2.10)$$

Here γ is supposed to be irrational but the strategy we shall use here is the same as that for the solution of the model equations (2.5), approximation of γ by rational numbers. The equations of motion for the chain are:

$$m\omega^2 u_n = \alpha_n(u_n - u_{n-1}) + \alpha_{n+1}(u_n - u_{n+1}). \qquad (2.11)$$

When $\gamma = L/N$, one may write $n=mN+j$ with $0 \leq j < N$ and the solutions u_{mN+j} satisfy $u_{n+N}=\exp(ik)u_n$. In this case the problem is reduced to the diagonalisation of an NxN matrix.

As an example we take $\gamma=(\sqrt{5}-1)/2$ which we approximate by a series F_n/F_{n+1}, where F_n are the Fibonacci numbers 1,2,3,5,8,13,.. The spectra for each of these rational approximations of γ are given in Fig.8. Numerical analysis leads to the conclusion that in the limit the spectrum is self-similar and is a Cantor set with an infinite number of gaps.

A second characteristic property of the spectra is the behaviour of the integrated density of states (IDS), $I(\Omega^2)$ which is the number of states with $\omega^2 < \Omega^2$ divided by the total number of states. For a one dimensional chain the density of states has singularities but the IDS remains finite. Since the IDS is constant in a gap the function has in the limit of irrational γ the property that its derivative is everywhere zero, but it increases nevertheless from zero to one. It is a Cantor function, sometimes also called a devil's staircase (Fig.9).

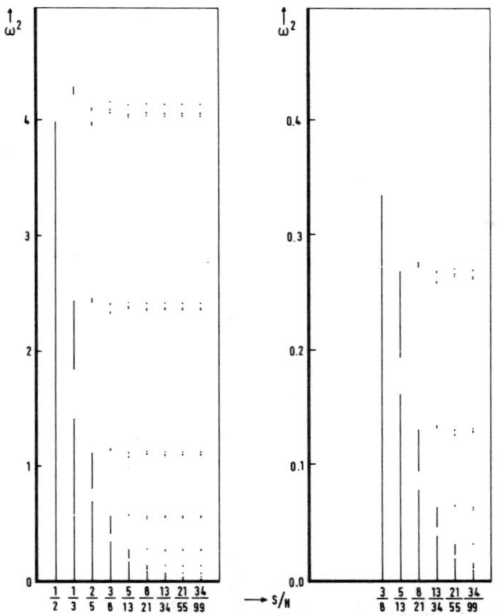

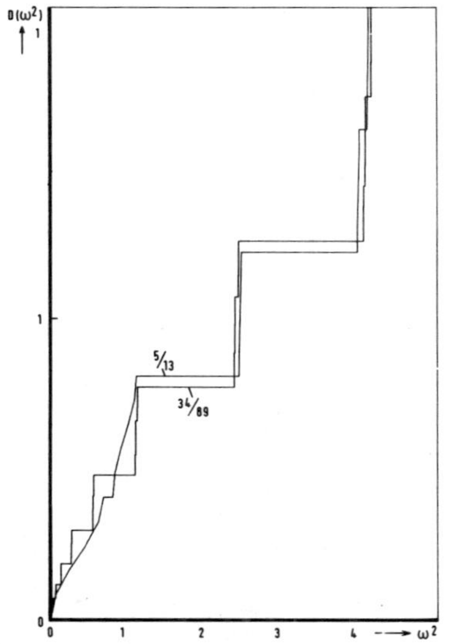

Fig.8 Spectra modulated spring model for $\gamma \rightarrow (\sqrt{5}-1)/2$.

Fig.9 Integrated density of states.

The self-similarity of the spectrum may also be seen if one plots the spectra for values of γ ranging from 0 to 1. In Fig.10 this has been done for a number of rationals L/N between 0 and 1 determined in the following way. The self-similarity may here be analysed using **Farey numbers** f(i,j). These are defined as follows. f(1,0)=0/1,f(1,1)=1/1. If $f(n,j)=r_1/s_1$ and $f(n,j+1)=r_2/s_2$ then $f(n+1,2j)=f(i,j)$ and $f(n+1,2j+1)=(r_1+r_2)/(s_1+s_2)$ for $j=0,1,..,2^{n-1}$. For example:

f(3,0)=0/1, f(3,1)=1/3, f(3,2)=1/2, f(3,3)=2/3, f(3,4)=1/1.

In Fig.10 the spectra are plotted for all rational numbers that are Farey numbers of degree 8. Numerical analysis then reveals that the spectra between f(n,j) and f(n,j+1) are (slightly distorted) scaled down versions of all spectra between 0 and 1. As an example look at the lower part of the spectra between 1/2 and 2/3. The spectra in this interval below ω=0.8 are a nearly exact copy of the whole set of spectra between 0 and 1.

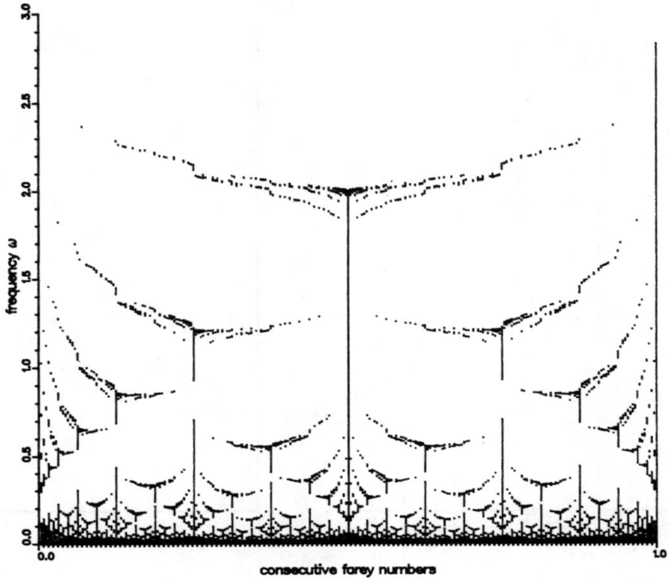

Fig.10 Spectra modulated spring model, for $\gamma=f(8,j)$, $j=0,..,128$.

The interesting properties of the spectra of incommensurate phases discussed here have not been proven rigorously, but follow from numerical analysis. It should be stressed that these properties only occur in this simple form because the modulation of the spring constants has been chosen as sinusoidal here. For modulation functions with higher harmonics the picture becomes much more complicated. Nevertheless it is to be expected that the normal-to-incommensurate phase transition gives rise to a change in the spectra with remarkable properties.

3. BIFURCATIONS AND SYMMETRY CHANGES.

3.1 Bifurcations in models.

In the preceding chapter we have considered a number of models for structural and magnetic phase transitions giving rise to a modulated phase. The models for **structural** transitions have as variables functions x_n on a d-dimensional lattice Z^d which may represent either positions or displacements (n Z^d). The potential energy is given by

$$V = \sum_n [\ \Phi_o(x_n) + \sum_s \Phi_s(x_n - x_{n-s})], \qquad (3.1)$$

where the first term gives the interaction with the background and the sum over s Z^d gives the interaction between particles a lattice vector s apart. The extrema of this potential energy are derived from $\partial V/\partial x_n = 0$. When we denote the set of parameters , for example the coupling strengths between particles such as the spring constants α, β and δ in the discrete ϕ^4-model, by μ this extremal condition may be written as

$$F(\mu,x) = 0. \qquad (3.2)$$

Then F is a mapping from the space of functions x on the lattice Z^d to itself. In the present case

$$F(\mu,x) = \Phi_o'(x_n) + \sum_s [\Phi_s'(x_n - x_{n-s}) - \Phi_s'(x_{n+s} - x_n)]. \qquad (3.3)$$

From the general theory one knows that solutions of (3.2) may show bifurcations if the Fréchet derivative, which is the linear operator in the function space defined by

$$F_x\, f = \lim_{t\to 0}\frac{F(\mu,x+tf) - F(\mu,x)}{t}, \qquad (3.4)$$

has a nontrivial kernel, or in other words if F_x has an eigenvalue zero. In this case

$$F_x f|_n = \sum_m \frac{\partial^2 V}{\partial x_n \partial x_m}\, f_m \qquad (3.5)$$

When one compares this to the equation of motion of small oscillations around a solution x^o of (3.2)

$$m\omega^2 \varepsilon_n = \sum_m \frac{\partial^2 V}{\partial x_n \partial x_m}\, \varepsilon_m \qquad \text{with } x_n = x_n{}^o + \varepsilon_n, \qquad (3.6)$$

it is obvious that a bifurcation may take place whenever there is an eigenmode with frequency zero. In solid state physics this is well known as a soft mode. For later reference we notice that the kernel $M := \mathrm{Ker}F_x$ carries a representation of the symmetry group of V (Sattinger[9]).

3.2 Symplectic mappings.

When d=1 the equation (3.2) leads to a recurrence relation if Φ_p is convex, because we can express then the most remote interacting neighbour x_{n+p} in terms of $x_{n+p-1}, x_{n+p-2}, \ldots, x_{n-p}$. In particular for the discrete ϕ^4-model one has

$$F(\mu,x) = (\alpha+2\beta+3\delta)x_n + x_n{}^3 + (\beta+2\delta)(x_{n-1}+x_{n+1}) + \delta(x_{n-2}+x_{n+2}). \qquad (3.7)$$

For $\delta=0$ one can introduce a vector v_n in R^2 with components x_n, x_{n-1} such that (3.7) becomes

$$v_{n+1} = \begin{pmatrix} x_{n+1} \\ x_n \end{pmatrix} = \begin{pmatrix} -\frac{\alpha+2\beta}{\beta}x_n - \frac{1}{\beta}x_n{}^3 - x_{n-1} \\ x_n \end{pmatrix} = S\, v_n. \qquad (3.8)$$

Eq.(3.8) defines a discrete, nonlinear mapping op R^2 onto itself. From the derivative

$$DS = \begin{pmatrix} -\frac{\alpha+2\beta+3x_n^2}{\beta} & -1 \\ 1 & 0 \end{pmatrix} \qquad (3.9)$$

it follows that the mapping S is area preserving since Det(DS)=1.

For the case that $\delta \neq 0$ one can define an analogous mapping in R^4 and in general it is a mapping in R^{2p}. Moreover, the mapping leaves a skew-symmetric nondegenerate form invariant and is therefore symplectic, and, a fortiori, volume preserving.

3.3 Orbits and crystal configurations.

To each orbit under the discrete mapping S in R^{2p} corresponds a crystal configuration for which the potential energy is an extremum. Periodic configurations (superstructures) correspond to periodic orbits, i.e. fixed points for the N-th iterate S^N of S defined by $S^n := S.S^{n-1}$ with $S^1 := S$. The character of the orbit is characterised by the linearised mapping $DS^N = \prod_{n=1}^{N} DS_n$, in particular its eigenvalues. If λ is an eigenvalue, also λ^{-1} and λ^* are eigenvalues, because S is symplectic and real. Hence for $\delta=0$ the 2 eigenvalues are on the unit circle or both on the real axis. The fixed point is then called elliptic, respectively hyperbolic. (When the eigenvalues are ± 1 it is called parabolic). For $\delta \neq 0$ there are 4 possibilities (Fig.11).

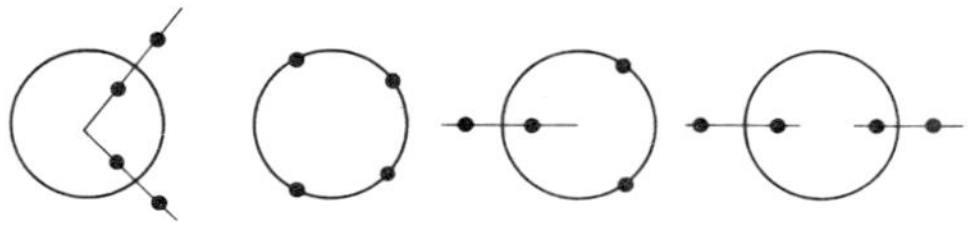

Fig.11 Eigenvalues of DS for $\delta \neq 0$.

When the orbit under S is not periodic it may be pseudo-periodic: the points are given by $x_n = f(\gamma n + \phi)$, where f is a periodic function $(f(x+1)=f(x))$ and γ is an irrational number. If the orbit is neither periodic nor pseudo-periodic it may be called chaotic. The distinction between the two types of infinite orbits is determined by the Ljapunov-exponent Λ.

$$\Lambda = \lim_{N\to\infty} \frac{1}{N} \log \left| \prod_{i=1}^{N} DS_i \xi \right|, \tag{3.10}$$

for almost every infinitesimal vector ξ in a point of the orbit. The Ljapunov exponent is zero for elliptic periodic and for pseudoperiodic orbits and non-zero otherwise. Examples of orbits in 2 and 4 dimensions are given in Figs.12 and 13.

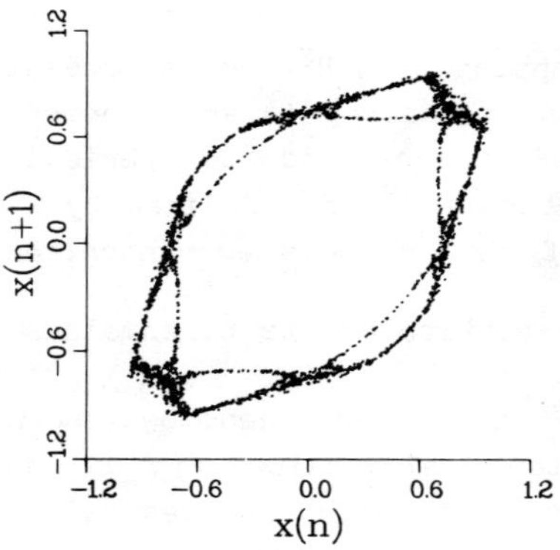

Fig.12 Orbit in R^2 (chaotic) (Ref.11)

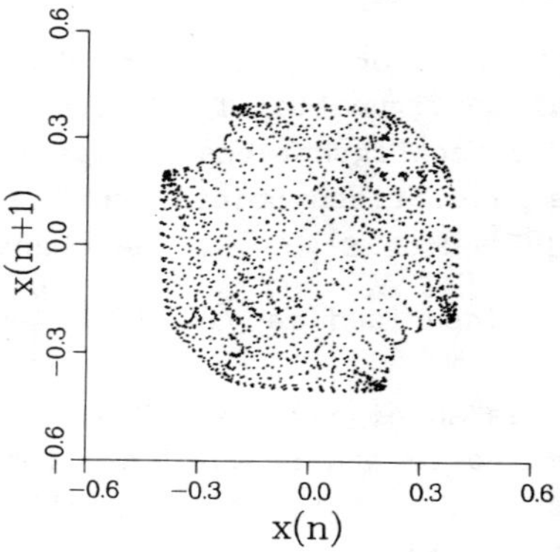

Fig.13 Projection of orbit in R^4 (Ref.11)

3.4 Stability.

One has to distinguish between the stability of the crystal configuration and that of the orbit under the symplectic mapping. The crystal is stable in a linear analysis if the eigenvalues ω^2 of (3.6) are all positive. For a periodic orbit the solutions of this equation satisfy the Bloch property: $\varepsilon_{n+N}=\varepsilon_n \exp(ik)$. Just as above, one can see eq.(3.6) also as a discrete (linear) mapping of R^{2p} onto itself. For $\omega=0$ this linear mapping is exactly the mapping DS. From this fact and the Bloch property follows that DS^N has an eigenvalue exp(ik) if and only if eq.(3.6) has an eigenvalue $\omega^2=0$.

Stability under the mapping S requires at least that the eigenvalues are at most of absolute value 1, i.e. that they are on the unit circle in the complex plane. Therefore, one may say that linear stability of the crystal configuration and stability of the nonlinear mapping are complementary: if the orbit is stable, the corresponding crystal is unstable (or marginally stable) and if the crystal is stable the corresponding orbit is unstable. Notice, however, that not every unstable orbit corresponds to a stable crystal.

The property mentioned above implies that it is hard to find the stable crystal configurations from the mapping by numerical methods.

The relation with the bifurcation discussed earlier implies that whenever the eigenvalues of DS^N approach the unit circle as a function of the parameters the corresponding crystal shows a soft mode. The motion of the eigenvalues of the trivial fixed point ($x_n=0$) are shown in Fig.14.

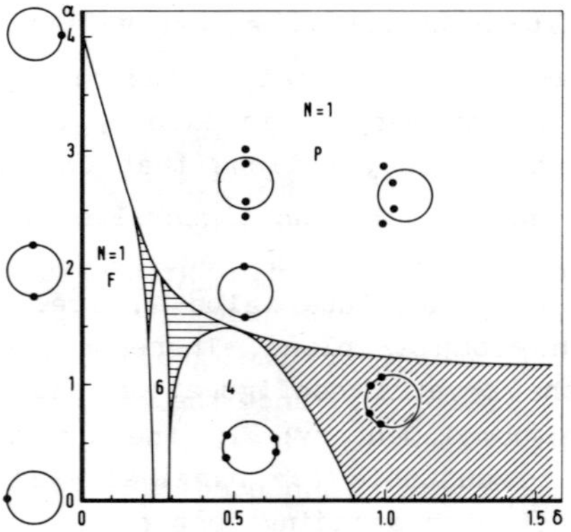

Fig.14 Phase diagram in parameter space (α-δ) and the change of the eigenvalues of DS for the trivial fixed point.

3.5 Phase diagram, bifurcations and ground state solutions.

If one determines the local minima by looking at those orbits that correspond to stable crystal configurations, one can compare these local minima to find the global minimum, i.e. the configuration of the ground state. This depends on the parameters. In Fig.14 a diagram is given indicating the structure of this ground state as a function of the parameters α and δ. Indicated is the period for a superstructure. There are 2 structures with period one: the trivial solution $x_n=0$ for sufficiently high value of α, and a nontrivial solution $x_n=c\neq0$ for low values of δ. The hatched regions contain pseudo-periodic structures and long-period superstructures.

If α decreases the trivial solution becomes unstable for $\alpha=\alpha_i$. There is a soft mode and a bifurcation takes place. For lower values of α the structure is periodic with period N if the eigenvalues reach the unit circle at $\exp(2\pi s/N)$ or pseudoperiodic if this occurs at $\exp(ik)$ with $k/2\pi$ irrational. The value of k is given by

$$\cos(k) = -\frac{\beta+2\delta}{4\delta}, \tag{3.11}$$

if this is smaller than one in absolute value. Otherwise $\cos(k)=\pm 1$.

In Fig.15 the orbits are given corresponding to the ground state for varying value of α, whereas $\beta=-1, \delta=0.23$. For low values of α this orbit is periodic with N=6 (crosses in the figure). On approaching the region where the N=6 superstructure is the ground state the structure of the ground state changes. The quasi-continuous curve breaks up and the points (black circles) cluster around the positions of the N=6 structure with the exception of some points that are not connected and which describe the discommensurations (open circles).

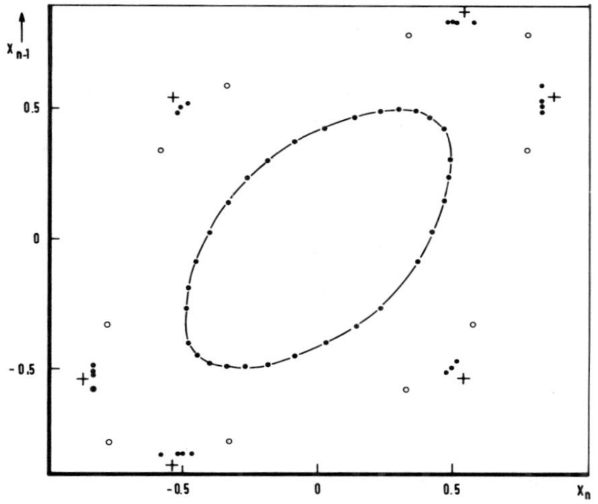

Fig.15 Orbits corresponding to the ground state for a number of values of α. Centre: trivial fixed point (high α). Ellipse:"incommensurate" phase. Crosses: superstructure with period N=6. Small circles: high-order commensurate orbit (N=32) with discommensurations.

The trivial solution always exists, but for $\alpha<\alpha_i$ there is a zero frequency mode at a wave vector that changes with α (Fig.16).

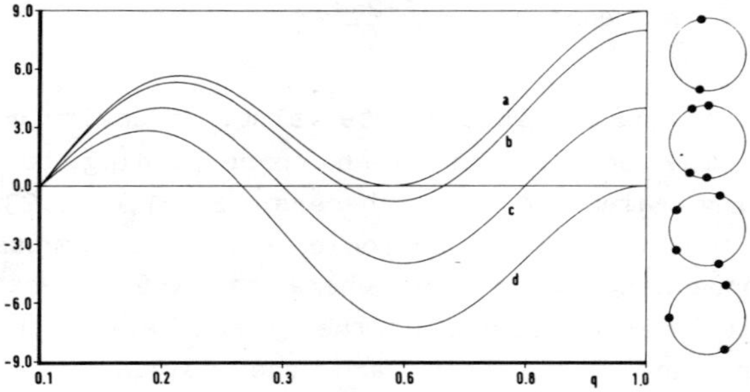

Fig.16 Dispersion curves trivial fixed point and corresponding eigenvalues of DS.
The wave vector is in units π and the 4 curves correspond to:
a) α=1.25, b) α=1, c) α=0 and d) α= -1.

Therefore, this solution stays bifurcating into new solutions until k=0 or π. Since there are 2 pairs of complex conjugate eigenvalues on the unit circle, one may have several bifurcations at the same time. For example, for δ=1 and β=-1 there is for α = -1 (d in Fig.16) a simultaneous bifurcation from the trivial solution to an N=6 and to an N=2 solution (Fig.17). [Moreover, there is in the neighbourhood a bifurcation to an N=23 orbit: the ring in Fig.17a condenses to a finite orbit in Fig.17b !] These bifurcations continue and also the bifurcated solutions bifurcate in turn. Then one can observe the well studied behaviour of Feigenbaum sequences also here. For physics the continuing bifurcation process means that also a large number of stationary points and local minima are formed, which has consequences for the thermodynamical behaviour.

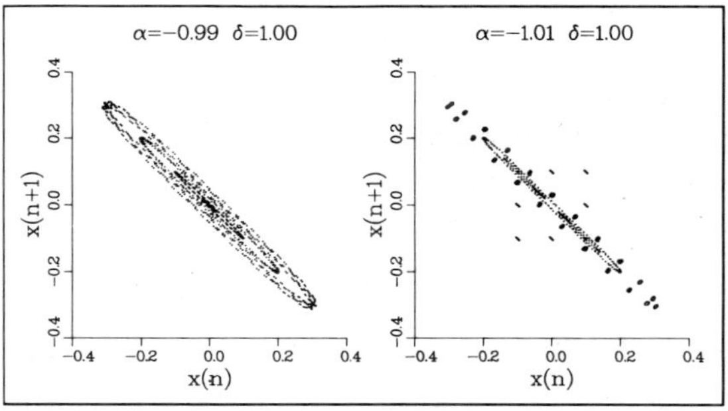

Fig.17 Simultaneous bifurcations. (Ref.11)

3.6 Symmetry of orbits.

As Sattinger[9] has shown one can reduce the bifurcation equations (3.2) in infinite-dimensional function space to a finite problem. To that end define the projection P on $M:=\mathrm{Ker}F_x$. An arbitrary function may be decomposed as

$$x = x_1 + x_2 \; ; \qquad \text{with } x_1:=Px, \text{ and } x_2:=(1-P)x. \tag{3.12}$$

Applying (1-P) to F one has the equation

$$(1-P)F(\mu, x_1 + x_2) = 0, \tag{3.13}$$

from which one may derive, using the implicit function theorem, an expression of x_2 in terms of $x_1 : x_2 = x_2(x_1)$. Substituting this in F and applying P gives

$$PF(\mu, x_1 + x_2(x_1)) = 0 \quad , \tag{3.14}$$

which is an equation in M, which is often finite-dimensional.

In the present model

$$M = \{z\cos(kn+\phi) \mid z,\phi \; R\} \tag{3.15}$$

This 2-dimensional space ($k\neq 0$ or π) carries a representation of the symmetry group G of the potential V. The group is the direct product of the one-dimensional space group of an equidistant array $p\overline{1}$ and , if the functions Φ_s are symmetric, a group of 2 elements. The generators of G are a,n and p given by their action:

$$\begin{aligned} a&: n \to n+1 \\ m&: n \to -n \\ p&: x_n \to -x_n \end{aligned} \tag{3.16}$$

The irreducible representations of G are labelled by k and ε with $0\leq k<2\pi$ and $\varepsilon = \pm 1$. The action of G in the representation space M is given by

$$\begin{aligned} a&: z \to z,\ \phi \to \phi+k \\ m&: z \to z,\ \phi \to -\phi \\ p&: z \to z,\ \phi \to \phi + \pi \end{aligned} \tag{3.17}$$

The function V(x) on the function space (3.1) determines a function $W(x_1)$ on M via $W(x_1)=V(x_1+x_2(x_1))$. A minimum of V that we look for corresponds to a minimum of W . The fixed points of the representation of G on M are certainly stationary points. If no other stationary points occur this implies that orbits under the symplectic mapping S have a symmetry determined by the fixed points of the representation. It follows that the orbits $\ldots x_n, x_{n+1}, x_{n+2}\ldots$ have the form

......c,b,a,b,c,....
......b,a,a,b,......
......b,a,-a,-b,....
......c,b,0,-b,-c,..

Hence the symmetry character of the orbits is determined: for example, it shows in 2 dimensions 2 mirrors (Fig.18).

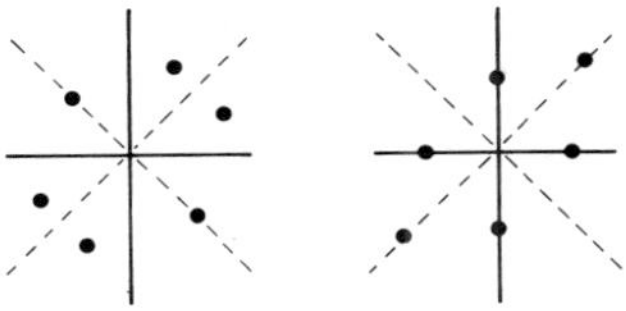

Fig.18 Stable and unstable N=6 orbit

The existence of these symmetry operations implies also symmetry elements for the crystal structure. If the orbit is given by $x_n=f(\gamma n + \phi)$ with periodic function f and with γ irrational or rational (in the latter case the crystal structure is a superstructure), the function f satisfies

$$f(x+1) = f(x); \quad f(x+1/2) = -f(x); \quad f(x+1/4) = f(-x+1/4). \quad (3.18)$$

3.7 Application to TMATC-metallates.

The model described above and the symmetry considerations about the orbits may be applied to a family of compounds with similar structure: the tetramethyl-ammonium-tetrachloro-metallates TMATC-M. They have chemical formula $[N(CH_3)_4]_2MCl_4$, where M=Zn,Co,Mn,Fe or Cu and show each a series of commensurate and incommensurate phases.

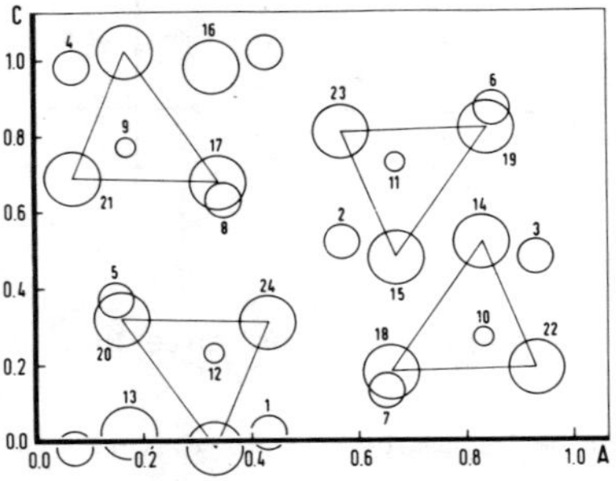

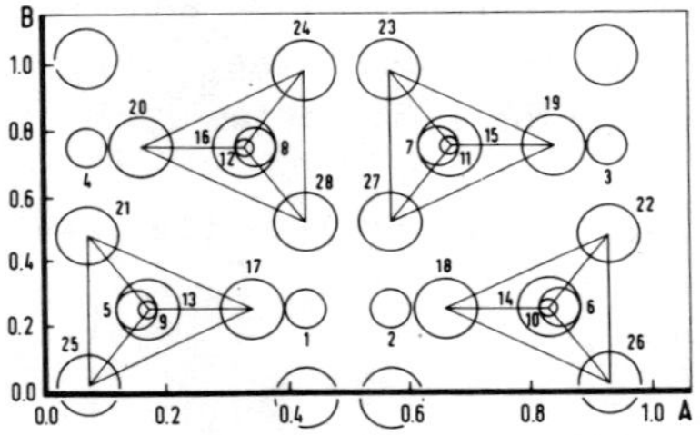

Fig.19 Structure TMATC-M

The structure may be considered as to exist of layers of (not fully regular) tetrahedra MCl_4 and TMA groups (Fig.19). The layers are perpendicular to the c-axis and are situated near $z=n+1/4$ and $z=n+3/4$ (integer n). Inside the layer there are electrostatic and sterical hindrance forces. One can show that in the layer one may have an instability and that when the instability sets in, the tetrahedra start to rotate (Fig.20). If one denotes the rotation angle of the tetrahedra at $z=n+1/4$ by x_{2n} and those at $z=n+3/4$ by x_{2n+1} the potential energy of one layer may be approximated by $Ax_n^2/2+x_n^4/4$. The interaction between layers may be approximated by the first (quadratic) nonvanishing terms. Then the potential energy in terms of the rotation angles x_n is

$$V = \sum_n \left[\frac{A}{2}x_n^2 + \frac{1}{4}x_n^4 + \frac{C}{2}(x_n-x_{n-1})^2 + \frac{D}{2}(x_n-x_{n-2})^2\right]. \tag{3.19}$$

This is the expression for the discrete ϕ^4 model. The phase diagram of this model has been discussed above.

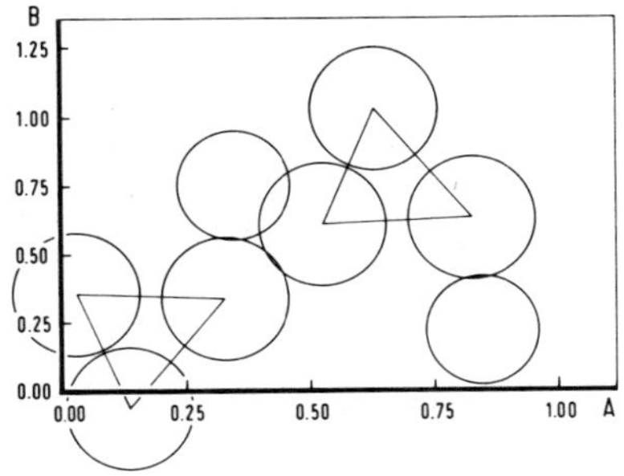

Fig.20 Rotation of tetrahedra (Ref.13)

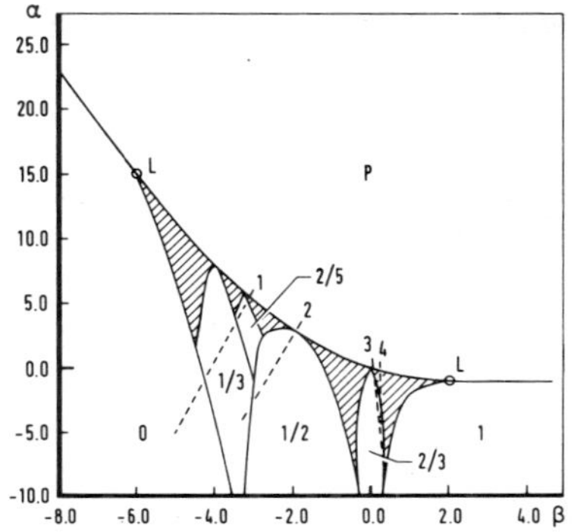

Fig.21 Phase diagram TMATC-M (Ref.13)

The basic space group of the structure is the orthorhombic group Pcmn. From the model it follows that the modulation wave vector $\vec{q}$ is $\gamma\vec{c}^*$. From the symmetry of the modulation function it may be derived that the (3+1)-dimensional space group is $P\frac{Pcmn}{1s\bar{1}}$. This higher-dimensional space group occurs irrespective of the value of γ. Hence in this model the group may be found both for commensurate and for incommensurate modulation. In the first chapter we have shown that for rational values of γ the 3-dimensional space group of the superstructure is determined by nominator and denominator of $\gamma=r/s$ and by the phase τ. The latter is the phase of the stable crystal configuration . In general, there are both a stable and an unstable solution with the same r/s (see Fig.18).

In Fig.21 the phase diagram for the model is given, in which are indicated the values of γ of the ground state solutions. The lines indicate the change of the parameters as a function of temperature for a number of TMATC-M compounds. In Table II the observed phase transitions are listed for the members of the family together with the 3-dimensional space group symmetry of the phase. These groups are exactly the groups one obtains from the (3+1)-

dimensional space group by the construction of section 1.5.

TABLE II

TMATC-Zn	TMATC-Co	TMATC-Mn	TMATC-Fe	TMATC-Cu
Pcmn	Pcmn	Pcmn	Pcmn	Pcmn
			IC	
IC	IC	IC	$\frac{3}{7}$:P112$_1$/n	IC
			IC	
$\frac{2}{5}$:Pc2$_1$n	$\frac{2}{5}$:Pc2$_1$n	$\frac{1}{2}$:P12$_1$/c1	$\frac{2}{5}$:Pc2$_1$n	
$\frac{1}{3}$:P112$_1$/n	$\frac{1}{3}$:P112$_1$/n	$\frac{1}{3}$:P112$_1$/n	$\frac{1}{3}$:P112$_1$/n	$\frac{2}{3}$:P2$_1$/c11
0:P2$_1$/c11	0:P2$_1$/c11		0:P2$_1$/c11	1:P112$_1$/n

The model thus gives a unifying view on the various members of the family of compounds.
There is one phase diagram and one (3+1)-dimensional space group for the whole TMATC-M family. A similar conclusion has been obtained by Hogervorst[10]

3.8 Concluding remarks.

The model discussed above gives a unified view of both the series of phase transitions and the space group symmetries of a family of TMA compounds. The space group symmetry is derivable from a space group in four dimensions which in turn is fully determined by the symmetry of the problem.

Acknowledgement.
Most of the results of chapters 2 and 3 have been obtained in stimulating collaboration with J.A. Tjon (Utrecht). We have profited also very much from collaboration and discussions with C. de Lange (Nijmegen) and J.C. Toledano (Paris).

References

[1] Landau, L. and Lifshitz, L., "Statistical Physics", Pergamon Press, London(1970).

[2] Wolff, P.M. de, Janssen, T. and Janner, A., "The superspace groups for incommensurate crystal structures with a one-dimensional modulation", Acta Cryst. A 37 625-636 (1981).

[3] Elliott, R.J., "Phenomenological discussion of magnetic ordering in the heavy rare-earth metals", Phys. Rev. 124 346 (1961).

[4] Janssen, T. and Tjon, J.A., "Model for a ferroelectric phase transition via an incommensurate phase", Ferroelectrics 36 285-288 (1981), "Microscopic model for incommensurate crystal phases", Phys. Rev. B 25 3767-3785 (1982).

[5] Bak, P. and Boehm, J. von, "An Ising model with solitons, phasons, and a floating modulated phase.", Phys. Rev. B 21 5297 (1980); Fisher, M.E. and Selke, W., "Infinitely many commensurate phases in a simple Ising model", Phys. Rev. Lett. 44 1502-1505 (1980).

[6] Aubry, S., Axel, F. and Vallet, F., "Devil´s staircases and Manhatten profile in an exact model for an incommensurate structure in an electric field", Preprint, (1985).

[7] Janssen, T. and Tjon, J.A., "Incommensurate crystal phases in mean-field approximation", J. Phys. C 16 4789-4810 (1983).

[8] Lange, C. de and Janssen, T., "Incommensurability and recursivity: lattice dynamics of modulated crystals", J. Phys. C 14 5269-5292 (1981).

[9] Sattinger, D.H., "Topics in stability and bifurcation theory", Springer, (1973).

[10] Hogervorst, A.C.R., Ph.D. thesis, TH Delft, The Netherlands (in preparation).

[11] Janssen, T. and Tjon, J.A., "Bifurcations in lattice systems", J. Phys. A 16 673-696 (1983).

[12] Janssen, T. and Tjon, J.A., "Incommensurate crystal phases in mean-field approximation", J. Phys. C 16 4789-4810 (1983).

[13] Janssen, T., "On the application of a frustration model to the phase diagram of tetramethylammonium-chlorometallates and other A_2BX_4 compounds", to appear in Ferroelectrics (1985).

APPLICABILITY OF THE LANDAU THEORY TO STRUCTURAL, INCOMMENSURATE, AND MAGNETIC PHASE TRANSITIONS

Pierre Tolédano

Groupe de Physique théorique, Faculté des Sciences d'Amiens
33, rue Saint-Leu, 80039 Amiens cedex
FRANCE

ABSTRACT

The essential features of the Landau theory of phase transitions are recalled. Its applicability to structural transitions between stricly periodic phases is discussed. Shortcomings of the current phenomenological description of incommensurate systems are pointed out, and a self-consistent method for these systems is proposed. The respective advantages of various approaches proposed for magnetic transitions, are illustrated in the case of a latent antiferromagnetic ordering.

1. INTRODUCTION

The elaboration of a theory for a given phase transition is a difficult problem, because of the variety and complexity of the experimental data which generally characterize real phase transitions. As an example, let us consider structural transitions which, stricly speaking, constitute only a fraction of the transitions involving a structural modification. In contrast to reconstructive transitions, structural transitions involve only small modifications in the length and orientation of the atomic bonds, i.e.no fundamental rearrangement of the lattice takes place. However, such a characterization of structural transitions is practically useless as, generally, no informations are available on the preservation of the atomic bonds, at least in the early stages of an experimental investigation.

Structural transitions are generally recognised when there exist a coherent set of observations among which a major part consists in anomalies of some physical quantities (e.g. susceptibility specific heat, spontaneous polar-tensors), and structural data denoting a symmetry relationship between the phases. The coherency of the observations implicitly refers to group-theoretical and thermodynamic concepts which constitute an intuitive phenomenological approach of the phase transition. A this point it can be stressed that the Landau theory is nothing else than the theoretical systematization of the preceding concepts. It thus constitute the more natural first-approach for the interpretation of a phase transition.

When considering the experimental data available for structural transitions, one can be struck by the variety of qualitative and quantitative situations found among their physical anomalies. For example, in TGS the main anomalies are connected with the dielectric properties of this material[1), whereas in LaP_5O_{14} it is the elastic properties which are mainly affected by the transition.[2)]
In KDP one finds strong dielectric and elastic anomalies[3)] while both types of anomalies are weak in NbO_2[4)]. For two compounds displaying major dielectric anomalies, namely TGS and GMO, there are respectively a divergence of the dielectric susceptibility $\chi(T)$ at the Curie point, and a small jump in $\chi(T)$ as the temperature is lowered[5)]. In different members of the boracite family, one has either a jump or a drop of the dielectric susceptibility[6,7)] at T_c. The same diversity can be verified in the temperature dependence of the elastic constants in compounds such as TeO_2[8)], LaP_5O_{14}[2)], $SrTiO_3$[9)] or V_3Si[10)]. Besides, in a given material different components of the same macroscopic tensor (e.g. the elastic constants in LaP_5O_{14}) have completely inequivalent behaviours[2)]. The understanding of a phase transition requires to relate the different anomalies in a given material within a single model, but also to explain why there exist other classes of behaviours.

As will be seen below, the first step for formulating a phenomenological theory for a given transition, consists in determining the transition's order-parameter, i.e. the phenomenological variable which accounts for the essential mechanism responsible for the transition. Such a determination is far from being obvious as many variables - which may display very similar temperature dependences - are potential candidates to be the order-parameter : polar tensors (e.g. polarization or strain), atomic displacements which are not connected to any macroscopic quantity, internal rearrangements in the molecules which modify the density of charges, etc... One of the main results of the Landau theory analysis of a phase transition is to provide the symmetry of its order-parameter.

From the knowledge of the order-parameter symmetry, a tentative model describing the transition anomalies can be constructed. Comparing the predictions of the model to the available experimental data allows verification of the correctness of the initial assumptions and if necessary to improve the model (e.g. by including higher degree terms or neglecting coupling terms in the Landau expansion, or by assuming some additional temperature-dependent coefficients). The preceding phenomenological approach should then be corrected and precised, using the Theory of Critical phenomena, and microscopic models. In this respect, let us note that these complementary methods of understanding phase transitions are not of much help in a first approach of a given transition. As for the Critical theory, it is out of its scope to account for non-universal quantities such as the magnitude of the anomalies, or the coupling strength between different physical quantities, as well as to describe the behaviour outside of the critical region. As for microscopic models, the complexity of the systems investigated, with many interesting degrees of freedom and large crystalline anisotropies, forbid them generally to account for the set of observed phenomena, unless they introduce these additional degrees of freedom in a phenomenological way. This removes their advantage which is being "first principles" explanations.

Two types of classifications underly the application of the Landau theory to a given phase transition. A first classification relates the observed point-group modification to the spontaneous macroscopic quantities arising at the transition. It allows to predict the domain pattern in the low-temperature phase, and the main type of transition anomalies. A second classification starts from the space-group change, and consists in determining the order-parameter symmetry, the coupling to macroscopic quantities, and the form of the relevant anomalies. In these lectures, we show how the preceding approach applies to structural, incommensurate and magnetic transitions.

At first, an intuitive approach of the Landau theory is given through particular examples of proper and improper transitions (§2). The general formulation of the theory is recalled (§3), and its applicability to structural transitions, between strictly crystalline phases, is discussed (§4). Then, the assumptions which constitute the starting point of the current Landau theory of incommensurate systems are underlined and the shortcomings of this theory are pointed out. Accordingly a "self-consistent" approach of incommensurate transitions is outlined (§5). Finally, the peculiar features of the Landau theory of magnetic systems are illustrated in the case of a "latent" antiferromagnetic transition (§6).

2. INTUITIVE APPROACH OF THE LANDAU THEORY

As a first example, let us consider a crystal undergoing a structural transition corresponding to the $4/m \rightarrow 4$ point-group change. As can be seen on Fig. 1, the phase having the lower point symmetry possesses two possible stable states, energetically equivalent, and differing by the orientation of their symmetry elements (e.g. the 4 axis) with respect to the higher symmetry phase. These states can coexist, in practice, in a given sample, and constitute the domains. With respect to a common frame of reference, the domains can be distinguished by the different sign of the spontaneous polarization component P_z (i.e. the component appearing below T_c and vanishing by symmetry in the high temperature phase). Let us note that one domain can be obtained from the other by applying a conjugated electric field E_z. In this paraelectric-to-ferroelectric transi-

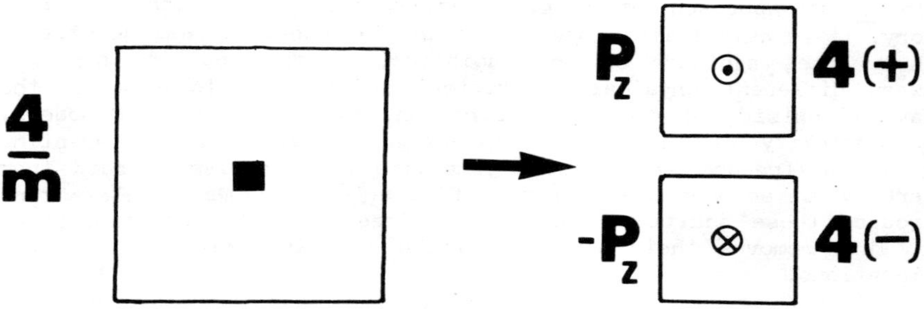

FIG.1 :Projection of the tetragonal unit-cells for the paraelectric-ferroelectric 4/m→ 4 transition.

tion, dielectric anomalies can be essentially expected, the main difference between the two phases being that the equilibrium values for P_z are :

$$P_z^e = 0 \text{ for } T \geqslant T_c, \text{ and } P_z^e = 0 \text{ for } T < T_c \qquad (1)$$

In an attempt to give a phenomenological picture of the preceding transition, let us introduce the non-equilibrium free-energy of the system F, which is assumed to be a function of the temperature T, the pressure P, and of the variational parameter P_z. Accordingly, minimization of $F(T,P,P_z)$ with respect to P_z should yield the equilibrium values :

$$F(T,P,0) \text{ for } T \geqslant T_c \quad , \quad \text{and } F(T,P,P_z^e = 0) \text{ for } T < T_c$$

Assuming P_z^e is sufficiently small near below T_c, i.e. it corresponds to atomic displacements which remain small with respect to the distances between atoms in the crystal, has the consequence that F can be developped as a polynomial expansion of F nearby $P_z^e = 0$:

$$F(T,P,P_z) = F(T,P,0) + aP_z + \frac{\alpha}{2} P_z^2 + \frac{b}{3} P_z^3 + \frac{\beta}{4} P_z^4 + \ldots \qquad (2)$$

Using a single expansion F for a description of the transition, implies to assume that F is invariant by the symmetry operations of the highest symmetry group 4/m. On a physical basis, this is justified by the fact that the interactions responsible for the transition have a symmetry which is at least as high as the observed phases. Application of the mirror reflexion σ_z to expansion (2) leads to :

$$F(T,P,P_z) = F(T,P,0) + \frac{\alpha}{2} P_z^2 + \frac{\beta}{4} P_z^4 + \ldots \qquad (3)$$

Minimization of (3) yields the following solutions :

$$P_z^e = 0 \text{ for } \alpha \geqslant 0, \text{ and } P_z^e = \pm\left(-\frac{\alpha}{\beta}\right)^{1/2} \text{ for } \alpha < 0 \text{ and } \beta > 0 \qquad (4)$$

Comparing (1) and (4) suggests the simplest requirements :
$\alpha = a(T-T_c)$ with a and β positive constants.

From (3), (4) and $E_z = \frac{\partial F}{\partial P_z}$, we deduce the temperature dependence of the dielectric susceptibility χ_{zz} in the two phases :

$$\chi_{zz}=\mathrm{Lim}\left(\frac{P_z}{E_z}\right)(E_z \to 0) = \frac{1}{a(T-T_c)} \text{ for } T \geqslant T_c \text{ and } \frac{1}{2a(T_c-T)} \text{ for } T < T_c \quad (5)$$

The temperature dependences found for P_z^e and χ_{zz}, which are represented on Fig. 2, are a test for our initial assumption that P_z is the order-parameter, i.e. the suitable quantity which determines the instability of the high symmetry phase. The experimental results show that there exist indeed cases of ferroelectric transitions (or other structural transitions) entering in the preceding scheme. These so-called proper transitions[11] are thus defined by the fact that the order-parameter identifies to the macroscopic polar tensor, which acquires a spontaneous value below T_c. However, we will show in the following example that proper transitions do not constitute the unique case. Before leaving the example under consideration let us represent a possible mechanism for the 4/m → 4 transition, by assigning the space-groups P4/m and P4 to the paraelectric and ferroelectric phases respectively. On Fig. 3 two unit-cells are shown in each case. One can see that the transition corresponds to the onset of microscopic dipoles, having the same magnitude and direction in each unit-cell.

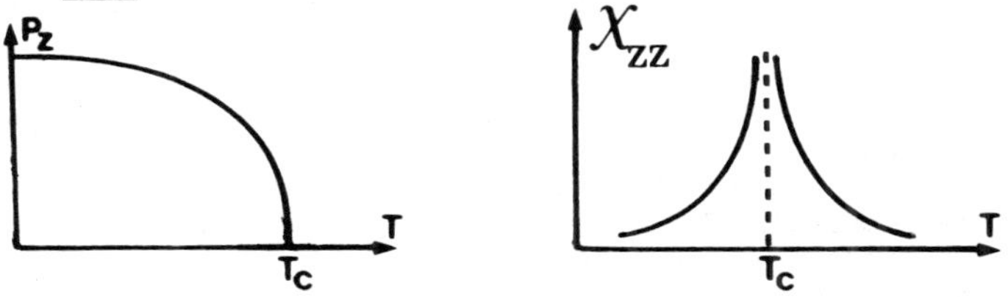

FIG.2 : Theoretical dependences for $P_z(T)$ and $\chi_{zz}(T)$ in a proper ferroelectric.

Each dipole corresponds to the displacement of a single positive ion outside its initial location, at the barycenter of the negative charges. The uniform values of the dipoles in all the unit-cells warrants that the primitive translations of the crystal are the same in both phases. Thus, the lower symmetry group can be considered as the intersection of the high-symmetry group (P4/m) and of the symmetry of the polarization component P_z. This model allows oneself to understand qualitatively, why the dielectric susceptibility is found to diverge at T_c : as the positive ions responsible for the onset of a dipole move spontaneously at T_c, they will be very loosely bound to their equilibrium location near T_c. Accordingly, their displacement along the c axis can be induced very easely by the conjugated field E_z, giving rise to a considerable increase of the dielectric susceptibility χ_{zz}. In a correlated way, one can say that the fluctuations ΔP_z of the polarization diverge at T_c.

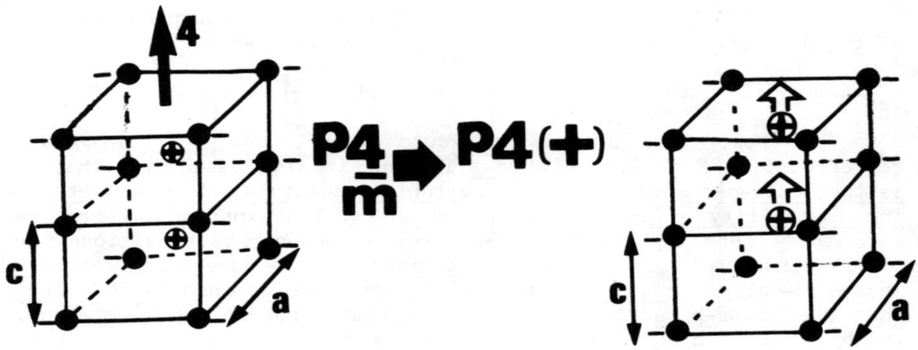

FIG.3 : Representation of two unit-cells in the paraelectric P4/m phase and in the ferroelectric P4(+) phase.

We now examine the more complex example of an improper ferroelectric transition corresponding to the $P4_2/m(V) \rightarrow P4_{1,3}(2V)$ space-group change. In the model represented on Fig.4 only the positive ions, all identical, are represented. During the transition, two groups of ions, labelled x and y, move independently, each one being likely to undergo two opposite sets of displacements. From Fig.4 it is obvious that the four displacements ±x and ±y are energetically equivalent and will lead to equally stable states. Their space groups can be easely identified as the enantiomorphous $P4_1$ and $P4_3$ groups respectively. In all cases, the primitive translation along the tetragonal axis is doubled in comparison to its value in the high symmetry phase. As in our first example, the change in point symmetry (4/m → 4) is a ferroelectric one. However, though this symmetry change is directly induced by the represented ionic displacements, these displacements do not generate any dipole in the unit-cell. A dipole, the existence of which is symmetry permitted, will eventually arise from secondary displacements of the ions in the structure. The resulting polarization will however be small, since it is a "side effect" of the transition.

This microscopic model, that was previously introduced in Ref. 12), illustrates a number of distinctive features of improper ferroelectric transitions, namely : 1) the symmetry of the ferroelectric phase cannot be obtained by retaining the symmetry elements common to the $P4_2/m$ phase and to a uniform polarization, as these elements do not account for the doubling of the crystal's periodicity along c.

2) the four stable states ± x and ± y correspond to two ferroelectric domains of opposite polarization (i.e. ± x can be associated to $+P_z$ whereas y corresponds to $-P_z$), and to two antiphase domains. These latter domains cannot be differentiated by any macroscopic quantity, e.g. −x is obtained from +x in the $P4_1$ phase, by a uniform shift equal to the non-primitive translation c.

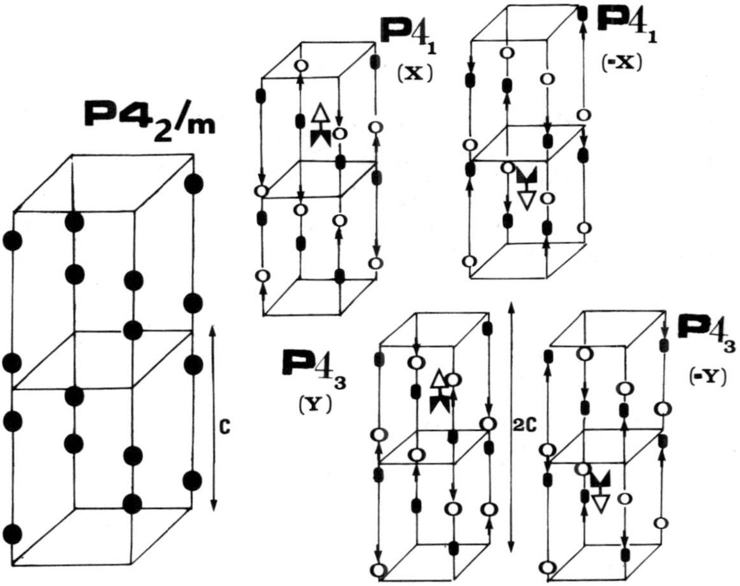

FIG.4 : Microscopic model of an improper ferroelectric transition (a) Two-unit-cells in the P4$_2$/m phase.(b) Four equivalently stable ferroelectric phases. The black arrows indicate the ions which move at the transition for each set of collective displacement.

3) no remarkable increase in the dielectric susceptibility of the crystal should be expected at the transition, as the ionic motions which become easier to induce at the transition are not directly related to the onset of a dipole. It is the collective displacements $\pm$ x and $\pm$ y that will display large fluctuations, and thus be associated to a large susceptibility of a non-macroscopic nature.

Thus, in the improper ferroelectric model under consideration, the most significant ordering parameter relative to the transition is not the polarization but the set of non-polar displacements $\pm$ x or $\pm$ y. It brings us to give a first list of the intrinsic properties characterizing the transition order-parameter, by contrast to the other quantities which acquire spontaneous values below T_c : i) the space-group of the low-symmetry phase is the intersection of the high-symmetry group and of the own symmetry of the order-parameter components ; ii) the susceptibility associated to the order-parameter diverges at T_c. Such properties are fulfiled by the polarization component P_z for a proper ferroelectric transition, where P_z is the order-parameter, but not in the present example of an improper transition. This will be made more clear now in an intuitive theoretical approach of the improper case.

Let us consider the non-equilibrium free-energy of the system as a polynomial expansion of the spontaneous variables x, y and P_z. One can write the most general expression :

$$F(T,P,x,y,P_z) = F(T,P,0,0,0) + \Sigma_{lmn} A_{lmn} x^l y^m P_z^n$$

The transformation properties of the variables x, y, P_z under the operations of the space-group $P4_2/m$, which are partially shown in Table I, supplies the following form for F :

$$F(T,P,x,y,P_z)=F(T,P,0,0,0)+ \frac{\alpha}{2} (x^2+y^2)+\frac{1}{2\chi_o} P_z^2+\delta P_z(x^2-y^2)+\frac{\beta_1}{4}(x^4+y^4) + \frac{\beta_2}{2} x^2y^2+ \frac{\gamma}{4} P_z^4+\mu P_z^2(x^2+y^2)+... \quad (6)$$

where the coefficients α, χ_o, δ, β_1, β_2, γ, μ are assumed a priori to be temperature and pressure dependent.

We first retain the quadratic terms in (6), which are the only relevant ones (if their use is conclusive) for small values of the variables near T_c. Minimization of F reveals that a necessary and sufficient condition for the state ($x^e=0$, $y^e=0$, $P_z^e=0$) to be stable, is that $\alpha > 0$ and $\chi_o > 0$. Therefore this condition must be verified for $T > T_c$ and should be violated below T_c. This may happen if α or (and) χ_o changes sign at T_c. As $\alpha(T,P)$ <u>and</u> $\chi_o(T,P)$ can only vanish <u>together</u> at a definite temperature <u>and</u> pressure, such a circumstance can occur only for a special type of transition, the character of which will change as soon as the pressure is varied. Therefore we have to choose between the conditions ($\alpha = 0, \chi_o > 0$) or ($\chi_o=0, \alpha>0$).

The latter condition must be discarded as it would imply, for small values of P_z below T_c, that the minimum of F with respect to x and y is $x^e=y^e=0$, in contradiction with the assumption of non-zero values for at least one of these variables below T_c. Thus the only condition adapted to our model is $\alpha= 0$, $\chi_o >0$. Accordingly, as for a proper transition we can assume $\alpha= a(T-T_c)$, $a > 0$.

<u>Table I</u>

Transformation properties of x, y and P_z under some operations of the space-group $P4_2/m$.

	$\vec{c}$	$(I/0,0,0)$	$(C_4/0,0,\frac{c}{2})$
x	$-x$	y	x
y	$-y$	x	$-y$
P_z	P_z	$-P_z$	P_z

Minimization of (6) with respect to P_z yields, in a first approximation :

$$P_z^e \simeq - \delta\chi_o (x^2 - y^2) \quad (7)$$

showing that for small x, y and P_z, P_z^e is of the order of the square of x^e and y^e. If we only keep in (6) the terms which are of order x^4, y^4 (such terms must be considered in order to stabilize the non-zero values for x^e or y^e below T_c), the free-energy gets the simplified form :

$$F(T,P,x,y,P_z)=F(T,P,0,0,0)+\frac{\alpha}{2}(x^2+y^2)+\frac{\beta_1}{4}(x^4+y^4)+\frac{\beta_2}{2}x^2y^2+\frac{P_z^2}{2\chi_o}+\delta P_z(x^2-y^2)$$

The absolute minima of F with respect to P_z, x and y determine <u>two couples</u> of possibly stable states for $T < T_c$:

I. $x^e = \pm(-\frac{\alpha}{\beta_1'})^{1/2}$; $y^e=0$, or $y^e = (-\frac{\alpha}{\beta_1'})^{1/2}$; $x^e = 0$, for $\beta_2' > \beta_1' > 0$,

II. $x^e = \pm y^e = (\frac{-\alpha}{\beta_1'+\beta_2'})^{1/2}$, for $\beta_1' > |\beta_2'|$,

with $\beta_1' = \beta_1 - 2\delta^2\chi_o$, $\beta_2' = \beta_2 + 2\delta^2\chi_o$ and P_z^e obeying to Eq(7) in states I and II. State I is non-ferroelectric (P_z=0) while in state II one has :

$$P_z = \pm \frac{\chi_o\delta a}{\beta_1'}(T_c - T) \qquad (8)$$

In each of the two ferroelectric domains of opposed polarization, definied by Eq(8), coexist the two antiphase domains corresponding respectively to x^e <u>or</u> y^e equal to $\pm(-\alpha/\beta_1')^{1/2}$. These four energetically equivalent stable states <u>actually correspond to the microscopic model represented on Fig.4</u>. Eq(8) shows that P_z has a <u>linear</u> dependence with temperature in the vicinity of T_c, thus differing from the $(T_c-T)^{1/2}$ variation law found for a proper ferroelectric. Furthermore, the dielectric susceptibility χ_{zz} is found to vary as :

$$\chi_{zz} = \chi_o \quad \text{for } T \geqslant T_c, \text{ and } \quad \chi_{zz} = \frac{\chi_o}{1-2\delta^2\chi_o/\beta_1'} \quad \text{for } T < T_c$$

so that an <u>upward</u> jump, but no divergence, is observed on cooling at T_c for χ_{zz}. The temperature dependences of P_z and χ_{zz}, which are represented on Fig.5, typify an improper ferroelectric transition.

The preceding discussion sheds light on two additional intrinsic properties of the transition order-parameter : iii) it is the degree of freedom with respect to which the high-symmetry phase becomes unstable ;

i4) it controls the onset of spontaneous values for the other (secondary) physical quantities.

The preceding examples have shown that the macroscopic characteristics of a structural transition (e.g. spontaneous quantities, symmetry change, form of the anomalies) can be deduced from a phenomenological treatment once that the order-parameter (P_z in our first example, the displacement (x,y) in the second one) is identified. One is then able to deduce informations about the

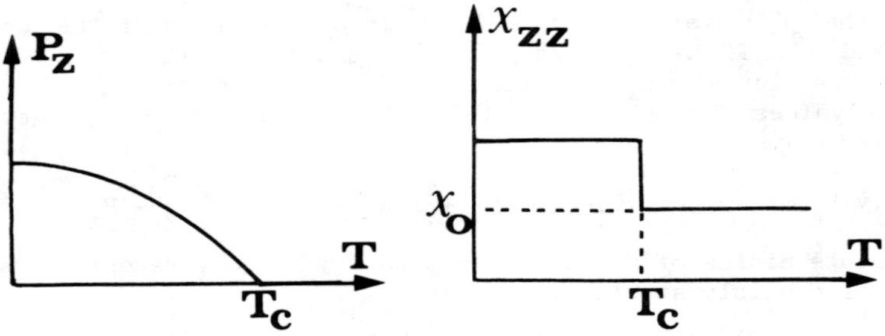

FIG.5 : Theoretical variation of $P_z(T)$ and $\chi_{zz}(T)$ for an improper ferroelectric transition.

other degrees of freedom (e.g. P_z when the order-parameter is (x,y)) by considering their coupling to the order-parameter. The form of the coupling term, as well as the form of the other invariants figuring in the free-energy, is entirely determined by the symmetry of the phases in presence. In the next section we are going to recall the basic arguments, first introduced by Landau[13)], which legitimate this approach.

3. THE LANDAU THEORY OF PHASE TRANSITIONS : GENERAL FORMULATION

In the phenomenological approach used in Sec. 2 two quantities play an essential part : the order-parameter, and the free-energy expansion. The most general description of a phase transition requires introduction of another crucial quantity : the set of probability density functions $\rho_i(\vec{r})$ describing the spatial distribution of the particles in the crystal. When considering the variation of the $\rho_i(\vec{r})$ and their symmetry, the Landau theory (at least in its initial formulation) applies only to the case where the $\rho_i(\vec{r})$ vary continuously at the transition with a change in their symmetry. However we will see a posteriori in Sec. 4 that the theory can be adapted to categories of transitions accompanied by a discontinous change of the $\rho_i(\vec{r})$ at the critical temperature T_c.

Let us consider a crystal, the symmetry of which is associated to the probability density $\rho(\vec{r})$. The initial assumptions of the Landau theory are :

1/ at T_c the symmetry group (i.e. the space-group) of the crystal changes from G_o (for $T \geqslant T_c$) to G (for $T < T_c$).

2/ $\rho(\vec{r})$ varies continously at T_c.

The thermodynamic potential at equilibrium is a function of P and T. However, we can construct a thermodynamic potential near equilibrium as : $F = F(T,P,\rho(\vec{r}))$ where F depends on the non-equilibrium configurational function $\rho(\vec{r})$. The equilibrium distribution of particles will thus be supplied by $\rho^e(\vec{r})$ such as $F(T,P,\rho^e(\vec{r}))$ corresponds to the absolute minimum of $F(T,P,\rho(\vec{r}))$ as ρ varies.

Near T_c, at fixed P, $\rho^e(\vec{r},T,P)$ varies continuously. Therefore, in order to determine $\rho^e(\vec{r})$ we can restrict the variation of $\rho(\vec{r})$ to functions which are close to $\rho_o^e(\vec{r})$ where $\rho_o^e(\vec{r})$ is the equilibrium value of $\rho(\vec{r})$ for $T = T_c$. In other terms :

$$\rho(\vec{r},T,P) = \rho_o^e(\vec{r},T_c,P) + \delta\rho(\vec{r},T,P) \qquad (9)$$

where $\delta\rho \ll \rho_o^e$. Accordingly, assuming $F(T,P,\rho)$ is non-singular at T_c, one can write :

$$F(T,P,\rho) = F(T,P,\rho_o^e) + F_1(\delta\rho) + F_2(\delta\rho) + \ldots \qquad (10)$$

where the F_1, F_2,... are functionals, respectively linear in $\delta\rho$, quadratic in $\delta\rho$, etc... We can note here that, as F is a scalar with respect to $\vec{r}$, the F_n functions should involve integrals $A_n \int [\delta\rho\ (\vec{r})\ ]^n\, d\vec{r}$.

$\delta\rho(\vec{r})$ can be written :

$$\delta\rho\ (\vec{r}) = \sum_{n,i} C_n^i\ \psi_n^i(\vec{r}) \qquad (11)$$

where, for n fixed, the $\psi_n^i(\vec{r})$ are normalized functions transforming into eachother like the basis vectors of an irreducible representation (IR) of the high-symmetry group G_o. The index n runs over the different IR's of G_o. However, since we are interested in $\delta\rho$ variations which will lead to a change in symmetry of $\rho = \rho_o + \delta\rho$, the identity representation Γ_1 (whose basis functions are invariant by G_o) is excluded from the summation (11). Accordingly $\delta\rho$ (and ρ) will be invariant under the symmetry operations of the low-temperature symmetry group G, the scalar coefficients C_n^i vanishing for all (n,i) when $\delta\rho$ goes to zero.

As $\delta\rho$ is real, either the C_n^i and ψ_n^i are real, or we can construct physically IR's composed by a complex IR and its complex conjugate so that the corresponding C_n^i are real numbers. Besides, instead of taking the C_n^i as fixed coefficients and the $\psi_n^i(\vec{r})$ as transforming functions, we can do the inverse and consider the ψ_n^i as a fixed basis and the C_n^i as transforming coefficients. This allows to avoid dealing with the action of G_o on vector functions but rather on scalar quantities. Thus the F_n(n = 1,2,...) can be considered as functions of the C_n^i, taken as the variational parameters. As the identity representation Γ_1 does not figure in the summation (11) we can deduce :

$$F_1(\delta\rho) = 0 \qquad (12)$$

Furthermore, as only physically IR's are considered, we have for a p-dimensional IR one and only one quadratic invariant of the C_n^i, which can be written for a given n :

$$\sum_{i=1}^{p} (C_n^i)^2$$

Accordingly, the form of the free-energy F, up to the second degree terms, is :

$$F(T,P,\rho) = F_o(T,P) + \sum_n \alpha_n(T,P)\left[\sum_{i=1}^{p} (C_n^i)^2\right] \qquad (13)$$

where α_n is the coefficient associated to the nth IR.

From a discussion, similar to the one performed in Sec.2, we obtain the following results :

i) <u>At T_c</u> the symmetry of the crystal is G_o, therefore all the equilibrium values $(C_n^i)^e = 0$, otherwise $\delta\rho^e$ would have a lower symmetry than ρ_o and so would the crystal. Then necessarily $\alpha_n \geq 0$ for all n. If $\alpha_n > 0$ (strictly), then $\alpha_n > 0$ in the neighborhood of T_c (by continuity) and $(C_n^i)^e = 0$, i.e. no symmetry change would take place at T_c. In consequence <u>at least one</u> $\alpha_n(T_c) = 0$.

ii) <u>only one</u> $\alpha_n(T_c) = 0$ (let say α_{no}), otherwise we would not have a <u>line</u> of continuous transitions but an isolated point in the pressure temperature diagram. Thus, at T_c : $\alpha_{no} = 0$ and $\alpha_n > 0$ for $n \neq n_o$. The set of C_{no}^i coefficients spanning the n_o^{th} IR of G_o constitute the <u>order-parameter</u> of the transition. The dimensionality of the order-parameter is equal to the number of ψ_{no}^i functions forming a basis for the preceding IR. The <u>non-zero</u> components of the order-parameter determine below T_c the symmetry of the crystal (i.e. the group G).

iii) As $\alpha_{no} \geq 0$ for $T \geq T_c$ and $\alpha_{no} < 0$ for $T < T_c$ we can assume in the vicinity of T_c a linear variation : $\alpha_{no}(T) = a(T-T_c)$ with $a > 0$. Above T_c, F_2 is sufficient to determine the minimum of F (Fig.6.a)). Below T_c the expansion (13) does not insure the stability of the solution $C_{no} \neq 0$ (Fig.6.b)). Thus, higher degree terms F_n must be taken into account. If we put :

$$C_{no}^i = C\,\gamma^i \quad \text{with} \quad C^2 = \sum_{i=1}^{p} (C_{no}^i)^2$$

the third degree term in F can be written :

$$F_3(C_{no}^i) = B(T,P)\, C^3 F_3(\gamma^i)$$

At T_c we have $F_2(C_{no}^i) = 0$. Accordingly :

$$F(T_c,P,\delta\rho) - F_o(T_c,P) \simeq F_3(C_{no}^i)$$

which has a saddle point for $C = 0$ (Fig.6c)). Therefore we must impose $F_3(C_{no}^i) = 0$. As the simultaneous conditions : ($\alpha_{no}(T_c,P)=0$, $B(T_c,P) = 0$) would define an isolated transition in the pressure-temperature diagram, we need :

$$F_3(\gamma^i) = 0 \qquad (14)$$

The absence of cubic invariants transforming as the n_o^{th} IR, is a <u>necessary condition (called the Landau condition)</u> to be imposed to the IR associated to the order-parameter, for a second-order transition to take place at T_c.

i4) If $F_3(\gamma^i) = 0$ it is (at least) the F_4 term which determines the absolute minimum of F below T_c :

$$F(T,P,\delta\rho)=F(T,P,0)+\alpha_{no}C^2+\beta(T,P)C^4F_4(\gamma^i) \qquad (15)$$

β must be such that $F-F_o$ be bounded from below (Fig.6d). The minimum of F will be reached only for certain $\{\gamma^i\}$ in the space of $\{C^i_{no}\}$, several direction corresponding eventually to the same group G. The symmetry G below T_c is that of the function

$$\delta\rho^e = \sum_{i=1}^{P'\leq P} (C_{no})^e \psi^i_{no} \qquad (16)$$

where the equilibrium values $(C_{no})^e = C\gamma^i$ are non-zero.

The main predictions of the Landau theory in its more general formulation can be summarized as follows : the instability of the crystal may take place at a second-order transition only if one IR (one irreducible degree of freedom) is involved at T_c. This IR is singled out by the fact that its quadratic contribution to the free-energy has a vanishing coefficient α_{no} at T_c. The symmetry properties of the C_{no} (the order-parameter components) determine the phase diagram, i.e. the symmetry of the low-temperature phase(s). The stability of these phases requires to take into account at least the fourth degree terms in F. As shown in Sec. 2, the preceding considerations must be completed by considering the $\{C^i_n\}$ corresponding to other degrees of freedom $n \neq n_o$, through coupling terms which are of degree higher than two. These couplings allow to describe the anomalies affecting the various physical quantities at the transition. We will now discuss the verification of this theoretical scheme for structural transitions.

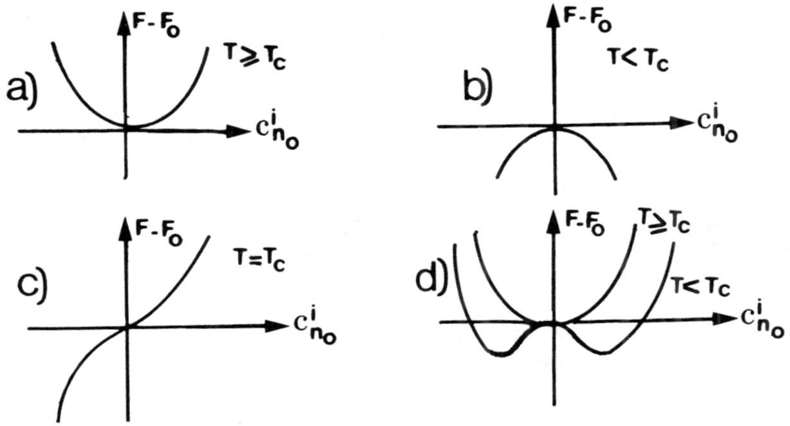

FIG.6: Form of the non equilibrium function $F(C^i_{no})-F_o$. a) and b) only F_2 is taken into account; c) $F_3 \neq 0$ at T_c; d) $F_3=0$ and F_4 is considered.

4. APPLICABILITY OF THE LANDAU THEORY TO STRUCTURAL TRANSITIONS BETWEEN STRICTLY CRYSTALLINE PHASES

4.1 The Lifshitz criterion

In the initial formulation of the Landau theory[13)] the assumption is made that the stable states of the crystal are characterized by definite C^i_{no} values of the order-parameter components corresponding to an <u>absolute minimum</u> of the free-energy F. Lifshitz has remarked[14)] that this latter condition did not test completely the stability of the phases at macroscopic level, as F has <u>also</u> to be minimum with respect to <u>variations</u> of the C^i_{no} which <u>are not homogeneous</u> (i.e. which vary from one point of the crystal to another).

Assuming for simplicity a one-coordinate dependence of the C^i_{no}, a non-homogeneous state will correspond to :

$$\delta\rho(\vec{r}) = \sum_{i=1}^{p} C^i_{no}(x)\ \psi^i_{no}(\vec{r})$$

Thus, instead of an homogeneous free-energy $F(C^i_{no})$ one has to consider a free-energy <u>density</u> $f[C^i_{no}(x)]$ with $F = \int_{crystal} f\,dx$,

where f depends on the $C^i_{no}(x)$ <u>and on their derivatives</u> with respect to x. If the $C^i_{no}(x)$ vary slowly at microscopic level, one can write

$$f(x) \simeq \frac{\alpha_{no}}{2} \sum_{i=1}^{p} (C^i_{no})^2 + f_4(C^i_{no}) + \sum_i a_i \frac{\partial C^i_{no}}{\partial x} + \sum_{i,j} b_{ij}\, C^i_{no} \frac{\partial C^j_{no}}{\partial x} \quad (17)$$

On the right-hand side of Eq.(17), the third term as well as the symmetric part of the fourth term contribute to F as <u>surface</u> terms and thus can be neglected. The only significant part of the derivative invariants is the anti-symmetric term :

$$\sum_{i,j} b_{ij}\left(C^i_{no} \frac{\partial C^j_{no}}{\partial x} - C^j_{no} \frac{\partial C^i_{no}}{\partial x}\right) \quad (18)$$

If such a term (so-called Lifshitz invariant) <u>is not</u> absent by symmetry, the homogeneous C^i_{no} do not determine the minimum of F as, for a suitable sign of the b_{ij}, the energy of the system <u>may be lowered.</u> Therefore <u>the Lifshitz criterion</u> imposes that : <u>in order to have a second-order transition between two macroscopically homogeneous phases</u> (i.e. strictly crystalline phases) <u>Lifshitz invariants of the type (18) should be absent by symmetry</u>.

From the preceding requirement, follows a drastic reduction of the IR's which are liable to induce continuous transitions between normal crystalline phases. Actually Lifshitz has shown[14)] that only should be considered IR's associated to k-vectors of the first Brillouin-zone, the invariance group of which is <u>non-polar</u>. This condition, which discards, in particular general points of the Brillouin zone, limits to a <u>finite number</u> (a few thousands) the IR's of interest. Actually, only the IR's corresponding to 84 distinct vectors of the 14 Brillouin zones (i.e. the centers and a few high-symmemetry points of the surfaces) must be considered[14-16]. Accordingly a systematic working out of the symmetry changes allowed by the Landau theory can be performed and compared to the available experimental data. This has been the subject of a number of comprehensive studies[15-19] the conclusions of which are summarized in the next section.

4.2. Phase diagrams for second and first-order transitions in the framework of the Landau theory

Once the space-group G_o associated to the high-temperature phase of a given crystal is known, one can deduce the possible symmetries for the low-temperature phases by applying to each IR of G_o a procedure which can be deduced from the theoretical considerations recalled in Sec.3 (for a practical application of the Landau approach see Refs 15-20). When restricting to the active IR's of G_o (i.e. the IR's fulfiling the Landau and Lifshitz conditions), only a limited number of cases have to be examined. Accordingly, the phase diagrams corresponding to the active IR's of the 230 space groups have been completed, and can be found under table forms in Refs 15-19). We will not analyse here the essential symmetries revealed by these studies[21] - (e.g. the possible dimensionalities of the order-parameters, the reduced number of distinct free-energies and order-parameter symmetries, etc...) but emphasize on the main conclusions that can be drawn about the applicability of the Landau theory to structural transitions in real systems. Such conclusions are deduced from a detailed comparison of the theoretical results with the available experimental data.

1.All known second-order structural transitions between strictly crystalline structures, can be unambiguously connected to a single active IR of the high-temperature space-group

This property can be verified in a large number of materials belonging to all classes of structural transitions, namely in pure ferroelectrics[15] (TGS, $LiTaO_3$, $Pb_5Ge_3O_{11}$), pure ferroelastics[17] (LaP_5O_{14}, $YNbO_4$, $SrTiO_3$, $KH_3(SeO_3)_2$), ferroelectric-ferroelastics[19] (KH_2PO_4, $NaKC_4H_4O_6.4H_2O$), secondary ferroics[16] (NH_4Cl) and non-ferroics ($Ag_2H_3IO_6$, $Bi_2Ti_4O_{11}$, $CsFeF_4$, $NaNO_3$)[18]. Actually, no counter-example is presently known to this essential prediction of the Landau theory, i.e. no second-order transition between crystalline phases has been found up to now, than can be clearly connected to more than one irreducible degree of freedom, or to an inactive IR.

2. A large fraction of the observed first-order structural transitions, involving a group-subgroup relationship, can also be associated to a single active IR.

This can be verified among pure ferroelectric materials (SbSI, $Sr_2Ta_2O_7$), pure ferroelastics (TeO_2, $RbCdCl_3$, VO_2, Mn_3O_4), ferroelectric-ferroelastics ($BaTiO_3$, KH_2AsO_4, $Gd_2(MoO_4)_3$), secondary ferroics

($LaCoO_3$, $AlPO_4$) and non-ferroics ($CsH_3(SeO_3)_2$, $(NH_4)_2H_3IO_6$, Mg_3Cd). The intriguing experimental fact that a majority of the structural transitions induced by a single active IR are first-order, has raised the question whether some fundamental property (fluctuations, defects) do not force certain transitions to be necessarily first-order. In this respect a tentative answer was suggested in the framework of the Renormalization-Group approach. More precisely, a number of authors[22,23] have hypothesized that transitions which do not possess a stable fixed point in their Landau-Wilson ε-expansion would necessarily be first-order. However this conjecture was shown to be non-predictive when the number n of components of the order-parameter is smaller than four (as for $n < 4$ there always exist a stable fixed point[24]). Furthermore, confirmed examples of second-order transitions were found recently in systems with $n \geqslant 4$ where no stable fixed points exist[25,26]. Accordingly, the large occurence of first-order transitions among structural systems, which are associated to active IR's, still remains unexplained. It denotes a possible incompleteness in the symmetry restrictions imposed by the Landau theory for a transition to be second-order.

3. The Landau condition is a very reliable sufficient condition for structural transitions to be first-order

When a cubic invariant of the order-parameter is allowed in the free-energy, the transition is expected to be first-order. The IR's (fulfiling the Lifshitz condition) which violate the Landau condition are the identity IR's of the 230 space-groups and a few-hundred of IR's at high-symmetry points of the rhombohedral, hexagonal and cubic Brillouin-zone surfaces[25,27]. Some isomorphous transitions (e.g. in SmS or Ce) and a large number of transitions involving a lowering of symmetry (as in V_3SI, InTl, KCN, Cu_3Au, $NiCr_2O_4$) are found to be respectively related to the two preceding categories of inactive IR's. These transitions are firmly established to be first-order (sometimes sligtly), thus confirming the validity of the Landau condition.

4. A number of first-order transitions are found to be associated to single IR's which do not satisfy the Lifshitz condition.

For the IR's which do not fulfil the Lifshitz criterion, it appears from the experimental data that a large majority of the corresponding transitions take place towards an inhomogeneous (incommensurate) state, in agreement with the meaning of this criterion (see also Sec.5). However, a small number of observed strongly discontinuous transitions (e.g. in $CsCuCl_3$, $Cd(NO_3)_2$, FeS, $LiNH_4SO_4$) between normal crystalline phases, are shown to be connected to such inactive IR's .

5. A small number of first-order transitions can be interpreted as being induced by more than one IR.

A few first-order transitions have been connected to a reducible representation[25]. This is the case for example in Benzil $(C_6H_5CO)_2$, Ag_2HgI_4, $Bi_4Ti_3O_{12}$, N_3AlF_6 and for some other materials for which a "triggering" mechanism involving two IR's was proposed, in order to interpretate the observed symmetry changes and transition anomalies.

In summary, although the Landau theory deals a priori with second-order transitions, it provides implicitly the framework for the interpretation of first-order transitions. Furthermore, the symmetry modifications observed at structural (second or first-order) transitions, are in very good agreement with the predictions of the theory. Let us stress however, that only a small fraction of the symmetry changes which are predicted on a theoretical basis[15-19] have been observed experimentally. Besides, most of the transitions which may occur continuously are actually first-order. These two facts suggest that the selection rules contained in the Landau theory are insufficiently restrictive, and that additional requirements could be needed in order to select the IR's which are liable to induce second and first-order structural transitions between normal crystalline phases.

4.3 Critical exponents

Another aspect of the Landau theory is the prediction of the external-variable dependence for the physical quantities which are mainly affected by the transition (i.e. the order-parameter and other spontaneous quantities, the susceptibilities, specific heat etc...). It is currently admitted that the theory does not account correctly, in general, for the critical behaviour of physical quantities, since it has the same range of validity as the mean-field approximation in microscopic theories. Accordingly, a number of works have been performed in the last decades, to extend the Landau theory to include spatial inhomogeneous fluctuations of the order-parameter in the region of the transition[28]. Recently, the advent of the Renormalization-group methods[29] has provided powerful techniques to study the influence of strong fluctuations on the features of systems undergoing a phase transition. Thus, on a theoretical ground, critical fluctuations have been shown to modify appreciably some predictions of the mean field theory, in particular the critical exponents[30].

In a recent work[31] the critical exponents predicted by the Landau and critical phenomena theories, have been compared to the experimental data available for structural transitions. Although these data concern a little number of real systems, one can draw the following conclusions.

1. For proper ferroelastic transitions in which the dimensionality of the soft acoustic directions in reciprocal space is m = 1, the experimental data are in a very good agreement with the theoretical values predicted by the Landau theory.

In this category of structural transitions, the order-parameter has the same symmetry than the spontaneous strain-tensor components.

Accordingly, the critical exponents and (corresponding respectively to the external-variable dependence of the spontaneous strain, and softening elastic constant) are found, with a great accuracy, to be $\beta = 1/2$ and $\gamma = 1$, for proper ferroelastic materials such as LaP_5O_{14}, Tanane, $TbVO_4$, $TmVO_4$, $BiVO_4$, $KH_3(SeO_3)_2$, $PrAlO_3$ or TeO_2.

Proper-ferroelastic transitions are assumed to display a particular critical behaviour among structural transitions, because of the strong anisotropies characterizing the corresponding elastic instabilities. The effect of these anisotropies is to confine the diverging fluctuations into limited regions of the reciprocal space, either in the vicinity of some special directions (m = 1) or along some planes (m = 2). As a consequence, the fluctuations will not disturb in an essential manner the behaviour of the tridimensional system and thus, the critical behaviour, as predicted by the Landau theory, will be satisfied[32,33]. However, although this analysis seems to be experimentally confirmed for materials were m = 1, the situation is still unclear for proper ferroelastics with m = 2[31]. In this latter case one finds either examples of materials, the critical behaviour of which illustrates the Landau theory predictions (e.g. KCN) or contradicts these predictions (e.g. NaN_3, $RbAg_4I_5$).

2. For all the other categories of structural transitions, a general disagreement can be verified between the critical behaviours predicted theoretically and the experimental results.

This is true not only for the Landau theory (mean-field) predictions but also for the critical phenomena theories which include the effect of fluctuations and take into account the specific spatial (d) and order-parameter (n) dimensionalities. Thus, for all the ferroelectric, or improper ferroelastic transitions in which the critical behaviour has been investigated carefully (i.e. K_2SeO_4, $BaMnF_4$, $KMnF_3$, $SrTiO_3$, $CsPbCl_3$, $NaNbO_3$,...) a systematic disagreement is found with the theoretical predictions[31]. A number of explanations have been proposed[34-37] in order to explain such a failure of the various theoretical frameworks, among which the influence of defects, which seems to increase the role of the fluctuations on the critical behaviour of most of the presently known structural transitions.

5. GENERALIZATIONS OF THE LANDAU THEORY TO INCOMMENSURATE SYSTEMS

In the study of incommensurate (INC) systems, as the wave-vector k varies with temperature in the modulated phase within an interval Δk one has to consider the variation of $\delta\rho_k(r)$ in this interval. Here the symmetry of the low temperature phase is expressed by the probability density :

$$\delta\rho_k(\vec{r}) = \sum_i \eta_i^k \psi_i^k(\vec{r}) \qquad (19)$$

In the current description of the standard high-temperature-INC-lock-in sequence of phases[38,39] the preceding variation is accounted by fixing the ψ_i^k functions in (19) at their values at k_c, where k_c is the rational vector associated with the lock-in commensurate (C) phase. The physical assumption underlying such a choice is that, within the INC phase, the system is asymptotically governed by a periodic potential possessing the symmetry of the lock-in phase, i.e. the INC phase is treated as a spatial modulation of the lock-in C phase. As a consequence, the order-parameter components transform as the single IR τ_{k_c} of the high-temperature (N) phase, but appear as slowly modulated functions of the actual order-parameter components η_i^k. It has the result that the free-energy F of the system is a continuous function of k through the whole sequence of phases and must be written as the sum, over the volume of the system, of a local density Φ. Φ depends on the $\eta_i^{k_c}(r)$ and of their derivatives with respect to the space coordinates.

In this section we briefly summarize the main predictions which can be obtained within the preceding description, that was first introduced by Dzialoshinskii[38] for magnetic systems, and adapted by Levanyuk and Sannikov[39] to structural incommensurabilities. Then we underline a number of shortcomings of such a generalization of the Landau theory, for the interpretation of real INC systems. An alternative approach is proposed, which is shown to be more adapted for the phenomenological description of complex INC systems.

1. Applicability of the Dzialoshinskii-Levanyuk-Sannikov approach

The specific standard properties of INC systems are :

a) a limited range of stability for the INC phase, which appears between two transitions at the temperatures T_I and T_L. T_I corresponds to a second-order transition from the high-temperature N phase, which is a normal crystalline phase. It marks the onset of the INC modulation with a definite wave-length and a vanishingly small amplitude. This onset is announced in the N phase by precursor effects. For a displacive type of modulation, these effects consist in an instability of the lattice (i.e. a soft mode). The transition at T_L is first-order and corresponds to the locking of the wavelength of the modulation on a simple multiple of the period of the underlying crystal structure.

b) within the range of the INC phase one can distinguish two regions. In the upper one, nearby T_I, the modulation is sinusoïdal i.e the wave-vector k is almost independent of temperature. In the lower one, on approaching T_L, it is expected to be constituted by crystalline regions (identical to the C phase below T_L) separated by walls, called discommensurations. These walls form a periodic array, the distance between two discommensurations being equal to the wavelength of the modulation. In this region k decreases more or less abruptly in the vicinity of T_L.

The preceding features can be observed in a number of real INC systems such as $(NH_4)_2BeF_4$, K_2SeO_4, Rb_2ZnCl_4, $NaNO_2$ or $[N(CH_3)_4]_2MnCl_4$.

They can be accounted within the current Landau theory of INC systems[38,39] by assuming the free-energy local density to be :

$$\Phi = \Phi_1(\eta_i(x_j)) + \Phi_2(\eta_i, \frac{\partial^n \eta_i}{\partial x \eta_j}) \qquad (20)$$

If we consider the particular case of a two-component order-parameter $\eta_1 = \rho \cos\theta$, $\eta_2 = \rho \sin\theta$, with a modulation along the x axis, one has for example :

$$\Phi_1 = \Phi_o + \frac{\alpha}{2}\rho^2 + \frac{\beta_1}{4}\rho^4 + \frac{\beta_2}{4}\rho^4 \cos 4\theta \qquad (21)$$

and $$\Phi_2 = -\delta\rho^2 \frac{\partial\theta}{\partial x} + \sigma[(\frac{\partial\rho}{\partial x})^2 + \rho(\frac{\partial\theta}{\partial x})^2] \qquad (22)$$

Neglecting the anisotropic term in (21) and assuming a constant amplitude for the order-parameter ($\rho = \rho_o$), minimization of the free-energy of the system $F = \int \Phi dv$, yields the equilibrium values :

$$k_o = \frac{|\delta|}{\sigma} \quad , \quad \rho_o = \frac{\delta^2/\sigma - \alpha}{\beta_1} \qquad (23)$$

which express a sinusoïdal variation of the η_i along the x direction, associated with an irrational k_o-vector. The N-INC transition takes place at T_I, given by : $\alpha = \delta^2/\sigma$, at a second-order transition. The lock-in transition can be predicted qualitatively by taking into account the anisotropic term in (21) which favours the low-temperature C phase, whereas the Lifshitz invariant in (22) favours the INC phase. Thus, it is the competition between the Lifshitz and anisotropic terms which allows a phenomenological description of the N-INC-C sequence of phases, as well as the existence of a region of discommensurations in which both terms influence the stability of the INC phase.

In its present state, the preceding approach provides above all a qualitative or semi-quantitative description of the standard features of INC systems. However some inedequacies of the theory appear when more than one reference wave-vector has to be considered. This is obviously the case when more than one lock-in transition takes place as it is observed in Thiourea, $NbSe_3$ or TTF-TCNQ. Even when only one lock-in is observed, the choice of the reference k-vector is not indisputable, in particular when the lock-in arizes at a point of special (larger) symmetry (e.g. the center of the Brillouin-zone). In this case the order-parameter is taken smaller than its actual value, as the number of branches of the star of k_c is smaller than the number of branches of the actual wave-vector k_c within the INC phase. As a consequence ,certain homogeneous couplings with macroscopic quantities are discarded, which are allowed when the actual order-parameter dimensionality is taken into account. Besides a number of non-standard behaviours of INC systems cannot be obtained within the preceding theoretical scheme, such as the absence of lock-in transition, or a succession of INC phases separated by partial locks-in at irrational k-values.

2. Self-consistent theory of INC systems

An alternative phenomenological description of INC systems can be used which avoids the preceding shortcomings. This approach which has been implicitly refered to by a number of authors[40-42] consists in retaining in (19) the set of ψ_i^k functions corresponding to each value of k. In other words, one always remain at the minimum of the k-dispersion curve (i.e. the Lifshitz invariants vanish identically at any temperature) in such a manner that the η_i^k remain spatially homogeneous. It follows that : a) when k varies, $\delta\rho_k$ varies continuously in the intervals of temperature corresponding to the INC phases and jumps discontinuously for given values of k associated to lock-in transitions. Thus, each value of k determines a set of order-parameter components η_i^k which transform as distinct IR's τ_k.

b) The free-energy F is a function of the homogeneous η_i^k and may take different forms following the symmetries of $\delta\rho_k$. Accordingly, the order-parameter dimension may change from one phase to another.

c) Variation of the wave-vector as a function of the external variable, must be introduced complementarily via the k-dispersion of the quadratic order-parameter invariant coefficient $\alpha(k)$.

As an illustration, let us consider the canonical example of a system undergoing a second-order transition to an INC phase, followed by a first-order transition to a C phase. The wave-vector is assumed to display a parabolic variation within the INC phase, which can be well approximated by a third-degree expansion of $\alpha(k)$. This point deserves justification. Indeed, the current theories predict a logarithmic decrease on cooling for the wave-vector k in the INC phase. However, experimentally the variation of k(T) is largely influenced by the first-order character of the lock-in transition and by thermal hysteresis effects which hinder a precise description. Thus, polynomial expansions for $\alpha(k)$ should provide a good first approximation for the equilibrium values of :

$$k_{INC} = \frac{\delta\alpha(k)}{\delta k} \qquad (24)$$

Let us assume that $k_{INC}(T)$ locks at a rational value $k_c = \frac{a^*}{n}$ where a* is a reciprocal lattice translation. If we consider a two-component order-parameter, it is well-known[17] that depending on the symmetry of the IR under consideration, two classes of Landau expansions can be constructed corresponding respectively to the images C_{nv} and C_n (i.e. the set of two-dimensional distinct matrices forming the IR, which are isomorphous to finite point-groups). Supposing here a C_{nv} image, the two expansions can be written:

$$F = F_o + \frac{\alpha}{2}\rho^2 + \frac{\beta}{4}\rho^4 + \ldots + \frac{\gamma}{2n}\rho^{2n}\cos 2n\theta$$

when n is even, and:

$$F = F_o + \frac{\alpha}{2}\rho^2 + \frac{\beta}{4}\rho^4 + \ldots + \frac{\gamma}{2n}\rho^n \cos n\theta$$

when n is odd.

It can be shown[43)] that the degree to which the isotropic invariants ρ^{2N} should be retained in the preceding expansions, depends on the number N/2 (for N even) or (N-1)/2 (for N odd) of INC phases displaying the same modulation, and of the order of the N-INC transition. As we assume here a second-order transition towards a single INC phase the sequence of phases can thus be described by the following expansions:

$$F_{INC} = F_o + \frac{\alpha}{2}\rho^2 + \frac{\beta}{4}\rho^4$$

for the INC phase, and

$$F_C = F_o + \frac{\alpha}{2}\rho^2 + \frac{\beta}{4}\rho^4 + \frac{\gamma}{2n}\rho^{2n}\cos 2n\theta \qquad (25)$$

for an even lock-in (i.e. $k_c = a^*/n$ with $n=2p$) , or

$$F_C = F_o + \frac{\alpha}{2}\rho^2 + \frac{\beta}{4}\rho^4 + \frac{\gamma}{n}\rho^n \cos n\theta \qquad (26)$$

for an odd lock-in ($k_c = a^*/n$ with $n=2p+1$), and by the $\alpha(k)$ expansion:

$$\alpha(k) = \alpha_o + \delta(k-k_c) + \frac{\sigma}{2}(k-k_c)^2 + \frac{\varepsilon}{3}(k-k_c)^3 \qquad (27)$$

where the coefficients $\beta, \gamma, \sigma, \varepsilon$ are taken constant, their sign being determined by the stability conditions: $0 < \gamma < \beta$, $\sigma < 0$, $\varepsilon > 0$.

In the INC phase the wave vector varies as:

$$k_i = k_c - \frac{\sigma}{2\varepsilon} + \frac{(\sigma^2 - 4\delta\varepsilon)^{1/2}}{2\varepsilon}$$

yielding a renormalized α-coefficient:

$$\alpha(k_i) = \alpha_o - \frac{1}{12\varepsilon^2}\left[(\sigma^2 - 4\delta\varepsilon)^{3/2} - \sigma(\sigma^2 - 6\delta\varepsilon)\right] \qquad (28)$$

$$= \alpha_o - \Delta(\delta)$$

with $\Delta(\delta) > 0$. Introducing (28) in F_{INC} provides the equilibrium values for ρ_{INC} and F_{INC} in the INC phase:

$$\rho^e_{INC} = \left(-\frac{\alpha(k_i)}{\beta}\right)^{1/2} \quad ; \quad F^e_{INC} = F_o - \frac{\alpha(k_i)^2}{4\beta}$$

In the C phase ($k=k_c$, $\theta=0$) the equilibrium equations for ρ, corresponding to (25) and (26) are respectively given by:

$$\alpha_o + \beta \rho^2 + \ldots - \gamma \rho^{2n-2} = 0 \tag{29}$$

and

$$\alpha_o + \beta \rho^2 + \ldots - \gamma \rho^{n-2} = 0 \tag{30}$$

Assuming ρ and γ are small, Eqs (29) and (30) can be approximated in the C phase, and provide the equilibrium values for F_C :

$$F_C^e = F_o - \frac{\alpha_o^2}{4\beta} - A(\beta,\gamma)(-\alpha_o)^n \tag{31}$$

and

$$F_C^e = F_o - \frac{\alpha_o^2}{4\beta} - B(\beta,\gamma)(-\alpha_o)^{n/2} \tag{32}$$

Thus, on the first-order transition line $F_{INC}=F_C$ one gets:

$$\alpha_o = -\varepsilon C(\beta,\gamma)\ \Delta(\delta)^{1/n-1} \tag{33}$$

or

$$\alpha_o = -\varepsilon D(\beta,\gamma)\ \Delta(\delta)^{2/n-2} \tag{34}$$

where $\varepsilon=+1$ for even n and $\varepsilon=-1$ for odd n;A,B,C and D being positive combinations of the constants β and γ.The curves $\alpha_o(\delta)$ corresponding to the N-INC and to the INC-C transition lines ,have for any n the shape represented on Fig.7 a).As noted in Ref.44, if α_o and $\delta$are linear functions of the temperature T and of another external variable (e.g. the pressure P),the conversion to the pressure-temperature phase diagram can be deduced from a linear transformation of Fig.7a).On the other hand,writing $\alpha_o=a(T-T_o)$ with $a>0$,where T_o is the Curie temperature corresponding to a virtual second-order transition occuring from the N to the C phase, one can deduce from Eqs. (28),(33) and (34) that $T_I>T_o$ and $T_L<T_o$,T_I and T_L being the actual transition temperatures to the INC and C phases respectively.

The preceding results apply directly when τ_{k_c} is associated to an image C_n (only the constants A,B,C,D are modified in Eqs. (29) to (34)),or when considering higher m-degree expansions for $\alpha(k)$.The case m=2 ($\varepsilon=0$, $\sigma>0$) reveals peculiar features.This case has been discussed by Indenbom and Loginov in the framework of a study devoted to the modulated ordering of smectic layers[44].Here,Eqs. (33) and (34) become respectively:

$$\alpha_o=-C\delta^{2/n-1} \quad \text{and} \quad \alpha_o=-D\ \delta^{4/n-2} \tag{35}$$

The curves corresponding to (35) are shown on Fig.7 b) for different significant values of n.

Let us stress that in the preceding example a single coefficient (α) is assumed to be temperature dependent ,the other coefficients in F_{INC} and $F_C$remaining fixed at their high-temperature values.This allows

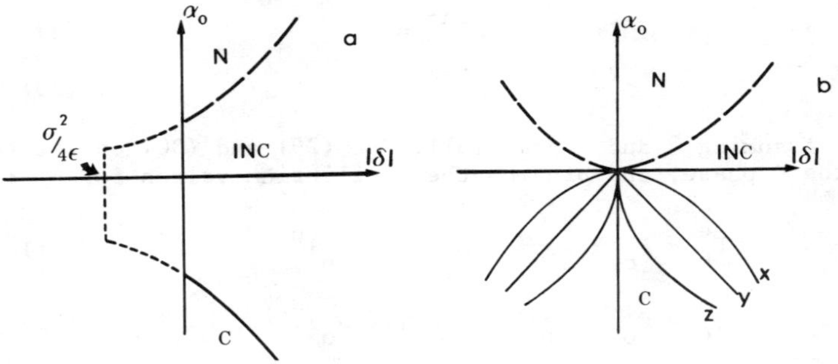

FIG.7 :Phase diagram in the $\alpha_o(\delta)$ plane.7 a):case corresponding to Eqs.(33) and (34).7 b):case corresponding to Eq.(35).The x,y and z curves on Fig.7 b) represent respectively the variation laws $\alpha_o \sim \delta^{2/n-1}$ with n=2,3 and $n \geq 4$, or $\alpha_o \sim \delta^{4/n-2}$ with n=(3,4,5),6 and $n \geq 7$.The N-INC and INC-C lines are respectively second-and first-order transition lines.

to introduce in the model,in agreement with the spirit of the Landau theory,the minimal number of phenomenological variables to be determined from experimental data.The applicability and advantages of this self-consistent approach to complex sequences of of INC and C phases, is discussed in Ref.43).

6. PECULIARITIES OF THE LANDAU THEORY OF MAGNETIC TRANSITIONS

The starting point for the phenomenological theory of magnetic transitions can be found in Landau's paper[13)],in which the author notices that the property of a magnetic crystal will depend not only on the symmetry of the density of charges ρ,but also on the symmetry of the mean current density $\vec{j}$ as well.Accordingly,1421 magnetic groups were constructed[45,46)] by combining the time-reflection symmetry R with the operations of the 230 cristallographic space-groups.The concept of co-representation introduced by Wigner[47)] allowed description of the irreducible degrees of freedom associated with a magnetic crystal.

However,the magnetic group of a crystal and its irreducible co-representations express only the weak relativistic interactions in the crystal (i.e.the anisotropic energy) and do not account for the exchange forces,wich generally play a fundamental role in the nature of the magnetic ordering.The terms figuring in the exchange energy are invariant by the isotropic group,and must be worked out separately by introducing special magnetic vectors,the form of which depend on the transformation properties of the mean spins located at the magnetic ions.

Using such vectors, a classification of the possible magnetic symmetries was recently performed by Andreev and Marchenko[48)] systematizing the approach used by Dzialoshinskii for the description of the various types of non-compensated antiferromagnets[49)]. As a particular example of the Dzialoshinskii method, we examine herebelow the case of Nickel-iodine boracite (Ni-I). Ni-I can be viewed as the first experimental example of a class of magnetic materials discussed on a theoretical basis by Dzialoshinskii and Man'ko[50] more than twenty years ago.

In the paramagnetic phase, the cubic primitive cell of Ni-I is a rhombohedron of volume V (Fig. 8) containing two formula units. Thus the conventional cell of symmetry $F\bar{4}3c1'$ contains eight formula units with twenty four magnetic ions Ni^{2+}. At T_c = 61.5 K the structure becomes monoclinic (Cc') the primitive monoclinic cell having the same volume as the conventional cubic cell. We can assume, withouth loss of generality, that the lowering of symmetry taking place at T_c is entirely connected with the displacement of the nickel ions. The twenty-four metallic ions will thus be distributed among twelve independent monoclinic sublattices. The positions of the ions forming each sublattice are given in Table II. In Fig. 9, the magnetic ions are represented in projection on the pseudo-cubic plane (00I). Having regard to their structural environment, the twelve independent nickel ions form three groups, denoted (1, 2, 3, 4), (5, 6, 7, 8) and (9, 10, 11, 12) in Fig. 9. The three groups of atoms lie respectively in planes perpendicular to the cubic z direction, and to the x direction (x = 0 and x = ½).

The magnetic structure of the crystal in the low temperature phase is completely determined if the spins of the ions belonging to each sublattice are given. Let us symbolize the spins by S_1, $S_2, \ldots, S_{12}$ and denote by $S_{13}, S_{14}, \ldots, S_{24}$ the spins of the ions respectively obtained by reflection in the monoclinic plane σ_{xz}. Assuming that the S_i are small close to T_c, the Landau free-energy F can be expanded as power series of the S_i components S_{iu} (u=x,y,z). We introduce the auxiliary vectors M^α and L_j^α (j, α = 1,2,3) defined by the equations :

$$
\begin{aligned}
L_1^1 &= S_1 + S_{13} + S_2 + S_{14} - S_3 - S_{15} - S_4 - S_{16} \\
L_2^1 &= S_1 + S_{13} - S_2 - S_{14} + S_3 + S_{15} - S_4 - S_{16} \\
L_3^1 &= S_1 + S_{13} - S_2 - S_{14} - S_3 - S_{15} + S_4 + S_{16} \qquad (36) \\
L_j^2 &= L^1 (S_{i+4}),\ L^3 = L^1(S_{i+8}) \quad (j = 1, 2, 3,\ i = 1\text{-}4 \text{ and } 12\text{-}16) \\
M^1 &= S_1 + S_{13} + S_2 + S_{14} + S_3 + S_{15} + S_4 + S_{16} \\
M^2 &= M^1(S_{i+4}),\ M^3 = M^1(S_{i+8}) \quad (i = 1\text{-}4 \text{ and } 12\text{-}16)
\end{aligned}
$$

It is obvious that the vector $M = M^1 + M^2 + M^3$ represents the total magnetization moment of the monoclinic cell, below T_c. It transforms as the tridimensional vector corepresentation of the $F\bar{4}3c1'$ paramagnetic group, at the Brillouin zone center. On the other hand, it can easely be found that the reducible corepresentation

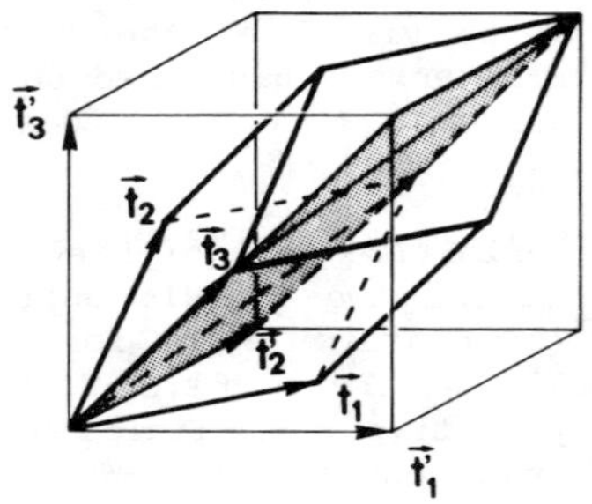

Fig.8: Lattice modification at the $F\bar{4}3c1' \to Cc'$ transition in Ni-I. The latent antiferromagnetic cell has the primitive translations: $\vec{t}'_1=\vec{t}_1-\vec{t}_2+\vec{t}_3$, $\vec{t}'_2=\vec{t}_1+\vec{t}_2-\vec{t}_3$, $\vec{t}'_3=-\vec{t}_1+\vec{t}_2+\vec{t}_3$.

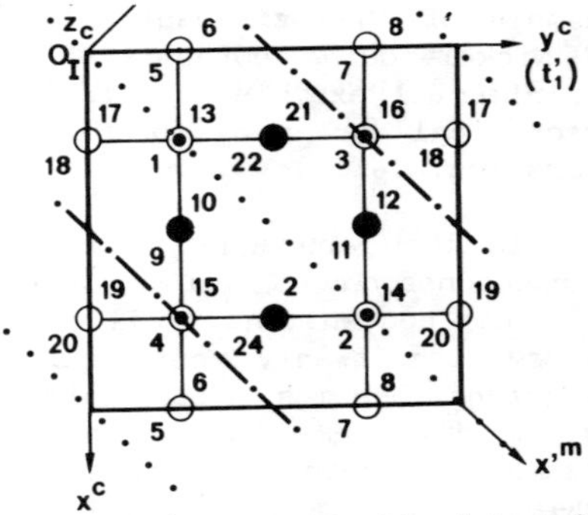

Fig.9: Position of the 24 nickel ions in the monoclinic phase of Ni-I. Projections on the xy cubic plane. Numbers below or above the metals indicate whether they are located at i) z=1/4 or z=3/4 for metals 5-12 and 17-24; ii) z=0 or z=1/2 for 1-4, 13-16.

given by the antiferromagnetic vectors L_j^α decomposes into two irreducible corepresentations of the $F\bar{4}3c1'$ group at the X point[53] of the face-centered Brillouin-zone. More precisely, $(L_{1x}^\alpha, L_{2y}^\alpha, L_{3z}^\alpha)$ transform as a three dimensional IC, whereas a six-dimensional irreducible corepresentation (denoted τ) describes the transformation properties of the components $(L_{1y}^\alpha, L_{1z}^\alpha, L_{2x}^\alpha, L_{2z}^\alpha, L_{3x}^\alpha, L_{3y}^\alpha)$, each of the sets of the L_j projections being distributed over the three arms of the star k* associated to the X point, namely $k_1 = (0, 0, \frac{\pi}{a})$, $k_2 = (0, \frac{\pi}{a}, 0)$ and $k_3 = (\frac{\pi}{a}, 0, 0)$.[51]

It can be shown[51] that it is τ which induces the cubic to monoclinic modification observed in Ni-I. We can thus express the expansion F in terms of the L^α, M^α, M_u^α and L_{ju}^α separating its exchange and anisotropic parts. For α = 1, 2, 3 one has, omitting the superscript α:

$$F_{exchange} = \frac{a}{2}\sum_j (L_j)^2 + \frac{B_1}{4}\{\sum_j (L_j)^2\}^2 + \frac{B_2}{4}\sum_j L_j^4 \qquad (37)$$
$$+ \frac{B_3}{4}\sum_{i\neq j}(L_iL_j)^2 + \frac{C}{2}M^2 + D\{(ML_1)L_2L_3 + (ML_2)L_1L_3 + (ML_3)(L_1L_2)\}$$
$$+ \frac{E_1}{2}(M^2)\{\sum_j (L_j)^2\} + \frac{E_2}{2}\sum_j (ML_j)^2$$

where the B_i, C, D, E_i are constant coefficients and $a \sim (T-T_c)$.

$F_{exchange}$ contains only terms which do not depend on the orientation of the vectors L_j^α and M^α, i.e. of the orientation of the spins with respect to the crystal axes. They represent exchange interactions. Minimization of (37) leads to the following results : at T_c = 61.5 K, where a vanishes, the components of L_i^α become non-zero. The L_i^α vary as $L_i^\alpha \sim (T_c-T)^{1/2}$. Since $F_{exchange}$ contains an

Table II

Coordinates of the nickel ions associated with the 24 average spins S_i (i=1-24) in the monoclinic cell of Ni-I, with respect to the cubic x,y,z axes.

S_1=1/4,1/4,0;S_2=3/4,3/4,0;S_3=1/4,3/4,0;S_4=3/4,1/4,0;S_5=0,1/4,1/4
S_6=0,1/4,3/4;S_7=0,3/4,1/4;S_8=0,3/4,3/4;S_9=1/2,1/4,1/4;S_{10}=1/2,1/4,3/4
S_{11}=1/2,3/4,1/4;S_{12}=1/2,3/4,3/4;S_{13}=1/4,1/4,1/2;S_{14}=3/4,3/4,1/2
S_{15}=3/4,1/4,1/2;S_{16}=1/4,3/4,1/2;S_{17}=1/4,0,3/4;S_{18}=1/4,0,1/4;S_{19}=3/4,0,3/4
S_{20}=3/4,0,1/4;S_{21}=1/4,1/2,3/4;S_{22}=1/4,1/2,1/4;S_{23}=3/4,1/2,3/4;S_{24}=3/4,1/2,1/4

invariant linear in M^α, a non-zero spontaneous magnetic moment will arise simultaneously with the appearance of the L_i^α, that will be of exchange origin. The expression for M^α is :

$$M^\alpha=-\frac{D}{C}[L_1^\alpha(L_2^\alpha L_3^\alpha)+L_2^\alpha(L_1^\alpha L_3^\alpha)+L_3^\alpha(L_1^\alpha L_2^\alpha)] \qquad (38)$$

which is proportional to $(T_c-T)^{3/2}$. The absolute minimum of $F_{exchange}$ associated with this non-zero magnetization corresponds to :

$$L_1^\alpha = \pm L_2^\alpha = \pm L_3^\alpha \qquad (39)$$

Introducing (39) in (36) leads to :

$$S_i+S_{i+12} = \frac{1}{4}(M^\alpha \pm 3L_1^\alpha) \text{ and } S_j+S_{j+12} = \frac{1}{4}(M^\alpha \pm L_1^\alpha)(j \neq i) \qquad (40)$$

where the number i in (40) is determined by the signs in Eq.(39) (e.g. for $L_1^\alpha = L_2^\alpha = L_3^\alpha$, i = 1, j = 2, 3, 4).

From Eqs.(40), it can be seen that the average spins of the ions can be divided into two groups, the absolute magnitude of the spins differing from one group to the other. Such a property is usual for ferrimagnets. However, here the spins are associated with one identical type of magnetic ions, which is found in equivalent crystallographic positions in the paramagnetic phase. This is in contrast with the situation found in standard ferrimagnets such as ferrites or garnets[52]. Dzialoshinskii and Man'ko[50] suggested to denominate this new type of uncompensated antiferromagnetism, latent antiferromagnetism. As noted by these authors, no confusion should be made with ferrimagnetism because of the peculiar temperature variation of the magnetization (i.e. $M \sim (T_c - T)^{3/2}$ in the vicinity of T_c). Another distinctive feature of latent antiferromagnetic materials that was partly overlooked by the authors of Ref.52), is that, despite its exchange origin, the magnetization must be expected to assume very weak values at any temperature below T_c. This is connected with the improper character of the transition, i.e. to the fact that M results from a coupling to the third power of the antiferromagnetic sublattices. In Ni-I boracite, the magnetization at 4.2 K is found to be about 0.9 Gauss[53] which represents 1 % of the nominal value. The other features distinguishing latent antiferromagnets, in particular the simultaneity of a structural modification at T_c which can be predicted from the symmetry of the order-parameter, are given in Ref. 51).

7.CONCLUDING REMARKS

In these lectures the applicability of the Landau theory to structural transitions has been discussed. The generalization of the theory to incommensurate and magnetic systems has been briefly outlined. In the recent years the Landau method has proved to be adaptable to a number of new experimental situations. In spite of its partial failure in the prediction of the critical behaviour, it remains an irreplaceable tool for the clarification of the macroscopic behaviour of systems possessing strong anisotropies and a large number of degrees of freedom.

ACKNOWLEDGMENT

This work has benefitted from very helpful discussions with Dr. J.C. Tolédano (CNET, Bagneux).

REFERENCES

1. Jona,F. and Shirane,G.,"Ferroelectric Crystals",Pergamon(New-York), 1962.
2. Errandonea,G.,Phys.Rev.B21,5221(1980).
3. Kobayashi,J.,Uesu,Y. and Enomoto,Y.,Phys.Stat.Sol.(b) 45,293(1971).
4. Shapiro,S.M.,Axe,J.D. and Raccah,P.M.,Solid State Commun.15,377(1974).
5. Cross,L.E.,Fouskova,A. and Cummins,S.E.,Phys.Rev.Letters 21,812(1968).
6. Smutny,F and Fousek,J.,Phys.Stat.Sol.(b) 40,K13(1970).
7. Shaulov,A.,Smith,W.A. and Schmid,H.,Ferroelectrics 34,219(1981).
8. Worlton,T.G. and Beyerlein,R.A.,Phys.Rev.B12,1899(1975).
9. Meeks,E.L. and Arnold,R.T.,Phys.Rev.B1,982(1970).
10. Menon,C.S. and Philip,J.,Solid State Commun.26,897(1978).
11. Dvorak,V.,Ferroelectrics 7,1(1974).
12. Tolédano,J.C.,J.of Solid State Chem.27,41(1979).
13. Landau.L.D.,Phys.Z.Sovjet 11,26 (1937),translated in L.D.Landau "Collected papers",Edited by D.ter Haar,Pergamon(1965).
14. Lifshitz,E.M.,Z.Eksp.Teor.Fiziki 11,255(1941).
15. Tolédano,P. and Tolédano,J.C.,Phys.Rev.B14,3097(1976).
16. Tolédano,P. and Tolédano,J.C.,Phys Rev.B16,386(1977).
17. Tolédano,J.C. and Tolédano,P.,Phys.Rev.B21,1139(1980).
18. Tolédano,P. and Tolédano,J.C.,Phys.Rev.B25.,1946(1982).
19. Tolédano,P. and Tolédano,J.C.,to be published.
20. Lyubarskii,G.Ya.,"The Application of Group Theory in Physics", Pergamon(New-York)1960.
21. Tolédano,J.C.,Ferroelectrics 35,31(1981).
22. Bak,P.,Krinsky,S. and Mukamel,D.,Phys.Rev.Letters 36,52(1976).
23. Brazovskii,S.A.,Dzialoshinskii,I.E. and Kuharenko,B.G.,Sov.Phys.JETP 43,1178(1976).
24. Brézin,E.,Le Guillou,J.C. and Zinn-Justin,J.,Phys.Rev.B10,892(1974).

25.Tolédano,P. and Pascoli,G.,in "Symmetries and Broken Symmetries in Condensed Matter Physics",231.Ed.N.Boccara(Idset,Paris)1981.
26.Depmeier,W. and Tolédano,P.,unpublished.
27.Tolédano,P. and Pascoli,G,Ferroelectrics 25,427(1980).
28.Ginzburg,V.L.,Sov.Phys.Solid State 2,1824(1960);Levanyuk,A.P.,Sov. Phys.Solid State 8,1294(1964).
29.Wilson,K.G.,Phys.Rev.B4,3184(1971).
30.Wilson,K.G. and Fisher,M.E.,Phys.Rev.Letters 28,240(1972).
31.Tolédano,J.C.,Annales des Télecommunications 39,277(1984).
32.Villain,J.Solid State Commun.8,295(1970).
33.Levanyuk,A.P. and Sobyanin,A.A.,JETP Letters 11,371(1970).
34.Lubensky,T.C.,Phys.Rev.B11,3573(1975).
35.Bergman,D.J.,Rice,T.M. and Lee.P.A.,Phys.Rev.B15,1706(1977).
36.Khmelnitskii,D.E.,Sov.Phys.JETP,41,981(1975).
37.Levanyuk,A.P.,Osipov,U.V.,Sigov,A.S. and Sobyanin,A.A.,Sov.Phys. JETP 49,176(1979).
38.Dzialoshinskii,I.E.,Sov.Phys.JETP 19,960(1964).
39.Levanyuk,A.P. and Sannikov,D.G.,Sov.Phys.Solid State 18,245(1976).
40.Moncton,D.E.,Axe,J.D. and Di Salvo,F.J.,Phys.Rev.Letters 34,734(1975).
41.Izumi.M.,Axe.J.D.,Shirane,G. and Shimaoka,K.,Phys.Rev.B15,4392(1977).
42.Golovko,V.A. and Levanyuk,A.P.,Sov.Phys.JETP 54,1217(1982).
43.Tolédano,P.,to be published.
44.Indenbom,V.L. and Loginov,E.B.,Sov.Phys.Crystallogr.26,526(1982).
45.Zamorzaev,A.M.,Sov.Phys.Crystallogr.2,10(1957).
46.Belov,N.V.,Neronova,N.N. and Smirnova,T.S.,Sov.Phys.Crystallogr.2, 311(1957).
47.Wigner.E.P.,"Group Theory and its application to the quantum mechanics of atomic spectra",Academic Press(New-York),1959.
48.Andreev,A.F. and Marchenko,V.I.,Usp.Fiz.Nauk,130,39(1980).
49.Dzialoshinskii,I.E.,Sov.Phys.JETP,5,1259(1957).
50.Dzialoshinskii,I.E. and Man'ko,V.I.,Sov.Phys.JETP 19,915(1964).
51.Tolédano,P.,Schmid,H.,Clin,M. and Rivera,J.P.,to be published.
52.Herpin.A.,"Théorie du Magnétisme",Presses Universitaires de France, (1968).
53.Zheludev,I.S.,Perekalina,T.M.,Smirnovskaya,E.M.,Fonton,S.S. and Yarmukhamedov,Yu.M.,JETP Letters 20,129(1974).

25. [illegible] "Symmetries and Broken Symmetries in Condensed Matter Physics" [illegible] 1.N.Boccara [illegible] Paris) 198[illegible]
26. Dombre [illegible] and [illegible] Phys [illegible]
27. Toledano [illegible] and [illegible] Surface [illegible]
28. [illegible]
[illegible]
29. [illegible] (198[illegible])
30. [illegible]
31. [illegible] Communications [illegible]
32. [illegible] (1970)
33. [illegible]
34. [illegible] 282 (1973)
35. [illegible]
36. [illegible] (1972)
37. [illegible]
[illegible] (1980)
38. [illegible]
39. [illegible] Phys. [illegible]
40. [illegible] Phys. [illegible]
41. [illegible]
42. [illegible]
43. Toledano [illegible]
44. [illegible] Phys. [illegible] (198[illegible])
45. [illegible] (198[illegible])
46. [illegible]
[illegible] (1973)
47. [illegible] "Group [illegible] application [illegible]
[illegible]
48. [illegible]
49. [illegible]
50. [illegible]
51. [illegible]
52. [illegible] Universite [illegible]
(1968)
53. [illegible]
[illegible]

COUPLING COEFFICIENTS FOR SPACE GROUP REPRESENTATIONS

Maciej Suffczyński
Institute of Physics, Polish Academy of Sciences
Lotników 32, Warszawa 02-668
POLAND

ABSTRACT

A method of calculation of Clebsch-Gordan coefficients from the small representations of the wave vector groups of the space group is described.

For the irreducible space group representation labelled by $\underline{k}l$, contained $m_{\gamma' \gamma'', l}$ times in the direct product of the irreducible representations $\underline{k}'l'$ and $\underline{k}''l''$, the basis functions $\Psi^{\underline{k}l\gamma}_{\underline{\sigma}\alpha}$ are linear combinations of the basis function products $\Psi^{\underline{k}'l'}_{\underline{\sigma}'\alpha'} \Phi^{\underline{k}''l''}_{\underline{\sigma}''\alpha''}$ with the coupling or Clebsch-Gordan coefficients:

$$\Psi^{\underline{k}l\gamma}_{\underline{\sigma}\alpha} = \sum_{\sigma'\alpha'} \sum_{\sigma''\alpha''} \left(\begin{matrix} \underline{k}'l' \\ \underline{\sigma}'\alpha' \end{matrix} \; \begin{matrix} \underline{k}''l'' \\ \underline{\sigma}''\alpha'' \end{matrix} \middle| \begin{matrix} \underline{k}l\gamma \\ \underline{\sigma}\alpha \end{matrix} \right) \Psi^{\underline{k}'l'}_{\underline{\sigma}'\alpha'} \Phi^{\underline{k}''l''}_{\underline{\sigma}''\alpha''} \qquad /1/$$

To calculate the Clebsch-Gordan /CG/ coefficients [1-7]

$$\left(\begin{matrix} \underline{k}'l' \\ \underline{\sigma}'\alpha' \end{matrix} \; \begin{matrix} \underline{k}''l'' \\ \underline{\sigma}''\alpha'' \end{matrix} \middle| \begin{matrix} \underline{k}l\gamma \\ \underline{\sigma}\alpha \end{matrix} \right) = U^{\gamma}_{\sigma'\alpha'\sigma''\alpha''\sigma\alpha} \qquad /2/$$

we decompose the space group G into left cosets with respect to the wave vector group $G(\underline{k})$

$$G = \sum_{\sigma=1}^{c_{\underline{k}}} \{R_\sigma | \underline{v}_\sigma\} \, G(\underline{k}) \qquad /3/$$

The coset representatives, i.e. the space group operations $\{R_\sigma | \underline{v}_\sigma\}$, remain fixed. The number $c_{\underline{k}}$ of the coset representatives, and thus of the arms $\underline{k}_\sigma = R_\sigma \underline{k}$ of the $\underline{k}$ wave vector star, equals $|\overline{G}|/|\overline{G}(\underline{k})|$, the order $|\overline{G}|$

of the point group of the space group G divided by the order $|\bar{G}(\underline{k})|$ of the point group of $G(\underline{k})$. We find $c_{\underline{k}'}$ and $c_{\underline{k}''}$ arms of the $\underline{k}'$ and $\underline{k}''$ wave vector stars, respectively. From the arms we compose all wave vector selection rules

$$R_{\sigma'}\underline{k}' + R_{\sigma''}\underline{k}'' = R_{\sigma}\,\underline{k} + \underline{K}_{\sigma'\sigma''} \qquad /4/$$

Here $\underline{K}_{\sigma'\sigma''}$ is a reciprocal-lattice vector which will not be written explicitly further. We choose one leading wave vector selection rule /LWVSR/

$$R_{\chi'}\underline{k}' + R_{\chi''}\underline{k}'' = \underline{k} \qquad /5/$$

with two space group operations $\{R_{\chi'}|\underline{v}_{\chi'}\}$ and $\{R_{\chi''}|\underline{v}_{\chi''}\}$.

The small representations $d^{\underline{k}'l'}$, $d^{\underline{k}''l''}$ and $d^{\underline{k}l}$, of dimension dim(l′) , dim(l″) and dim(l), respectively, are tabulated for the first arm of the wave vector star. In the necessary transformation

$$d^{R_{\chi'}\underline{k}'l'}(\{R_S|\underline{v}_S\}) = d^{\underline{k}'l'}(\{R_{\chi'}|\underline{v}_{\chi'}\}^{-1}\{R_S|\underline{v}_S\}\{R_{\chi'}|\underline{v}_{\chi'}\}) \qquad /6/$$

$$d^{R_{\chi''}\underline{k}''l''}(\{R_S|\underline{v}_S\}) = d^{\underline{k}''l''}(\{R_{\chi''}|\underline{v}_{\chi''}\}^{-1}\{R_S|\underline{v}_S\}\{R_{\chi''}|\underline{v}_{\chi''}\}) \qquad /7/$$

primitive translation vectors from the translational subgroup T of G may arise in the space group operations on the right-hand side of eqs. /6-7/.

The principal, or $\sigma'=\chi'$, $\sigma''=\chi''$, $\sigma=1$, block of CG coefficients is calculated from the tabulated [8] small representations $d^{\underline{k}'l'}$, $d^{\underline{k}''l''}$ and $d^{\underline{k}l}$

$$U^{\gamma}_{\chi'a'\chi''a''1a} = \left\{\sum_S d^{R_{\chi'}\underline{k}'l'}(S)_{b'b'}\, d^{R_{\chi''}\underline{k}''l''}(S)_{b''b''}\, d^{\underline{k}l}(S)^{*}_{bb}\right\}^{-1/2}$$

$$\times\left\{\frac{\dim(l)}{|\bar{G}(\underline{k})|}\right\}^{1/2} \sum_S d^{R_{\chi'}\underline{k}'l'}(S)_{a'b'}\, d^{R_{\chi''}\underline{k}''l''}(S)_{a''b''}\, d^{\underline{k}l}(S)^{*}_{ab} \qquad /8/$$

by performing summations over the space group operations belonging to the intersection of the wave vector groups,

$$S = \{R_S|\underline{v}_S\} \in G(R_{\chi'}\underline{k}') \wedge G(R_{\chi''}\underline{k}'') \wedge G(\underline{k}) / T \qquad /9/$$

The indices $\gamma\ b''b$ in eq./8/ have to be chosen such that the sum with diagonal indices yields a nonvanishing value.

Using the decompositions

$$\dot{G}(\underline{k}) = \sum_{\Sigma}\{R_{\sigma'\Sigma}|\underline{v}_{\sigma'\Sigma}\}C\,G(R_{\sigma'}\underline{k}')\wedge G(\underline{k})) \qquad /10/$$

of the wave vector groups, for each wave vector selection rule of eq./4/ we find one space group operation $\{R_{\Sigma}|\underline{v}_{\Sigma}\}$ which rotates the principal block into the $\sigma'\sigma''\sigma$ block

$$R_{\Sigma}R_{\chi'}\underline{k}' = R_{\sigma'}\underline{k}', \quad R_{\Sigma}R_{\chi''}\underline{k}'' = R_{\sigma''}\underline{k}'', \quad R_{\Sigma}\underline{k} = R_{\sigma}\underline{k} \qquad /11/$$

The $\sigma'\sigma''\sigma$ block is now computed from the principal block by matrix multiplication

$$U_{\sigma'\alpha'\sigma''\alpha''\,\sigma\alpha} = \sum_{a'=1}^{\dim(l')}\sum_{a''=1}^{\dim(l'')}\sum_{a=1}^{\dim(l)} d^{\underline{k}'l'}(\{R_{\sigma'}|\underline{v}_{\sigma'}\}^{-1}\{R_{\Sigma}|\underline{v}_{\Sigma}\}$$

$$\times\{R_{\chi'}|\underline{v}_{\chi'}\})_{\sigma'a'}\; d^{\underline{k}''l''}(\{R_{\sigma''}|\underline{v}_{\sigma''}\}^{-1}\{R_{\Sigma}|\underline{v}_{\Sigma}\}\{R_{\chi''}|\underline{v}_{\chi''}\})_{\alpha''a''}$$

$$\times U_{\chi'a'\chi''a''\,1a}\; d^{\underline{k}l}(\{R_{\sigma}|\underline{v}_{\sigma}\}^{-1}\{R_{\Sigma}|\underline{v}_{\Sigma}\})^{-1}_{a\alpha} \qquad /12/$$

In case of multiplicity $m_{l'l'',l} > 1$, the multiplicity index is $1 \leq \gamma \leq m_{l'l'',l}$, thus a suitable choice of the column indices $b'b''b$ in eq./8/ is needed to find the CG coefficients for all γ [9], and one has to secure orthogonalization of columns and rows of the CG square matrix of dimension

$$c_{\underline{k}'}c_{\underline{k}''}\dim(l')\dim(l'') = c_{\underline{k}}\sum_{l} m_{l'l'',l}\dim(l) \qquad /13/$$

REFERENCES

[1] F. Iachello, Phys. Rev. Lett. 44 (1980) 772.

[2] Gard, P., J. Phys. A6, 1837 /1973/.

[3] Berenson, R. and Birman, J.L., J. Math. Phys. 16, 227 /1975/.

[4] Berenson, R., Itzkan, I. and Birman, J.L., J. Math. Phys. 16, 236 /1975/.

[5] van den Broek, P.M., phys. stat. sol./b/ 94, 487 /1979/.

[6] Dirl, R., J. Math. Phys. 20, 671 /1979/.

[7] Kunert, H. and Suffczyński, M., J. Physique 41, 1361 /1980/.

[8] Cracknell, A.P., Davies, B.L., Miller, S.C. and Love, W.F., Kronecker Product Tables /Plenum Press, New York, 1979/.

[9] Davies, B.L. and Dirl, R., Proceedings of 12th Int. Coll. on Group Theor. Methods in Physics, eds. Denardo, G., Ghirardi, G. and Weber, T. /Springer, Berlin, 1984/ p. 373.

SOME ASPECTS OF THE SYMMETRY BREAKING IN THE 2-D ISING MODEL AND IN THE ONE COMPONENT COULOMB SYSTEM

D. Merlini

Mathematisches Institut Ruhr Universität
4630 Bochum
GERMANY

and

Dipartimento di Fisica Università di Milano
Via Celoria 16, 20133 Milano
ITALY

ABSTRACT

We review and discuss some results and problems concerning the symmetry breaking in the two models.

INTRODUCTION

In this lecture we review and discuss some results and questions concerning the symmetry breaking in the 2-d Ising model and in the 2-d one component Coulomb model(OCP), i.e. a discrete and a continous model of great interest in statistical mechanics.

For the Ising model described by the Hamiltonian $H_\Lambda = -\sum_{i,j} J_{ij}\sigma_i\sigma_j$ + boundary conditions, where $-J_{ij}$ is the nearest neighbor interaction, we first discuss in Section 1 the translation invariance of the state at low-temperature (non-existence of a symmetry breaking of the translation group) ; in Section 2, in connection with the symmetry breaking of the symmetry group and related phase transition we analyze the grand canonical surface tension of the model, its convergence to the canonical one and to that of the

SOS (solid on solid) model for $T \leq T_c$. For the second model, sometimes called jellium , we devote Section 3 and Section 4. The model consists of an assembly of N classical point charges interacting with the long range Coulomb logarithmic potential together with a neutralizing charge background. Its study is related to one of the basic problem of statistical mechanics, namely, the possible existence of a crystalline state in a classical model.

The Hamiltonian is given by $H_\Lambda = -\sum \phi_{x_i x_j} + \rho \cdot e^2 \sum_i \int dx\, \phi_{x x_i} + - \frac{1}{2} \iint dx\, dy\, \phi_{xy}$, where ϕ_{xy} is the Coulomb potential and ρ the constant charge density of the background. In Section 3 we first discuss some relevant boundary conditions of Dobrushin type for the model and the uniqueness of thermodynamic funtions. In Section 4 we discuss the problem of absence or presence of the symmetry breaking of the translation group in connection with the decay property of the correlation functions, i.e. a condition for the absence of a crystalline state in the system.

1. TRANSLATION INVARIANCE OF THE EQUILIBRIUM STATE STATE AT LOW TEMPERATURE

The first aspect we want to discuss concerns the non-existence of a symmetry breaking of the translation group in the 2-d Ising model. In 3-d the situation is different and one of the open problems is the possibility of the existence of a roughening transition; in fact it is only in the 3-d case that non-translation invariant equilibrium states exist(at sufficiently low temperatures).The existence of such states was proven by Dobrushin [1] who showed that the +- mixed boundary condition yields, at low enough temperatures, a non -translationally invariant equilibrium state. This fact is related to and expresses the fact that the interface between the two pure phases is rigid, in contrast to the 2-d case where the oscillation of the interface is relatively large [2]. We recall that the roughening temperature is defined as the lowest temperature above which the state obtained by means of the +- boundary condition is translation invariant, i.e. $\mu_{+-} = \frac{1}{2}(\mu_+ + \mu_-)$ if $T \geq T_r$. Numerical studies [3] gives $T_r \sim 0.57\ T_c$, a roughening temperature very close to the critical temperature of the two-dimensional model; the Van Beijeren inequality [4] yields $T_c(\ d=2) \leq T_r \leq T_c(\ d=3)$ but presently it has not been proven that T_r is strictly less then T_c. In two-dimension, $T_r=0$ since the state +- is translation invariant 5): So, some workers investigated the problem of non-existence of a stable phase coexistence in 2-d for $T > 0$; [6,7,8,9]. The problem was solved and the state is always translation invariant. In this discussion we present a simple construction which, combined with the cluster property and standard inequalities, give a simple proof of the translation invariance of the equilibrium state at low-temperatures for the 2-d Ising model in cylindrical geometry (this was the first proof of the insensitivity of the even correlation function to the whole set of boundary conditions at low temperatures for the domain considered) ; the results of applying the cluster property is suggestive in two-dimension since the cardinality of the boundary may grow at most as the linear dimension of the lattice. For the model on a square box, the same method applies, and the uniqueness of the correlations was proven in the case that the strength of the boundary fields is smaller than the bulk contribution [6].

To proceed we first treat general boundary conditions for an auxiliary model; the uniqueness of the even correlation functions (symmetric state) for the domain, then follows simply by a trivial partial trace transformation technique. The auxiliary model is defined as follows: let $\tilde{\Lambda}$ be a simply connected domain on the surface of a 3-d sphere containing the south pole x_o. On the surface of the sphere we consider a set of ℓ_1 latitudes and ℓ_2 longitudes; with each intersection point of a latitude with a longitude (or a longitude with a longitude) in $\tilde{\Lambda}$, we associate a spin variable $\sigma = \pm 1$ and ferromagnetic interactions $-J_{ij}$ between nearest neighbors; we thus obtain an Ising model on $\tilde{\Lambda}$ in which each spin interacts with its 4 nearest neighbors, except the spin at the south pole x_o which interacts with all the ℓ_2 nearest neighbors. We assume that the spin at x_o interacts either with ferromagnetic or antiferromagnetic coupling between its ℓ_2 neighbors, so that the set of bonds with antiferromagnetic interaction at x_o will be denoted by b_o . We need to consider all the boundary conditions on the boundary $\partial\tilde{\Lambda}$, which is chosen as the ℓ_1-ten latitude counted from x_o .(notice that a general boundary condition b on $\partial\tilde{\Lambda}$ is the set of points of $\partial\tilde{\Lambda}$ which are fixed in the configuration $\sigma = -1$. The model may be represented on the plane: then ℓ_2 resp. ℓ_1 is the number of bonds on each concentric circle (latitude) resp. the number of bonds on each radius (longitude) around x_o and $|\tilde{\Lambda}| = \ell_1 \cdot \ell_2$. The Hamiltonian is given by $H_{\tilde{\Lambda}} = -\sum_{i,j \in \tilde{\Lambda}} J_{ij}\sigma_i\sigma_j - \sum_{i_o \notin \tilde{\Lambda}} \sigma_{i_o}\sigma_i J_{i i_o}$, $i_o \in \partial\tilde{\Lambda}$.

Let us now consider a low-high temperature dual model to $\tilde{\Lambda}$, obtained by drawing a perpendicular bond between each pair of nearest neighbors (boundary point included); it is defined on $\tilde{\Lambda}^*$, a square lattice on the surface of a three dimensional sphere as before with north pole instead of south pole and with open boundary conditions (selfduality). We shall now prove that as $|\tilde{\Lambda}|$ or $|\tilde{\Lambda}^*|\to\infty$ the even correlation functions $\langle\sigma_X\rangle_{\tilde{\Lambda},b_0,b}$, $X\subset\tilde{\Lambda}$ become independent of b_0 , b and converge to $\langle\sigma_X\rangle_{\tilde{\Lambda},+,+}$ (++ denotes + boundary conditions on $\partial\tilde{\Lambda}$ i.e. b $=\phi$ and $b_0=\phi$, i.e. all interactions at x_0 are ferromagnetic and all boundary fields on $\partial\tilde{\Lambda}$ are positive). The fact that we consider the dual lattice is that we need the cluster expansion for closed trajectories on $\tilde{\Lambda}^*$ as given by the Sherman theorem [10] which reads:

$$Z_{\tilde{\Lambda}^*}=\underset{\sigma}{\mathrm{Tr}}\, e^{-\beta H_{\tilde{\Lambda}^*}}=C_0\cdot e^{\sum_C W(C)}=C_0\cdot e^{\sum_C \frac{(-1)^{N_C}}{\mu_C}\cdot\prod_{B\in C}(e^{-2K_B})^{n_B}} \tag{1}$$

C_0 is an immaterial constant and C denotes any cycle or closed connected path on $\tilde{\Lambda}^*$ by weight $W(C)$; μ_C is the multiplicity of C, N_C the number of selfcrossing of C and n_B the number of times a bond B occurs in C. Further $\sum_{B\in C} n_B = \ell(C)$ is the lenght of C; exp (-2K) is the low-temperature parameter associated with any bond B. With $\xi=\langle\sigma_X\rangle_{\tilde{\Lambda},b_0,b}/\langle\sigma_X\rangle_{\tilde{\Lambda},+,+}$ the quantity of interest, we then have the following inequality:[6]

$$1-(3e^{-2K})^{2d(b,b_0,X)}\cdot e^{4K(|b|+|b_0|)}\leq\xi\leq 1 \tag{2}$$

The left hand side may be proven with the help of the cluster expansion above and the right hand side simply follows from standard inequalities. Notice that d $(b\ ,b_0,X)$ is the minimal distance between any point of X and any point of $\partial\tilde{\Lambda}$ (b) or $x_0(b_0)$. ($|b|,|b_0|\leq\ell_2$).The application of the above inequality allows us to control explicitely the speed at which boundary effects vanish as the thermodynamic limit is approached. Clearly of interest are subsets X in the central region of $\tilde{\Lambda}$ or $\tilde{\Lambda}^*$ such that d $(b\ ,b_0,X)$ $\sim\ell_1/2$. We are also free to choose a well defined sequence of domains $\tilde{\Lambda}_i$ or $\tilde{\Lambda}_i^*$ increasing to infinity in both directions in such a way that $\ell_2<\ell_1/8-o(\ell_1)$; with the above inequality we then obtain:

$$\xi>1-(3e^{-2K})^{\ell_1}\cdot e^{8K\cdot\ell_2}\geq 1-e^{-8\cdot o(\ell_1)} \tag{3}$$

where $o(\ell_1)\to\infty$ as $\ell_1\to\infty$ and if $e^{-2K}<1/9$; thus at low-temperatures such that $e^{-2K}<1/9$ we obtain $\xi=1$ as $|\tilde{\Lambda}|\to\infty$ and we have:

<u>Property 1</u>

$$\lim_{|\tilde{\Lambda}|\to\infty}\langle\sigma_X\rangle_{\tilde{\Lambda},b_0,b}=\lim_{|\tilde{\Lambda}|\to\infty}\langle\sigma_X\rangle_{\tilde{\Lambda},+,+} \tag{4}$$

This proves the insensitivity of the even correlation functions in the central region of $\tilde{\Lambda}$ to all boundary conditions on $\partial\tilde{\Lambda}$.

We then have:

Property 2

$$\lim_{|\tilde\Lambda|\to\infty} \langle \sigma_X \rangle_{\tilde\Lambda,+,+} = \langle \sigma_X \rangle_{\mathbb{Z}^2,+} \tag{5}$$

To prove this, let us apply GKS inequalities:we introduce an additional infinite magnetic field at the south pole x_o as well as at all points on a suitable longitude running from x_o to $\partial\tilde\Lambda$. Then $\langle\sigma_X\rangle_{\tilde\Lambda,+,+} \le \langle\sigma_X\rangle_{\Lambda,+}$ where Λ, + refer to the model defined on a $\Lambda \subset \mathbb{Z}^2$ rectangular box Λ, with + boundary conditions and with the set X lying near the center of Λ. In the same way we may neglect all ferromagnetic interactions of the spin at x_o with its neighbors and all boundary fields as well as all ferromagnetic interactions between two longitudes. Then $\langle\sigma_X\rangle_{\tilde\Lambda,+,+} \ge \langle\sigma_X\rangle_{\Lambda,op}$ where Λ is as before and open means open boundary conditions. Application of the equality $\langle\sigma_X\rangle_{\mathbb{Z}^2,+} = \langle\sigma_X\rangle_{\mathbb{Z}^2,op}$ yields then property 2. Thus all the boundary conditions for the auxiliary model on $\tilde\Lambda$ yields the same translation invariant correlation function as for the model on $\mathbb{Z}^2$ if $\exp(-2K) \le 1/9$.

Finally, uniqueness of the symmetric state for the model in cylindrical geometry $\tilde\Lambda_c$ (where the ferromagnetic or antiferromagnetic interactions of the spin at x_o are replaced by positive resp. negative boundary fields, giving thus rise to general boundary conditions on $\tilde\Lambda_c$ is obtained by performing a partial trace [11] on the spin at x_o for the auxiliary model. With $\tilde\Lambda_c = \tilde\Lambda / x_o$, and $\kappa = Z_{\tilde\Lambda_c,\bar b_o,b} / Z_{\tilde\Lambda_c,b_o,b}$:

$$\langle \sigma_X \rangle_{\tilde\Lambda,b_o,b} = \frac{\langle \sigma_X \rangle_{\tilde\Lambda_c,b_o,b} + \langle \sigma_X \rangle_{\tilde\Lambda_c,\bar b_o,b} \cdot \kappa}{1+\kappa} \tag{6}$$

where b_o, $\bar b_o$ are the two complementary boundary conditions ($|\bar b_o| + |b_o| = \ell_2$) at the bottom of $\tilde\Lambda_c$ introduced by the partial trace transformation at x_o. $Z_{\tilde\Lambda_c,\bar b_o,b}$ denotes the partition function for the model on $\tilde\Lambda_c$ with b resp. $\bar b_o$ (or b_o) boundary condition in the upper, resp. the lower boundary of $\partial\tilde\Lambda_c$. Notice $\kappa \in [0,\infty]$. If $\kappa = 0$ or $\kappa = \infty$ in the thermodynamic limit, then we obtain at once that

$$\lim_{|\tilde\Lambda|\to\infty} \langle \sigma_X \rangle_{\tilde\Lambda,b_o,b} = \lim_{|\tilde\Lambda_c|\to\infty} \langle \sigma_X \rangle_{\tilde\Lambda_c,b_o,b} = \langle \sigma_X \rangle_{\mathbb{Z}^2,+} . \tag{7}$$

If on the other hand, κ remains finite in the thermodynamic limit, then the above inequality may be written as:

$$\kappa\big(\underbrace{\langle\sigma_X\rangle_{\tilde\Lambda,b_o,b} - \langle\sigma_X\rangle_{\tilde\Lambda_c,++}}_{\varepsilon\to 0} + \underbrace{\langle\sigma_X\rangle_{\tilde\Lambda_c,++} - \langle\sigma_X\rangle_{\tilde\Lambda_c,\bar b_o,b}}_{\ge 0}\big) =$$

$$= \underbrace{\langle \sigma_x \rangle_{\tilde{\Lambda}_c, b_0, b} - \langle \sigma_x \rangle_{\tilde{\Lambda}_c, ++}}_{\leq 0} + \underbrace{\langle \sigma_x \rangle_{\tilde{\Lambda}_c, ++} - \langle \sigma_x \rangle_{\tilde{\Lambda}, b_0, b}}_{\varepsilon \to 0} \quad (8)$$

Thus, we obtain:

Property 3

$$\lim_{|\tilde{\Lambda}_c| \to \infty} \langle \sigma_x \rangle_{\tilde{\Lambda}_c, b_0, b} = \lim_{|\tilde{\Lambda}_c| \to \infty} \langle \sigma_x \rangle_{\tilde{\Lambda}_c, ++} = \langle \sigma_x \rangle_{\mathbb{Z}^2, +} . \quad (9)$$

and thus all boundary conditions on $\partial \tilde{\Lambda}_c$ yields the same translation invariant equilibrium state $\langle \sigma_x \rangle_{\mathbb{Z}^2}$ at low temperatures. The result proves the nonexistence of a stable phase coexistence for the 2-d Ising model in the geometry we have considered.

2. SYMMETRY BREAKING AND GRAND CANONICAL SURFACE TENSION

A second aspect of spontaneous symmetry breaking in the model we want to discuss concerns the phase transition defined by means of a surface tension (canonical or grand canonical).For a large class of spin lattice models it is known that the surface tension is zero at high temperature and non zero or not defined at low temperature [12].

The surface tension may be introduced as a definition of a phase transition related to a spontaneous symmetry breakdown of the internal symmetry group, i.e. one can say that there exists a phase transition associated with a surface tension τ^{12} between two phases ω^1 and ω^2 if there exists a critical temperature T_c such that

$$\begin{aligned} \tau^{12} &= 0 \qquad T > T_0 > T_c \\ \tau^{12} &\neq 0 \text{ or not defined for } T < T_c \end{aligned} \quad (10)$$

For specific models like the 2-d model, many calculations, treatements and more results are known [13,14]. For example, in the inequality above, T_0 may be shown to coincide with T_c, i.e., $\tau^{12} = 0$ for $T \geq T_c$ and $\tau^{12} \neq 0$ for $T < T_c$, where T_c is such that the spontaneous magnetization vanishes above T_c.In fact, $\tau^{12} = 2(K - K^*)$ (Onsager), where $K = \beta \cdot J$ and $K^* = -\frac{1}{2} \ln \operatorname{th} \beta J$ is the dual interaction ; $K = K^*$ for $T = T_c$. From now on we shall concentrate on the relation between the canonical and the grand canonical surface tension for the model. Using some inequalities it will be shown that the two coincide in the whole temperature range; second, using a generalization to open trajectories of the cluster expansion mentionned above it will be shown that the SOS limit is exact for $T \leq T_c$ as pointed out some years ago by Temperley. A new derivation of the surface tension (grand canonical) for $T \leq T_c$ is also given and coincides with Onsager formula. To define τ^{12} we consider the system confined to a rectangular box $\Lambda \subset \mathbb{Z}^2$ of length L + 1 and height 2M, centered at the origin and consider the mixed boundary condition +- between the two phases ω_1 and ω_2. (here ω_1 and ω_2 are related by internal symmetry).Then

$$\tau^{12} = -1/L \cdot \ln\left((Z^+ \cdot Z^-)^{1/2} / Z_{+,-}\right) \tag{11}$$

where Z^+ , Z^- are the partition functions with + resp. - boundary condition. We want to remark that by a duality transformation, τ [11] is related to the expectation value of a product of characters on the dual model with open boundary condition, i.e. the expectation value of a non local observable associated with the dividing surface between the two phases, i.e.

$$\Phi = e^{-\tau_\Lambda \cdot L} = Z^{+-}/Z^+ (K) = \langle \sigma_{i_1} \cdot \sigma_{i_2} \rangle (K^*) \tag{12}$$

where i_1 and i_2 are the two extreme points on the dual open lattice in the middle of the box (Fig. 1).

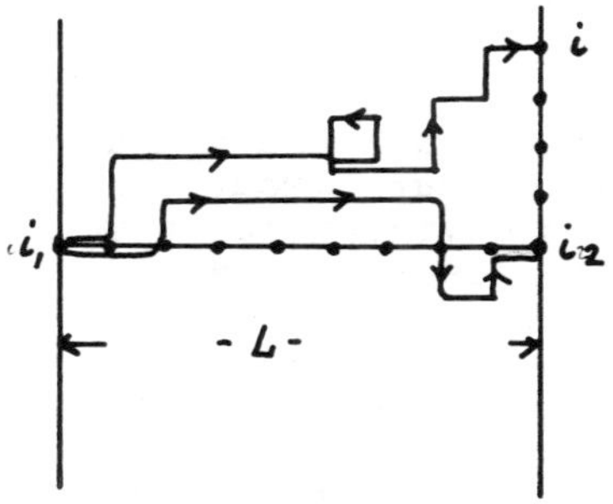

Fig.1 Paths in the dual lattice.

By a generalization of the Sherman theorem on path to open trajectories [15], we have in particular that:

$$\langle \sigma_{i_1} \sigma_{i_2} \rangle (K^*) = \sum_{P_{1,2}} (-1)^{N_{P_{1,2}}} \cdot e^{-2K_1 \cdot n_2} \cdot e^{-2K_2 \cdot n_1} \tag{13}$$

where n_i, i=1,2 is the number of interactions in $P_{1,2}$ along the i-ten direction and $P_{1,2}$ is any path from i_1 to i_2; $N_{P_{1,2}}$ is the number of selfcrossing of $P_{1,2}$. So , Φ is the sum of all open random trajectories starting at i_1 and ending up in i_2. To make contact with the SOS limit one should consider the grand canonical surface tension as introduced by Gallavotti [16] and proven to coincide with τ at sufficiently low temperatures; it is defined as:

$$\bar{\Phi} = e^{-\bar{\tau} \cdot L} = \lim_{M \to \infty} \sum_{i \in \mathbb{Z}^1} \left(Z^{i,+-}_{L,M} / Z^{i,+}_{L,M} \right) = \sum_{i \in \mathbb{Z}^1} \langle \sigma_{i_1} \sigma_i \rangle_{L,\infty} \tag{14}$$

In Eq. (14), $Z^{i+-}_{L,M}$ is defined with +- boundary conditions but where on one side the separation line between + and - is at the height i and where the limit $M \to \infty$ has been taken. $\overline{\Phi}$ is the sum of all paths starting at $i_1 = (-\frac{L}{2}, 0)$ and ending up at any boundary points $i = (\frac{L}{2}, i)$, $i \in \mathbb{Z}'$ (See Fig. 1). We now establish the equivalence between τ and $\overline{\tau}$ in the whole range of temperature. We have the

Property 4

$$\tau = \overline{\tau} \quad \forall\, T. \tag{15}$$

In fact let us consider the triplet (i_1, i_2, i), Applying standard inequalities [5] we have that, for all $\varepsilon > 0$

$$<\sigma_{i_1}\sigma_i>^{1-\varepsilon} \leq <\sigma_{i_1}\sigma_{i_2}>^{1-\varepsilon}, \quad <\sigma_{i_1}\sigma_i>^{\varepsilon} \leq <\sigma_{i_2}\sigma_i>^{\varepsilon} \tag{16}$$

thus
$$\Phi \leq \overline{\Phi} = \sum_{i \in \mathbb{Z}'} <\sigma_{i_1}\sigma_i> \leq \sum_{i \in \mathbb{Z}'} <\sigma_{i_1}\sigma_{i_2}>^{1-\varepsilon} \cdot <\sigma_{i_2}\sigma_i>^{\varepsilon} \tag{17}$$

Since $\forall\, T < T_c$, i.e. $\forall\, K^* < K_c$, $\exp(-2K_c) = \sqrt{2} - 1$,

$$\sum_{i \in \mathbb{Z}'} <\sigma_{i_2}\sigma_i>^{\varepsilon} \leq \sum_{i \in \mathbb{Z}'} e^{-|i-j| \cdot m \cdot \varepsilon} = (1/1 - e^{-m\varepsilon}) \tag{18}$$

and we obtain $\tau \leq \overline{\tau} \leq \tau^{1-\varepsilon}$. . As $\varepsilon \to 0$, $\tau = \overline{\tau}$ $\forall\, T$.

Property 5

For the 2-d Ising model the grand canonical surface tension coincides with that computed in the SOS limit for $T \leq T_c$. It was noted some years ago by Temperley that bylneglecting all trajectories having at least four bonds at some point and/or coming back (SOS limit) one obtains the Onsager formula and thus that in 2-d the SOS limit is exact. The miracle of this delicate cancellation between overhang contributions is hard to be proven looking directly into the geometrical structure of the trajectories whose set is very big [17]; moreover the series expansion for $\overline{\Phi}$ indicates that should be true as it may easily be checked; we recall that the SOS limit of an interface is obtained by letting some couplings in one direction go to infinity, i.e.

$$\Phi_{SOS} = e^{-2K_2 \cdot L} \cdot \widetilde{\Phi} \tag{19}$$

where $\widetilde{\Phi} = \lim_{K_2 \to \infty} \overline{\Phi} / e^{2K_2 \cdot L} (K_1, K_2)$. Thus only the contributions of trajectories containing L interactions in the 1-direction survive. To prove property 2, we then use the combinatorial method which takes into account geometrically the local structure of the two point interactions on the lattice. The counting and thus the problem of cancellation will appear as properties of the eigenvalues of the matrix propagator for trajectories [15]. In Fourier space it reads:

$$M(\vec{K},x) = \begin{pmatrix} xe^{-iK_1} & xe^{-iK_2-i\frac{\pi}{4}} & 0 & xe^{iK_2+i\frac{\pi}{4}} \\ xe^{-iK_1+i\frac{\pi}{4}} & xe^{-iK_2} & xe^{iK_1-i\frac{\pi}{4}} & 0 \\ 0 & xe^{-iK_2+i\frac{\pi}{4}} & xe^{iK_1} & xe^{iK_2-i\frac{\pi}{4}} \\ xe^{-iK_1-i\frac{\pi}{4}} & 0 & xe^{iK_1+i\frac{\pi}{4}} & xe^{iK_2} \end{pmatrix}$$

For simplicity we restrict to the isotropic case such that $K_1 = K_2$; it may then be shown that:

$$\bar{\Phi} = \int_0^{2\pi} dK_1 \sum_{i=1}^{4} \frac{\lambda_i^{L-1}}{1-\lambda_i} e^{iK_1(L-1)} = \int_0^{2\pi} dK_1\, f(\lambda_i, x, K_1) / \mathrm{Det}(1 - M(K_2=0)) \qquad (20)$$

and $\vec{k} = (k_1, k_2)$. Det $(I - M) = (1+x^2)^2 - 2x(1-x^2)(1+\cos K_1)$; there $x = \exp(-2K)$. The eigenvalues of the matrix propagator $M = \lambda I$ is given by the solution of the equation:

$$(\lambda^2 + x^2)^2 - 2\lambda x(\lambda^2 - x^2)(1+\cos K_1) = 0 \qquad (21)$$

Similarly, the matrix propagator of trajectories for the SOS model may be constructed and M for this model is such that the corresponding eigenvalue equation is given by:

$$\lambda(\lambda - x)(\lambda^2 - \lambda x(1+e^{-iK_1}) - x^2 e^{-iK_1}) = 0 \qquad (22)$$

Introducing the variable $\frac{\lambda}{x} = \xi$ and $e^{iK_1} = z$, then the above equations reduces to

$$\left(z + \frac{\xi(1-\xi)}{1+\xi}\right) \cdot \left(z + \frac{1+\xi}{\xi(1-\xi)}\right) = 0 \qquad (23)$$

$$\xi(\xi - 1) \quad \cdot \left(z + \frac{1+\xi}{\xi(1-\xi)}\right) = 0 \qquad (24)$$

For the SOS model $\xi_1 = 0$ and $\xi_2 = 1$ don't give any contribution to $\bar{\Phi}$; for the Ising model the contribution of ξ_1 and ξ_2 to $\bar{\Phi}$, solutions of the equation $z + \xi(1-\xi)/(1+\xi)$, reduces to the integral on a circle of a meromorphic function with no pole inside the circle for $T < T_c$ and thus vanishes also; it should now be noted that ξ_3 and ξ_4 are equal for both models and this establishes property 5. So the delicate cancellation of trajectories in $\bar{\Phi}$ is reflected here by the property of the eigenvalues for trajectories we have considered. It is expected that this result may be established for a large class of spin models by different techniques, but even for the Ising model considered here, no other proof has appeared.

Finally, we may recover Onsager formula for $\bar{\Phi}$ by noting that $\bar{\Phi}$ is the integral of a meromorphic function of z with a simple pole at $z_c = \frac{x(1+x)}{(1-x)}$ for $T \leq T_c$, with the result

$$\bar{\tau} = \lim_{L\to\infty} \frac{-\ln \bar{\Phi}}{L} = 2(K - K^*) \qquad T \leq T_c\,. \qquad (25)$$

Thus $\bar{\tau} = \tau$ as computed by Onsager and others. In conclusion $\bar{\tau}$ coincides with the SOS limit for $T \leq T_c$.

3. CRYSTALLINE BOUNDARY CONDITIONS AND UNIQUENESS OF THE FREE ENERGY DENSITY

The first aspect related to the problem of symmetry breaking in the 2-d one component Coulomb system (OCP plasma) we want to discuss concerns the uniqueness of the free energy density in the thermodynamic limit in presence of Dobrushin boundary conditions (somewhat like in the Ising model), espected to characterize different possible phases in the system (fluid, solid); there is a large body of literature on computation for thermodynamic functions.Monte Carlo as well as molecular dynamic experiments indicate the existence of a melting transition (first order) for the 2-d OCP at sufficiently low temperatures i.e. at a value $\gamma_0 \sim 140$ of the plasma parameter $\gamma = \beta \cdot e^2$, $\beta = (KT)^{-1}$. These methods compare the free energy resulting from integrating internal energy, with the free energy of the harmonic crystal, in a simulation with a few hundred of particles confined to some domain Λ of $\mathbb{R}^2$. The two branches of the energy curve intersect in the low temperature region near γ_0 [18-21]. So, in a theoretical context, it is tempting to introduce some relevant boundary conditions and analyze the problem of uniqueness of the free energy density. We first notice that the model possesses the H-stability property, i.e. there exists B independent of N, the number of particles, such that [22]

$$H_N \geq -B \cdot N \tag{26}$$

B is in fact very close to the energy per particle of the regular crystalline configurations on Λ :the triangular, the square and the hexagonal lattice; the energy of the triangular lattice is the lowest one with a value extremely close to B. [23]

The problem of symmetry breaking may then be pursued like in the Ising model either by means of the infinite BBGKY hierarchy or by introducing relevant boundary conditions.These are the Dobrushin boundary conditions for the model and have been recently introduced [24].They are of the type of a fixed crystalline configuration of point charges outside a bounded region with neutralizing homogeneous charge background, i.e. we put fixed charges $+e$ in a crystalline configuration $Y \subset \Lambda_0 / \Lambda$, in a domain Λ_0 surrounding the domain Λ . One then considers the lim $(\Lambda_0, \Lambda) \to \infty$ in a suitable manner. In fact under a suitable boundedness condition on the one particle correlation function, it is shown that the thermodynamic limit of the corresponding free energy for these Dobrushin boundary conditions is the same as the one wishout Dobrushin boundary conditions.

Many forms of domains may be considered. In all the cases the proof of uniqueness consists in controlling energy averages with respect to the canonical Gibbs measure, using explicit computations, the symmetry property of the model and the property of the Coulomb potential together with estimates on one particle correlation function. Here we discuss the method for the case of mixed boundary conditions (cylindrical domain).

In general, let Λ be a bounded domain and H_Λ be the classical interaction energy given by:

$$H_\Lambda(\mathbf{x}) = \sum_{i<j} \varphi(x_i, x_j) - \sum_i \varrho \int_\Lambda \varphi(x, x_i)\,dx + \tfrac{1}{2}\varrho^2 \int_\Lambda\int_\Lambda \varphi(x,y)\,dx\,dy \qquad (27)$$

where $\mathbf{x} = (x_1, \ldots x_i, \ \ldots x_N)$, x_i, i=1,2,..N being the coordinates of N positive unit charges in Λ ,with volume $|\Lambda| = \frac{N}{\varrho}$; ϱ is the constant background density (so that $\varrho|\Lambda|$ is the negative charge in Λ) balancing exactly the positive N charges in Λ; φ is the 2-d Coulomb potential, i.e. the kernel of Δ^{-1} , Δ being the Laplacian. The canonical free energy density is given by

$$f_\Lambda = -(1/\beta|\Lambda|)\cdot \ln Q_\Lambda \ , \quad Q_\Lambda = (N!)^{-1}\int_{\Lambda^N} e^{-\beta H_\Lambda(\mathbf{x})}\,d\mathbf{x} \qquad (28)$$

Let now Λ_0 be a larger bounded domain such that $\varrho|\Lambda_0 - \Lambda|$ is again an integer. We place charges $+1$ at each point of some discrete subset Y of $\Lambda_0 - \Lambda$ (crystalline configurations). This is a Dobrushin boundary condition for the configuration inside Λ. With $H_{\Lambda,\Lambda_0} = H_\Lambda + V_{\Lambda,\Lambda^0}$ where V_{Λ,Λ^0} is the interaction between charges in Λ and in Y (we do not consider in V_{Λ,Λ_0} the selfenergy of the system in $\Lambda_0 - \Lambda$), then with $f_{\Lambda,\Lambda_0} = -(\beta/|\Lambda|)\cdot \ln Q_{\Lambda,\Lambda_0}$ the free energy given by H_{Λ,Λ_0} and using some inequalities we have that:

$$f_{\Lambda,\Lambda_0} - f_\Lambda \leq |\Lambda|^{-1} \langle V_{\Lambda,\Lambda_0}\rangle_0 \ , \quad f_{\Lambda,\Lambda_0} - f_\Lambda \geq |\Lambda|^{-1}\langle V_{\Lambda,\Lambda_0}\rangle_D \qquad (29)$$

where $\langle\ \rangle_0$, $\langle\ \rangle_D$ are expectations with respect to μ_0, μ_D and the two measures μ_0, μ_D refer to the system without and with boundary conditions. In order to control the limit of the right hand side as $\Lambda, \Lambda_0 \uparrow \mathbb{R}^2$, we consider the one particle correlation functions g_Λ and g_{Λ,Λ_0}. Then

$$\langle V_{\Lambda,\Lambda_0}\rangle_0 = \int_\Lambda (g_\Lambda(x) - \varrho)\, F_{\Lambda,\Lambda_0}(x)\,dx \qquad (30)$$

where $$F_{\Lambda,\Lambda_0}(x) = \sum_{x_\ell \in Y} \Big[\varphi(x, x_\ell) - \varrho \int_{C_\ell} \varphi(x,y)\,dy\Big] \qquad (31)$$

C_ℓ denotes the Wigner-Seitz cell with center x_ℓ. Similarly

$$\langle V_{\Lambda,\Lambda_0}\rangle_D = \int_\Lambda (g_{\Lambda,\Lambda_0}(x) - \varrho)\cdot F_{\Lambda,\Lambda_0}(x)\,dx \qquad (31)$$

Let now $\Lambda = \Lambda_N$ be the surface of the cylinder $S^1 \times [-N-1/2, N+1/2]$ for some integer N , S^1 being a circle of radius $R = M/2\pi$, for some M . Let $\Lambda_0 = \Lambda_{N_0}$ be $S^1 \times [-N_0 - 1/2, N_0 + 1/2]$ for some $N_0 > N$. In this case

$$\varphi(x,y) = M^{-1}\int_{\mathbb{R}^1} dq_1 \sum_{q_2 = \frac{2\pi n}{M}} 2\pi (q_1^2 + q_2^2)^{-1} \qquad (32)$$

For symmetry reason, $\lim_{\Lambda_0 \uparrow \mathbb{R}^2} V_{\Lambda,\Lambda_0} = 0$.
We then get

Property 6

$$\lim_{\Lambda_0 \uparrow \mathbb{R}^2} f_{\Lambda,\Lambda_0} - f_\Lambda \leq 0 \tag{33}$$

Moreover, we obtain

$$\langle V_{\Lambda,\Lambda_0} \rangle_D = -1/2 \int g_{\Lambda,\Lambda_0}(x) \cdot \sum_{y \in Y} \ln\left(1+\xi^2 - 2\xi \cos \tfrac{2\pi}{M}(x_2-y_2)\right) \tag{34}$$

With $\xi = e^{-\frac{2\pi}{M}(x_1-y_1)}$, where we used the reflection symmetry of g_{Λ,Λ_0} in the x_1 coordinate to cancel the contribution of the term $-\pi/M(|x_1-y_1|)$ in the potential. We now compute the sum over Y as $\sum_{j=1}^{M} \sum_{i=N-N_0}^{N_0-N}$. We remark that, calling ξ_i the quantity defined as ξ with $y_1 = N+i$, we have

$$\sum_{j=1}^{M} -\tfrac{1}{2} \ln\left(1+\xi_i^2 - 2\xi_i \cos \tfrac{2\pi}{M}(x_2-y_{2,j})\right) \geq -\xi_i^M \tag{34}$$

Introducing the bound above we get

$$\langle V_{\Lambda,\Lambda_0} \rangle_D \geq -\int_\Lambda dx\, g_{\Lambda,\Lambda_0}(x) \sum_{i=1}^{N_0-N} \left[e^{-2\pi|x_1-(N+i)|} + e^{-2\pi|x_1+(N+i)|}\right] \geq$$

$$\geq -\int_\Lambda g_{\Lambda,\Lambda_0}\, e^{-2\pi N}\left(e^{2\pi x_1} c_1 + e^{-2\pi x_1} c_2\right). \tag{35}$$

Let now assume the bound $|g_{\Lambda,\Lambda_0}| < C(N)$ then

$$|\Lambda|^{-1} \langle V_{\Lambda,\Lambda_0} \rangle_D \geq -C'(N) e^{-2\pi N} \int_\Lambda dx \left(c_1 e^{2\pi x_1} + c_2 e^{-2\pi x_1}\right), \quad C'(N) = \frac{C(N)}{M(2N+1)}.$$

and thus $\lim_{N\to\infty} \lim_{N^0\to\infty} |\Lambda|^{-1} \langle V_{\Lambda,\Lambda_0} \rangle_D = 0$, with $C(N) = o(N)$.
This then yields similarly as before the

Property 7

$$\lim_{N\to\infty} \lim_{N_0\to\infty} f_{\Lambda,\Lambda_0} - f_\Lambda \geq 0 \tag{36}$$

Finally we obtain the result $\lim_{N\to\infty} \lim_{N_0\to\infty} f_{\Lambda,\Lambda_0} = \lim_{\Lambda\to\infty} f_\Lambda$ and the proof extends to the case of rectangular and circular domains.

4. DECAY OF THE PAIR CORRELATION FUNCTION AND ABSENCE OF A CRYSTALLINE STATE

The second point we want to discuss concerns the connection between the behavior of the pair correlation function and the absence of a symmetry breaking of the translation group in the model. To break the translation invariance of the Hamiltonian at finite volume one may consider one of the relevant boundary conditions discussed above. Here we are concerned with the case of doubly periodic boundary conditions which has previously not been discussed.

In presence of this boundary condition, the way to break the translation invariance consists in introducing an external localizing one-body potential $\alpha\varphi_{ext}(x)$, $\alpha<0$, with peaks at the side of a given lattice and letting $\alpha\to 0$ afterwards. This is also the analogon to the other method known for the Ising model, where in order to prove or disprove a symmetry breakdown of the state at low temperature (existence of a spontaneous magnetization) one can introduce a small external magnetic field h at every lattice point, compute the magnetization and remove the field afterwards. Here we use the Mermin argument and show that if the net pair correlation function decays faster or with equal speed as $1/r^2$ [25], $r\to\infty$, then no positional long range order will persist in the system. This is the content of the following result, which in the present form does not exclude the possibility of the existence of an orientational ordering in the system. Let (a_1,a_2) be the generators of a Bravais lattice $\mathbb{L}^2$ and let $\Lambda=\{x\in\mathbb{R}^2;\ x=Lx_1a_1+Lx_2a_2,\ 0\le x_1\le 1,\ 0\le x_2\le 1\}$. Let us consider in Λ a system of N classical charges and a neutralizing charge background described by a uniform charge density $\varrho=\frac{N}{|\Lambda|}\cdot e$, interacting through the two-body potential $\varphi_\Lambda(x,y)$ where φ_Λ is the kernel of the inverse of $-\Delta$ on $\tilde{L}^2(\Lambda)$, with periodic boundary condition and where $\tilde{L}^2(\Lambda)$ is the orthogonal complement in $L^2(\Lambda)$ of the constant functions In the Fourier representation we have

$$\varphi_\Lambda(x,y)=|\Lambda|^{-1}\cdot\sum_{k\neq 0,\,k=2\pi L^{-1}(\mathbb{L}^{2*})} e^{ik(x-y)}\cdot\frac{1}{k^2} \qquad (37)$$

where $\mathbb{L}^{2*}$ is the reciprocal lattice. The Hamiltonian of the system is given by $H_\Lambda=e^2\sum_{i<j}\varphi_\Lambda(x_i,x_j)$.
We want to consider perturbations of H_Λ by means of an external localizing potential $\alpha\varphi_{ext}(x)$, then

$$H_\Lambda^\alpha(x_1,\dots x_N)=e^2\sum_{i<j}\varphi_\Lambda(x_i,x_j)+\alpha\sum_i\varphi_{ext}(x_i)\,. \qquad (38)$$

Here φ_{ext} is assumed to be a smooth positive function invariant under translations of the lattice $\mathbb{L}^2$ and well localized around the sites in $\mathbb{L}^2$, where it obtains its maximum; α is an arbitrary negative constant. We then have the

Property 8

Let H_Λ be defined as above and let $h(r)$ be the pair correlation function related to the structure factor $S(p)$ defined in the usual way ,i.e. $\varrho h(r)=\widehat{F}(h(p))=S(p)-1$. Suppose that there exist positive constants r_0, C_1, C_2 independent of L for r sufficiently large, such that

$$|h_\Lambda(r)|\le C_1\ \forall |r|\le r_0,\ |h_\Lambda(r)|\le C_2/r^2\ \forall |r|>r_0\,.$$

Then with ϱ_K the Fourier component of the one-body density ϱ , one has, using standard method [25], the inequality

$$S(K+k)\ge\frac{[(K+k)\cdot e_t]^2\cdot|\varrho_K|^2}{(k\cdot e_t)^2+D_{tt}(k)} \quad K\in 2\pi(\mathbb{L}^2)^*,\ |e_t|=1 \qquad (39)$$

where $S(K+\kappa)=\frac{1}{N}\left\langle\left|\sum_{j=1}^{N}\left(e^{i(K+\kappa)x_j}-\delta_{K+\kappa}\right)\right|^2\right\rangle_\alpha$, $\rho_K=\left\langle\frac{1}{N}\sum_{j=1}^{N}e^{iK\cdot x_j}\right\rangle_\alpha$.

$$D_{tt}(\kappa)=\frac{1}{N}\left\langle\sum_{i,j}e^{i\kappa(x_i-x_j)}\nabla_{t_i}\nabla_{t_j}H_\Lambda\right\rangle+\alpha\langle\nabla_t\nabla_t\varphi_{ext}\rangle_\alpha=I_{1,\Lambda}+I_{2,\Lambda}$$

We now note that $I_{2,\Lambda}$ vanishes if we take the limit $\Lambda\uparrow\mathbb{R}^2$ and $\alpha\to 0$ in the stated order. The first term is given by

$$I_{1,\Lambda}(\kappa)=\beta\cdot\rho\int_\Lambda d^2r\, h(r)(1-\cos\kappa\cdot r)\,\nabla_t^2\varphi_\Lambda(r) \qquad (40)$$

where we have chosen $e_t\cdot\kappa=0$ and ∇_t denotes the gradient $t\cdot\nabla_x$. It may now be seen that for r independent of Λ, where C_0 is the unit cell of $(\mathbb{L}^2)$, we have

$$|\nabla_t^2\varphi_\Lambda(r)|=|L^{-2}\,\nabla_t^2\varphi_{\Lambda=C_0}(r/L)|\le C_3/r^2+C_4/L^2 \qquad (41)$$

The first equality follows from the definition of $\varphi_\Lambda(r)$ and the inequality from the fact that $\varphi_\Lambda=c_0(r)-\ln r$ is a harmonic function in C_0. Using the assumption above we may now easily estimate $\lim_{\Lambda\uparrow\mathbb{R}^2}I_{1,\Lambda}(\kappa)$ by $\lim_{\Lambda\uparrow\mathbb{R}^2}I_{1,\Lambda}(\kappa)\le -C_5\,\kappa^2\ln\kappa$,

for $|\kappa|$ sufficiently small, $C_5>0$. Then with the above inequality we have $\forall\kappa\in(2\pi/L)(\mathbb{L}^2)^*$ that $S(K+\kappa)\ge\frac{(\kappa\cdot e_t)^2\cdot|\rho_K|^2}{-\beta C_5\cdot\kappa^2\ln\kappa}$ (42)

In $S(K+\kappa)$ we have already taken the $\lim_{\alpha\to 0}$. The final step follows the line of the original argument given by Mermin. Both sides of (42) are multiplied by a positive Gaussian function $f(|K+\kappa|)$, divided by the volume of Λ and summed on all $\kappa\in 2\pi/L\cdot(\mathbb{L}^2)^*$. Taking the thermodynamic limit $\Lambda\uparrow\mathbb{R}^2$ we get

(43): $$F(0)=\rho\int(h(r)+1)\cdot F(r)\,d^2r\ge\int d^2\kappa\,|\kappa\cdot e_t(\kappa)|^2\,|\rho_K|^2\,f(|K+\kappa|)(-\beta C_5\,\kappa^2\ln\kappa)^{-1}$$

where $F(r)$ is the Fourier transform of f . By restricting the region of integration on the right-hand side of (43) to the values of κ such that $|\kappa\cdot e_t(\kappa)|\ge\delta>0$ and using the divergence $\int d^2\kappa/-\kappa^2\ln\kappa$, we get that $\rho_K=0\ \forall K\neq 0,\ K\in 2\pi(\mathbb{L}^2)^*$. We remark here that using our assumption on $h(r)$ the left-hand side of (43) is finite.

The decay in $1/r^2$ should be the borderline for the possible occurrence of long range order in the system, and this result is peculiar to the 2-d case; similar results may be obtained in the ν dimensional case using the BBGKY hierarchy.[26] To conclude, we want to remark that in recent computer experiments, at low temperature, the radial correlation function has been found to decay much slower than in the fluid case and shows typical solid-like structures, with peak positions corresponding to characteristic distances of the triangular lattice. Unfortunately, the tail of the function has not been investigated. Concerning the pair correlation function in the fluid phase, another peculiarity of the 2-d case is the existence of some well defined BBGKY truncation scheme, which may be solved exactly.[27]

REFERENCES

1. Dobrushin R.L., Theory Appl. 17, 582 (1972)
2. Gallavotti G., Comm. Math. Phys. 27, 103 (1972)
3. Weeks J.D., Gilmer G.H., Leamy H.J., Phys. Rev. Lett. 31, 549 (1973)
4. Van Beijeren H., Commun. Math. Phys. 40, 1 (1975)
5. Messager A. , Miracle-Sole S., J. Stat. Phys. 17, 245 (1977)
6. Merlini D. , J. Stat. Phys. 21, 739 (1979)
7. Merlini D. , presented at the XLI Statistical Mechanics Meeting, Rutgers, may 1979 and Lett. Nuovo Cimento 30, 474 (1981)
8. Higuchi Y., Proceeding of the Estergom conference on random fields (1979)
9. Aizenman M. , Proceeding of the ICMP, Lausanne (1979)
10. Sherman S., J. Math. Phys. , 8,399 (1960) and Sherman S., J. Math. Phys. , 4,1213 (1963); see also Burgoyne P.N., J. Math. Phys., 4 , 1320 (1963)
11. Gruber C.,Hintermann A. , Merlini D., Lectures Notes in Physics, 60, Springer Verlag (1977)
12. Fontaine J.R., Comm. Math. Phys., 70, 243 (1979) Gruber C.
13. Abraham D.B., Reed P., Comm. Math. Phys., 49,35(1976)
14. Bricmont J., Lebowitz J.L., Pfister C.,J. Stat. Phys. 26, 313 (1981)
15. Calheiros F.J.,Johannesen S.,Merlini D.,preprint (1984), presented at the Köszeg ICRF (1984)
16. Gallavotti G., Comm. Math. Phys. 27,103 (1972)
17. We acknowledge a discussion on the problem with J. Groeneveld.
18. De Leeuw S.W.,Perram J.W., Physica 113A, 546 (1982)
19. Caillol J.M., Levesque D., Weiss J.J., Hansen J.P. J. Stat. Phys. 28, 324 (1982)
20. Johannesen S., Merlini D., Phys. Lett. 93A,21(1982)
21. Choquard Ph., Clerouin J.,Phys.Rev. Lett.50,2086(1983)
22. Sari R.,Merlini D., J. Stat. Phys. 14,91 (1976)
23. Sari R.,Calinon R.,Merlini D.,J.Phys. A9,1539, (1976)
24. Albeverio S.,Dürr D.,Merlini D.,J.Stat.Phys., 31,389 (1983)
25. Martinelli F.,Merlini D.,J. Stat. Phys. 34,313 (1984)
26. Gruber Ch.,Martin Ph.A., Oguey Ch.,Comm. Math. Phys. 84,55 (1982)
27. Calinon R.,Golden K.,Kalman G.,Merlini D., Phys.Rev. A 20,327 (1979).

RACAH ALGEBRA FOR PERMUTATION REPRESENTATIONS OF FINITE GROUPS

T. Lulek

Institute of Physics, A. Mickiewicz University, Poznań
POLAND

1. INTRODUCTION

The expression "Racah algebra" is a frequently used in a physical literature abbrevation for an ensamble of notions and computational techniques of quantum mechanics, called "Wigner-Racah calculus", or "quantum theory of angular momentum" (cf. e.g. Fano and Racah [1], Biederharn and van Dam [2], Biederharn and Louck [3,4]). Typical notions associated with Racah algebra in the above meaning are Clebsch--Gordan coefficients, 3jm Wigner symbols, recoupling matrices and 6j, or, more generally, 3nj Wigner symbols, irreducible tensorial sets, reduced matrix elements, coefficients of fractional parentage etc. Racah algebra for the case of spherical symmetry, i.e. for the group SO(3), or for its covering group SU(2), has been widely applied in the theory of a multielectron atom, and its undobtful success is a satisfactory explanation of the observed spectroscopic properties of spectra of all known atoms (i.e. N-body systems, with $N \approx 10^2$), such as selection rules for quantum transitions, splittings of spectral lines in external fields, distances of energy levels, intensity and width of a spectral line etc. (cf. e.g. Wybourne [5]). These methods are also thoroughly applied in nuclear theory (cf. e.g. Moshinsky [6], or Vanagas [7]), as well as in the condensed matter physics, e.g. for an interpretation of spectra of paramagnetic ions admixed in a diamagnetic host (Griffith [8,9], Sugano, Tanabe, and Kamimura [10], Butler [11]). In the last case the Racah algebra is associated with point groups. A full description and some standardization of principal notions of Racah algebra for an arbitrary groups is provided in a review article of Butler [12]. Sometimes the phrase "Racah algebra" is used in a narrower, more exact mathematical meaning, as the associative algebra of irreducible tensor operators (Kibler [13]).

These spectacular successes of Racah algebra in the above mentioned cases suggest a natural extension to the case of multicentre multielectron systems like molecules or crystals. Then one has to deal with groups of permutations of the centres, and with permutation representations describing actions of several permutation groups on sets of identical physical objects. In particular, permutation representations have been used to a unique classification of symmetric coordinates of a molecule or a crystal, and to a classification of one-electron states in the method of molecular orbitals (Flurry [14], Fieck [15], Lulek [16], Kuplowski, Kuźma, and Lulek [17], Newman [18], Chan and Newman [19], Chan and Newman [20], Michel and Mozrzymas [21], Litvin [22], Butler, Ford, and Reid [23]). All these remarks justify, in our opinion, a program of construction of Racah algebra for permutation representations (Newman [18], Lulek and Lulek [24], Lulek, Lulek, Riel, and Chatterjee [25]), where the role of an irreducible representation is replaced by a transitive one. There is an evident way of construction of Racah algebra, since the direct product of two permutation representations of a group is also a permutation representation (acting on the set of appropriate pairs), which legitimates introduction of corresponding Clebsch-Gordan coefficients. Moreover, it is useful to study some quantities related to decomposition of a transitive representation into its irreducible components.

In general, Racah algebra is intended to provide a standardization of quantum-mechanical calculations for many-body systems using the symmetry of the problem. It requires some standardization of the description of appropriate mathematical objects, e.g. enumeration of elements of a set, a choice of some representative, a choice of phases for standard irreducible bases etc. From the other hand, however, the physical theory has to be independent of e.g. the labelling of permuted objects. More specificaly, in aproperly formulated theory the essential properties of a physical system should be reflected only in invariant properties of the corresponding mathematical structure, e.g. a group, a linear space, an algebra etc. but not in technical detail of description of this structure, like labelling of its elements. Thus the condition of invariance of a theory under the way of enumeration of elements of appropriate sets, or, more generally, under all automorphisms of the corresponding algebraic structure, can be considered as an analogue of the relativity postulate (Weyl [26], p. 138, cf. also Mozrzymas [27], p. 35). Racah algebra can be therefore interpreted as a compromise between a manifestly covariant theory operating only with algebraic structures, and avoiding any reference systems in these structures, and calculational needs of quantum mechanics, where the choice of definite reference system, like a complete system in a Hilbert space (Fourier transforms in the case of a free particle) proves to be quite useful, since it allows to simplify calculations and compare them with experiment. Racah algebra should provide an implicitly covariant description, by a determination of appropriate transformation rules under admissible changes in labelling of objects.

We intend to describe in this article some elements of Racah algebra for permutation representations in terms of notions associated with "the action of a group on a set". In Sec. 2 we give the principal definitions in a way of introducing of the notation, and formulate a general definition of equivalence of permutation representations. In Sec. 3 we propose a standard form for an arbitrary transitive representation, and in particular, we demonstrate a fibration of the orbit of this representation by means of imprimitivity sets generated by an intermediate subgroups between a given group and the stability group of an element of the orbit. In Sec. 4 we propose a decomposition of the carrier set of an arbitrary permutation representation into strata associated with the assumed definition of equivalence, together with a further decomposition into substrata associated with a narrower definition. In Sec. 5 we describe the Clebsch-Gordan decomposition for transitive representation as the level of bases, as a particular case of the theorem of Mackey for induced representations.

2. EQUIVALENCE OF PERMUTATION REPRESENTATIONS

Let G be a finite group, and $\tilde{P} = \{p|p = 1, 2, \ldots, |\tilde{P}|\}$ - a set ($|\tilde{P}|$ is the number of elements of the set $\tilde{P}$). A permutation representation P of the group G on the set $\tilde{P}$ is a homomorphism $P: G \to \Sigma_{|\tilde{P}|}$, where $\Sigma_{|\tilde{P}|}$ is the symmetric group on the set $\tilde{P}$. The permutation representation P associated thus each element $x \in G$ with a permutation

$$P(x) = \begin{pmatrix} p \\ P(x|p) \end{pmatrix}, \quad x \in G,\ p \in \tilde{P}, \tag{1}$$

so that $P(x|p) \in \tilde{P}$ is the image of $p \in \tilde{P}$ under the permutation $P(x) \in \Sigma_{|\tilde{P}|}$. It can be said briefly, that the group G acts on the set $\tilde{P}$.

The kernel of the mapping P,

$$\text{Ker } P = \left\{x \in G | P(x) = \begin{pmatrix} p \\ p \end{pmatrix},\ p \in \tilde{P}\right\} \triangleleft G, \tag{2}$$

is a normal subgroup of the group G, and consists of all elements of G acting trivially on $\tilde{P}$. The image of P,

$$\text{Im } P \equiv \Sigma(P) = \{P(x) | x \in G\} \subset \Sigma_{|\tilde{P}|}, \tag{3}$$

is called the constituent group of G on $\tilde{P}$, and acts effectively on $\tilde{P}$. Evidently, we have

$$\Sigma(P) \cong G/\text{Ker } P. \tag{4}$$

Let $P: G \to \Sigma_{|\tilde{P}|}$ and $P': G' \to \Sigma_{|\tilde{P}'|}$ be two permutation representations of the group G and G' on the set $\tilde{P}$ and $\tilde{P}'$, respectively. Representations P and P' are permutationally equivalent if there exist

(i) such an isomorphism τ: $G \to G'$, given by

$$\tau = \begin{pmatrix} x \\ \tau(x) \end{pmatrix}, \quad x \in G , \tag{5}$$

(ii) such a bijection ρ: $\tilde{P} \to \tilde{P}$, given by

$$\rho = \begin{pmatrix} p \\ \rho(p) \end{pmatrix}, \quad p \in \tilde{P}, \tag{6}$$

that

$$P'(\tau(x)|\rho(p)) = \rho(P(x|p)), \quad x \in G,\ p \in \tilde{P}. \tag{7}$$

Such a definition of permutational equivalence, assumed e.g. by Kurosh [28], Hall [29], Vilenkin [30], Naimark [31], assures a complete independence of a theory of the way of labelling both the carrier set $\tilde{P}$ and the group G. One frequently uses a narrower definition, consisting in taking into account only all bijections (ii), with the restriction to the case of $G = G'$ (Burnside [32], Gorenstein [33], Palais [34], Kirillov [35]). The narrower definition reduces effectively to equivalence with respect to the subgroup $\mathrm{Int}\ G \triangleleft \mathrm{Aut}\ G$, where Int G and Aut G are respectively groups of internal and all automorphisms of G.

There is still another definition of equivalence, the so-called weak equivalence, proposed by Michel [36]. Permutation representations P and P' of two groups G and G' acting both on the set $\tilde{P} = \tilde{P}'$ are weakly equivalent if their consituent groups are conjugated in the group $\Sigma_{|\tilde{P}|}$, i.e. if there exists such $\rho \in \Sigma_{|\tilde{P}|}$, that

$$\Sigma(P') = \rho\Sigma(P)\rho^{-1}. \tag{8}$$

The weak equivalence is thus associated with the effective action of the group G on the set $\tilde{P}$: the representations P and P' can be weakly equivalent even in cases when the groups G and G are not isomorphic (for reason of different kernels, Ker P $\neq$ Ker P').

The definition (5) - (7) allows to define, for the case $G = G'$

and $\tilde{P} = \tilde{P}'$, the transform P^{ρ} of permutation representation P by the bijection $\rho\in\Sigma_{|\tilde{P}|}$ as

$$P^{\rho}(x) = \rho P(x)\rho^{-1} = \begin{pmatrix} \rho(p) \\ \rho(P(x|p)) \end{pmatrix}, \quad x\in G,\ p\in\tilde{P}, \tag{9}$$

and the transform P_{τ} by the automorphism $\tau\in\mathrm{Aut}\ G$ as

$$P_{\tau}(x) = P(\tau(x)) = \begin{pmatrix} p \\ P(\tau(x)|p) \end{pmatrix}, \quad x\in G,\ p\in\tilde{P}. \tag{10}$$

The transforms P^{ρ} and P_{τ} of P consist therefore of a change of labelling of elements of the carrier set $\tilde{P}$ and the group G, imposed by the bijection ρ and the automorphism τ, respectively. It is easy to prove that

$$(P^{\rho})_{\tau} = (P_{\tau})^{\rho} = P^{\rho}_{\tau}, \quad \rho\in\Sigma_{|\tilde{P}|},\ \tau\in\mathrm{Aut}\ G, \tag{11}$$

so that the set

$$\tilde{\pi}[P] = \left\{P^{\rho}_{\tau} \middle| \rho\in\Sigma_{|\tilde{P}|},\ \tau\in\mathrm{Aut}\ G\right\} \tag{12}$$

of all transformsof the permutation representation P consitutes the orbit of a transitive representation $\pi^{[P]}: \Sigma_{|\tilde{P}|}\times\mathrm{Aut}\ G \rightarrow \Sigma_{|\tilde{\pi}[P]|}$, defined by the formula

$$\pi^{[P]}(\rho,\tau) = \begin{pmatrix} P' \\ P'^{\rho}_{\tau} \end{pmatrix}, \quad P'\in\tilde{\pi}[P]. \tag{13}$$

3. THE CANONICAL FORM OF A TRANSITIVE REPRESENTATION

Let

$$G = \bigcup_{r\in\tilde{R}(G:H)} g_r H, \tag{14}$$

where

$$\tilde{R}(G:H) = \{r | r = 1,2,\ldots,|G|/|H|\} \tag{15}$$

is the set of indices, be the decomposition of the group G into left cosets with respect to a subgroup $H\subset G$. The set of these left cosets,

$$\tilde{R}(H) = \{g_r H | r\in\tilde{R}(G:H)\}, \tag{16}$$

consitutes the orbit of the transitive representation $R^{G:H}$ of the group G, with the stability group H and the action determined by the formula

$$R^{G:H}(x) = \begin{pmatrix} g_r H \\ x g_r H \end{pmatrix}, \quad r\in\tilde{R}(G:H),\ x\in G. \tag{17}$$

Eqs. (14) - (17) define the canonical form of a transitive representation. This form is determined by choosing the labels for left cosets in decomposition (14).

A natural convention of partial labelling of elements of the canonical orbit $\tilde{R}(H)$ of the transitive representation $R^{G:H}$ is associated with a coarsening of this orbit, i.e. a decomposition into imprimitivity sets according to an intermediate subgroup K in the chain $H\subset K\subset G$. Let

$$K = \bigcup_{r_2\in\tilde{R}(K:H)} g_{r_2} H \tag{18}$$

be the decomposition of the subgroup $K\subset G$ into left cosets with respect to $H\subset K$. The decomposition (14) can be now rewriten in a form

$$G = \bigcup_{r_1\in\tilde{R}(G:K)} \ \bigcup_{r_2\in\tilde{R}(K:H)} g_{r_1} g_{r_2} H, \tag{19}$$

so that

$$\tilde{R}(H) = \bigcup_{r_1\in\tilde{R}(G:K)} g_{r_1} \tilde{R}^K(H) \tag{20}$$

is the decomposition of the orbit $\tilde{R}(H)$ into imprimitivity sets $g_{r_1}\tilde{R}^K(H)$, where

$$\tilde{R}^K(H) = \{g_{r_2} H | r_2\in\tilde{R}(K:H)\} \tag{21}$$

is the canonical orbit of the transitive representation $R^{K:H}$ of the intermediate subgroup K.

Thus one can define a bijection ψ: $\tilde{R}(K)\times\tilde{R}^{K}(H)\to\tilde{R}(H)$ by means of the formula

$$g_{r_1}g_{r_2}H = g_rH, \quad r_1\epsilon\tilde{R}(G:K),\ r_2\epsilon\tilde{R}(K:H),\ r\epsilon\tilde{R}(G:H), \tag{22}$$

which creates a temptation to interpret the orbit $\tilde{R}(H)$ as the cartesian product of orbits $\tilde{R}(K)$ and $\tilde{R}^{K}(H)$. One has to note, however, that the bijection ψ is not determined uniquely by the intermediate subgroup K, but is also dependent upon the choice of left coset representatives in the decomposition (19). In particular, the decomposition (20) of the orbit $\tilde{R}(H)$ into imprimitivity sets $g_{r_1}\tilde{R}^{K}(H)$ depends solely on the subgroup K, and is independent of the choice of representatives g_{r_1}, $r_1\epsilon\tilde{R}(G:K)$, whereas the labelling of cosets g_rH within a particular imprimitivity set $g_{r_1}\tilde{R}^{K}(H)$ evidently depends on the choice of the representative g_{r_1}. Such a situation can be adequately described in terms of a finite analogue of the notion of a fibre bundle, which can be interpreted as a generalization of the cartesian product.

We consider the following definition: let E, B, W be finite sets, and p: $E\to B$ - a surjection. The tetraiad

$$\xi = (E,B,W,p) \tag{23}$$

is called a fibration, if there exist such bijections Φ_b: $W\to p^{-1}(b)$, $b\epsilon B$, that

$$p(\Phi_b(w)) = b, \quad w\epsilon W,\ b\epsilon B. \tag{24}$$

The sets E, B, W, and the counterimage

$$p^{-1}(b) = \{e\epsilon E|p(e) = b\} \tag{25}$$

are then called the fibre bundle, the base, the standard fibre, and the fibre upon the element b of the base, respectively, whereas the surjection p is referred to as projection of the fibre bundle onto its base. This definition is evidently a finite analogue of the standard definition of a fibre bundle (cf. e.g. Spanier [37], p. 90), with neglection of the notion of neighbourhood, which is irrelevant in the case of finite sets. The definition of the fibration can be illustrated by the following pattern,

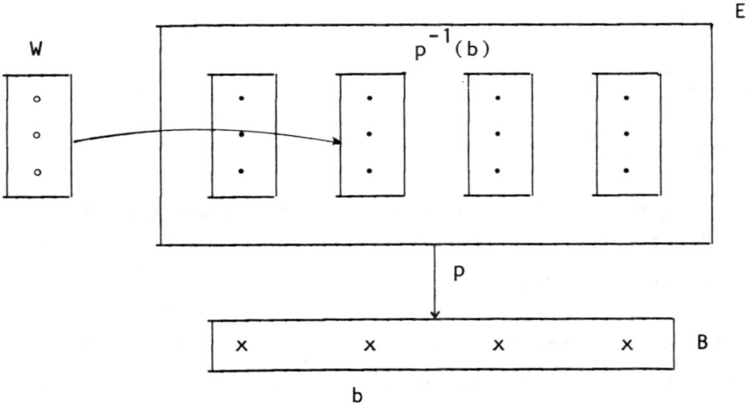

where the set E (the fibre bundle) is presented as a collection of fibres upon of the base B, each fibre $p^{-1}(b)$, $b \in B$, being a copy of the standard fibre W.

With this definition, it is easy to observe that putting

$$E = \tilde{R}(H) = \{g_{r_1} g_{r_2} H \mid r_1 \in \tilde{R}(G:K),\ r_2 \in \tilde{R}(K:H)\}, \qquad (26)$$

$$B = \tilde{R}(G:K) = \{r_1 \mid r_1 = 1,2,\ldots,|G|/|K|\}, \qquad (27)$$

$$W = \tilde{R}^K(H) = \{g_{r_2} H \mid r_2 \in \tilde{R}(K:H)\}, \qquad (28)$$

$$p(g_{r_1} g_{r_2} H) = r_1, \quad r_1 \in \tilde{R}(G:K),\ r_2 \in \tilde{R}(K:H), \qquad (29)$$

one obtains a fibration

$$\xi = (\tilde{R}(H), \tilde{R}(G:K), \tilde{R}^K(H), p) \qquad (30)$$

of the canonical orbit $\widetilde{R}(H)$ of the transitive representation $R^{G:H}$, imposed by the intermediate subgroup K. Namely, we can introduce bijections $\Phi_{r_1} : \widetilde{R}^K(H) \to p^{-1}(r_1)$, where

$$p^{-1}(r_1) = \{g_{r_1} g_{r_2} H \mid r_2 \in \widetilde{R}(K:H)\}, \quad r_1 \in \widetilde{R}(G:K), \tag{31}$$

by the formula

$$\Phi_{r_1}(g_{r_2} H) = g_{r_1} g_{r_2} H, \quad r_2 \in \widetilde{R}(K:H). \tag{32}$$

We have

$$p(\Phi_{r_1}(g_{r_2} H)) = p(g_{r_1} g_{r_2} H) = r_1, \quad r_1 \in \widetilde{R}(G:K),\ r_2 \in \widetilde{R}(K:H), \tag{33}$$

satisfying thus Eq. (24).

It is evident that the base (27) of the fibration (30) is uniquely determined by the subgroup K, whereas the position of an element of the canonical orbit $\widetilde{R}(H)$ within its fibre is determined by the bijection Φ_{r_1}, dependent on the choice of the representative g_{r_1}. This choice can be done independently for each left coset of G with respect to the intermediate subgroup K, so that a comparison of projections of elements of two different fibres $p^{-1}(r_1)$ and $p^{-1}(r_1')$, $r_1 \neq r_1'$, onto the standard fibre $\widetilde{R}^K(H)$, is evidently meaningless. Borrowing a terminology from kinematics, we can say that the base is "absolute", like the time in Galilean theory, whereas the fibre is "relative", like the space distance, which can be well defined only for simultaneous events, i.e. for elements of the space-time belonging to one fibre, determined by a definite moment of time.

The fibration $\xi = (G, \widetilde{R}(G:H), H, p)$ is associated with the presentation of an arbitrary element g of the group G in a form

$$g = g_r h, \quad r \in \widetilde{R}(G:H),\ h \in H, \tag{34}$$

and the projection p can be written in a form

$$p(g) \equiv p(g_r h) = r, \quad g \in G. \tag{35}$$

The action of the regular representation $R^{reg} \equiv R^{G:\{e\}}$ on the group manifold of G can be written with this presentation as

$$R^{reg}(x) = \begin{pmatrix} g_r h \\ x g_r h \end{pmatrix} = \begin{pmatrix} g_r h \\ g_{r''} h_r(x) h \end{pmatrix}, \tag{36}$$

where r'' is the label of the coset enclosing the element $xg_r h$, and

$$h_r(x) = g_{r''}^{-1} x g_r \in H \tag{37}$$

is called the subelement of the element $x \in G$ under the representative g_r. Therefore, the regular representation (i) permutes the while fibres of ξ according to the transitive representation $R^{G:H}$ (the so called ground representation in the terminology of Altmann [38]), (ii) permutes elements within a fibre in a way determined by subelements, (iii) does not exchange individual elements between different fibres (one can say that there is no "diffusion" between fibres). It can be written symbolically as

$$R^{reg}(x) \cong \begin{pmatrix} r & h \\ R^{G:H}(x|r) & R^{H:\{e\}}(h_r(x)|h) \end{pmatrix}, \tag{38}$$

where

$$R^{G:H}(x|r) = r'', \quad R^{H:\{e\}}(h_r(x)|h) = h_r(x)h. \tag{39}$$

4. THE STANDARD FORM OF AN ARBITRARY PERMUTATION REPRESENATION

Let $\tilde{L}(G)$ be the lattice of subgroups of the group G and let C: Aut $G \to \Sigma_{|\tilde{L}(G)|}$ be the permutation representation realising the natural action of the group Aut G on the set $\tilde{L}(G)$ according to the formula

$$C(\tau) = \begin{pmatrix} H \\ \tau(H) \end{pmatrix}, \quad H \in \tilde{L}(G), \ \tau \in \text{Aut } G. \tag{40}$$

Let

$$\tilde{L}(G) = \bigcup_{H\in\tilde{I}(G)} \tilde{C}[H] \tag{41}$$

be the decomposition of the carrier set $\tilde{L}(G)$ of the representation C into orbits

$$\tilde{C}(H) = \{\tau(H)|\tau \text{ Aut } G\}, \tag{42}$$

consisting of all these subgroups of G, which are conjugated with H by an arbitrary automorphism of G, so that $\tilde{I}(G)$ is the set of representatives of these orbits, and the restriction

$$C|_{\tilde{C}[H]} = R^{\text{Aut } G:J(H)} \tag{43}$$

is a transitive representation of the group Aut G, with the stability group

$$J(H) = \{\tau\in\text{Aut } G|\tau(H) = H\}. \tag{44}$$

Further, let

$$\Phi(H) = \langle J(H),\text{Int } G\rangle \tag{45}$$

be the group generated by J(H) and Int G. Then it can be easily shown that the transform $(R^{G:H})_\tau$ of the transitive representation $R^{G:H}$ by the automorphism $\tau\in$Aut G is equal to the transform $(R^{G:H})^\rho$ by a bijection $\rho\in\Sigma_{|G|/|H|}$ if and only if $\tau\in\Phi(H)$. Let, moreover

$$\text{Aut } G = \bigcup_{t_1\in\tilde{R}(\text{Aut } G:\Phi(H))} \bigcup_{t_2\in\tilde{R}(\Phi(H):J(H))} \tau^{t_1}\tau^{t_2}J(H) \tag{46}$$

be the decomposition of Aut G into left cosets with respect to its subgroup J(H), with the coarsening into imprimitivity sets associated with the intermediate subgroup Φ(H) in the chain

$$J(H)\subset\Phi(H)\subset\text{Aut } G. \tag{47}$$

Using this notation, one can present an arbitrary permutation representation P of the group G in a standard ordered form. In the first step one can decompose the carrier set $\tilde{P}$ into subsets consisting of orbits of equivalent representations, called strata, and classified by means of elements of the set of representative $\tilde{I}(G)$, i.e.

$$\tilde{P} = \bigcup_{H \in \tilde{I}(G)} S[H], \tag{48}$$

where

$$S[H] = \{p \in \tilde{P} \mid \exists\ \tau \in \mathrm{Aut}\ G,\ G^{(p)} = \tau(H)\}, \tag{49}$$

and

$$G^{(p)} = \{x \in G \mid P(x|p) = p\} \tag{50}$$

denotes the stability group for $p \in \tilde{P}$. In the second step each stratum $\tilde{S}[H]$ can be decomposed into substrata consisting of orbits of those representations whose stability groups are conjugated by internal automorphisms of G, i.e.

$$S[H] = \bigcup_{t_1 \in \tilde{R}(\mathrm{Aut}\ G:\Phi(H))} S[H,t_1], \tag{51}$$

where

$$S[H,t_1] = \{p \in S[H] \mid \exists\ \tau \in \mathrm{Int}\ G,\ G^{(p)} = \tau(\tau^{t_1}(H))\}. \tag{52}$$

In the third step each transitive representation entering P can be put in a canonical form, determined by the choice of an appropriate bijection on this orbit.

5. THE CLEBSCH-GORDAN DECOMPOSITION FOR TRANSITIVE REPRESENTATIONS

Let us consider two transitive representations $R^{G:H}$ and $R^{G:D}$. The direct product $R^{G:H} \otimes R^{G:D}$ of these representation acts on the set

$$\tilde{R}(H) \times \tilde{R}(D) = \{(g_{r_1}H, g_{r_2}D) \mid r_1 \in \tilde{R}(G:H),\ r_2 \in \tilde{R}(G:D)\}, \tag{53}$$

i.e. on the Cartesian product of orbits of the consituent representations. In general, the set (53) is not an orbit, so that the direct product $R^{G:H} \otimes R^{G:D}$ decomposes onto a direct sum of transitive representations of the group G. This decomposition can be written in a form

$$R^{G:H} \otimes R^{G:D} \cong \sum_{\omega\in\Omega(D,H)} \oplus R^{G:L_\omega} \tag{54}$$

where $\Omega(D,H)$ is the set of indices of double cosets of the group G with respect to the pair (D,H) of subgroups, determined by the decomposition

$$G = \bigcup_{\omega\in\Omega(D,H)} Dg_\omega H, \tag{55}$$

and

$$L_\omega = D \cap g_\omega H g_\omega^{-1}. \tag{56}$$

The Clebsch-Gordan decomposition (54) for transitive representations is essentially a particular case of Mackey's theorem for induced representations (Mackey [39]; cf. also Bradley [40,41], Altmann [38]).

The decomposition (54) is formulated at the level of representations. Sometimes one need a more exhaustive formulation, on the level of standard bases (cf. Edwards [42]). To this aim, we define the Clebsch-Gordan coefficients for transitive representations by means of the formula

$$|R_1R_2R_3\omega r_3\rangle = \sum_{\substack{r_1\in\tilde{R}(G:H)\\ r_2\in\tilde{R}(G:D)}} \begin{vmatrix} R_1 & R_2 & R_3 & \omega \\ r_1 & r_2 & r_3 & \end{vmatrix} |R_1r_1\rangle|R_2r_2\rangle, \tag{57}$$

where we use a traditional Dirac notation $|R_1r_1\rangle$, $|R_2r_2\rangle$, and $|R_1R_2R_3\omega r_3\rangle$ for elements of orbits of the constituent representations $R_1 = R^{G:H}$, $R_2 = R^{G:D}$, and the resultant representation $R_3 = R^{G:L_\omega}$. The symbol in rectangular brackets in Eq. (57) is the Clebsch-Gordan coefficient for transitive representations. By a comparison of Eq. (57) with the corresponding definition for the case of irreducible representations (cf. e.g. Butler [11,12]) we observe that the label ω of a double coset plays a role of the repetition index distinguishing equivalent representations R_3. Note that the repetition index in the transitive case is uniquely determined by the structure of the group G, i.e. by Eq. (55), whereas it can be chosen rather arbitrarily in the irreducible case.

In order to evaluate the Clebsch-Gordan coefficients occuring in Eq. (57), we observe that the direct product of transitive representations is an induced representation, which can be written e.g. in a form

$$R^{G:H} \otimes R^{G:D} \cong (R^{G:H} \downarrow D) \uparrow G, \tag{58}$$

so that it is a representation induced from the subgroup $D \subset G$, with the restriction $R^{G:H} \downarrow D$ of the first constituent representation $R^{G:H}$ to the stability group D of the second in the role of inducing representation. The orbit $\tilde{R}(H)$ decomposes under this restriction into orbits with respect to the subgroup D according to a formula

$$\tilde{R}(H) = \bigcup_{\omega \in \Omega(D,H)} O_D[g_\omega H], \tag{59}$$

where

$$O_D[g_\omega H] = \{d g_\omega H | d \in D\} \equiv \{d_s g_\omega H | s \in \tilde{R}(D:L_\omega)\}. \tag{60}$$

Correspondingly, the direct product (53) can be written in a form

$$\tilde{R}(H) \times \tilde{R}(D) \cong \{(\omega, s, r_2) | \omega \in \Omega(D,H),\ s \in \tilde{R}(D:L_\omega),\ r_2 \in \tilde{R}(G:D)\} \tag{61}$$

where the orbit ω of the resultant representation $R^{G:L_\omega}$ is subjected to a fibration according to the chain of subgroups

$$L_\omega \subset D \subset G, \tag{62}$$

and the indices s and r_2 label respectively elements of a fibre and the base of this fibration. The labels ω and s are determined by r_1 and r_2 through the condition

$$g_{r_2}^{-1} g_{r_1} H = d_s g_\omega H, \quad s \in \tilde{R}(D:L_\omega), \omega \in \Omega(D,H), \tag{63}$$

and the standard index r_3 of the resultant representation - from

$$g_{r_2} d_s L_\omega = g_{r_3} L_\omega, \quad r_3 \in \tilde{R}(G:L_\omega). \tag{64}$$

Thus we have

$$\begin{vmatrix} R_1 & R_2 & R_3 & \omega \\ r_1 & r_2 & r_3 & \end{vmatrix} = \begin{cases} 1, & \text{if Eqs. (63) and (64) are satisfied,} \\ 0 & \text{otherwise.} \end{cases} \quad (65)$$

6. FINAL REMARKS

We have presented in this article the problem of standardization of Clebsch-Gordan decomposition for permutation representations. Another problem which arises naturally is a decomposition of a transitive representation $R^{G:H}$, or, more exactly, its linear equivalent, into irreducible representations of the group G. This problem has been treated in detail in papers of Lulek and Lulek [24], using the Frobenius reciprocity theorem at the level of irreducible bases (Edwards [42], Kirillov [43]). In particular, there are given analytic expressions for irreducible bases of induced representations in terms of irreducible bases for representations subduced to appropriate stability subgroups. These expressions do not involve the summation over the group G, avoiding thus an annoying calculational difficulty.

Up to our knowledge, there is still no papers dealing explicitly with multiple Clebsch-Gordan decompositions for transitive representations, which would be a natural continuation of development of Racah algebra. It seems that some results are implicit in the representation theory for symmetric groups (James and Kerber [44]), especially with connection with the wreath product (Kerber [45]).

The methods of Racah algebra can be also applied in solid state theory, in particular in crystallography and for a classification of states (Lulek [46]).

7. REFERENCES

[1] Fano, U. and Racah, G.,"Irreducible Tensorial Sets", Academic Press, New York 1959.

[2] Biederharn, L.C. and van Dam, H., "Quantum Theory of Angular Momentum", Academic Press, New York 1965.

[3] Biederharn, L.C. and Louck, J.D., "Angular Momentum in Quantum Physics", Encycl. Math. Vol. 8, Addison-Wesley, Reading, Mass. 1981.

[4] Biederharn, L.C. and Louck, J.D., "The Racah-Wigner Algebra in Quantum Theory", Encycl. Math. Vol. 9, Addison-Wesley, Reading, Mass. 1981.

[5] Wybourne, B.G., "Spectroscopic Properties of the Rare Earths", Wiley, New York 1965.

[6] Moshinsky, M., "Group Theory and the Many-Body Problem", Gordon and Breach, New York 1968.

[7] Vanages, V., "Algebraic Methods in Nuclear Theory", Mintis, Vilnius 1971.

[8] Griffith, J.S., "The Irreducible Tensor Method for Molecular Symmetry Groups", Prentice-Hall 1962.

[9] Griffith, J.S., "The Theory of Transition-Metal Ions", Cambridge Univ. Press, Cambridge 1964.

[10] Sugano, S., Tanabe, Y. and Kamimura, H., "Multiples of Transition--Metal Ions in Crystals", Academic Press, New York 1970.

[11] Butler, P.H., "Point Groups Symmetry Applications", Plenum Press, New York 1981.

[12] Butler, P.H., Phil. Trans. Roy. Soc. (London) 227, 545-85 (1975).

[13] Kibler, M., Int. J. Quantum Chem. 10, 87-111 (1976).

[14] Flurry, R.L., Int. J. Quantum Chem. 65, 455-8 (1972); Theor. Chim. Acta 31, 221-30 (1973).

[15] Fieck, G., Theor. Chim. Acta 44, 279-91 (1977); 49, 187-98, 199-210, 211-22 (1978); Physica 105A, 577-92; 106A, 521-38 (1981).

[16] Lulek, T., Acta Phys. Pol. A57, 407-14 (1980).

[17] Kuplowski, I., Kuźma, M. and Lulek, T., Acta Phys. Pol. A57, 415-28 (1980).

[18] Newman, D.J., J. Phys. A14, 3143-51 (1981); A15, 3395-404 (1982); A16, 2375-85 (1983).

[19] Chen, S.C. and Newman, D.J., J. Phys. A15, 331-41 (1982).

[20] Chan, K.S. and Newman, D.J., J. Phys. A15, 3383-93 (1982); A16, 2389-403 (1983); A17, 253-65 (1984).

[21] Michel, L. and Mozrzymas, J., C.R. Acad. Sc. Paris 295, s. II, 435-7 (1982).

[22] Litvin, D.B., J. Math. Phys. 23, 337-44 (1982).

[23] Butler, P.H., Ford, A.M. and Reid, M.F., J. Phys. B16, 967-74 (1983).

[24] Lulek, B. and Lulek, T., Acta Phys. Pol. A66, 149-65 (1984); J. Phys. A17, 3077-89 (1984).

[25] Lulek, B., Lulek, T., Biel, J. and Chatterjee, R., Can. J. Phys., submitted for publication.

[26] Weyl, H., "Symmetry", Princeton Univ. Press, New Jersey 1982.

[27] Mozrzymas, J., "Application of Group Theory in Physics"(in Polish), PWN, Warsaw 1977.

[28] Kurosh, A.G., "Theory of Groups", Chelsea, New York 1955.

[29] Hall, M., "The Theory of Groups", Macmillan, New York 1959.

[30] Vilenkin, N.Ya., "Special Functions and the Theory of Group Representations", Providence, Rhode Island, 1968.

[31] Najmark, M.A., "Theory of Group Representations" (in Russian), Nauka, Moscow 1976.

[32] Burnside, W., "Theory of Groups of Finite Order", Dover Publ., New York 1911.

[33] Gorenstein, D., "Finite Groups", Harper and Row, New York, 1968.

[34] Palais, R.S., Mem. Am. Math. Soc. 36, 1 (1960).

[35] Kirillov, A.A., "Elements of the Theory of Representations, Springer-Verlag, Berlin, 1976.

[36] Michel, L., Lectures at the present School.

[37] Spanier, E.H., "Algebraic Topology", McGraw-Hill, New York 1966.

[38] Altmann, S.L., "Induced Representations in Crystals and Molecules", Academic Press, London 1977.

[39] Mackey, G.W., Ann. Math. 55, 101-39 (1952); 58, 193-221 (1953).

[40] Bradley, C.J., J. Math. Phys. 7, 1145-52 (1966).

[41] Bradley, C.J., "The Mathematical Theory of Symmetry in Solids", Clarendon, Oxford 1972.

[42] Edwards, S.A., J. Phys. A13, 1563-73 (1980).

[43] Kirillov, A.A., "Elements of the Theory of Representations", Springer-Verlag, Berlin 1976.

[44] James, G.D. and Kerber, A., "The representation Theory of the Symmetric Group", Encycl. Math. Vol. 16, Addison-Wesley, London 1981.

[45] Kerber, A., "Representations of Permutation Groups" I, II, Lecture Notes in Math. Vols. 240 and 495, Springer-Verlag, Berlin 1971 and 1975.

[46] Lulek, T., J. Physique, 45, 29-34 (1984).

SYMMETRY CHANGES AT A TRICRITICAL POINT

Jerzy Kociński

Institute of Physics, Warsaw Technical University
Koszykowa 75, 00-662 Warszawa
POLAND

ABSTRACT

Landau type theory of symmetry changes at a tricritical point in two-sublattice metamagnets has been formulated and applied to $FeCl_2$ crystals.

1. INTRODUCTION

Magnetic crystals which exhibit field-induced phase transitions are in general either highly anisotropic or isotropic (weakly anisotropic). The phase transitions in the highly anisotropic crystals are characterized by simple reversals of the local spin directions in contrast with transitions in the isotropic (weakly anisotropic) crystals in which there appears a rotation of the local spin directions. An example of the first-type crystal is $FeCl_2$ [1,2]. As this crystal is cooled in a zero magnetic field, it undergoes a continuous phase transition at a certain temperature T_N, and it orders antiferromagnetically below T_N with a two-sublattice structure in which spins are constrained to point either parallel or antiparallel to the easy axis. Below T_N but at high temperatures in a magnetic field parallel to the easy axis there again appears a continuous phase transition from antiferromagnetic (AF) to paramagnetic (P) state. This behaviour persists in a certain interval of temperatures below T_N, and, consequently, we deal with a line of critical points. While the magnetic field is applied at low temperatures the crystal behaves differently. The transition from the AF state to the P state is discontinuous. This situation persists for an interval of temperatures down to $T = 0$, and, consequently, we deal with a line of phase transition points. The lines of critical points for the discontinuous and for the continuous transitions meet with the same slope at a point which has been called the tricritical point (TCP). The crystal spins are constrained by the strong anisotropy to lie along the easy axis in the AF phase as well as in the P phase. The phenomenological theory of phase transitions at a TCP has been discussed by Boccara [3].

2. A METHOD OF CALCULATING SYMMETRY CHANGES AT A TRICRITICAL POINT OF A TWO-SUBLATTICE METAMAGNET

The situation at the TCP of a metamagnet differs in the following respects from that at a critical point at which takes place a continuous phase transition. First of all we deal with a phase transition in a magnetic field, from a magnetically ordered phase P to another magnetically ordered phase. This implies an application of corepresentations. Next, when the tricritical point is approached along the line of critical points for discontinuous transitions then the discontinuity in the behaviour of the customary order-parameter $(M_A - M_B)/2$, where M_A and M_B are the sublattice magnetizations, diminishes to zero. The convergence of the thermodynamic potential series expansion in the vicinity of the TCP can therefore be assumed. Consequently, we can expect that a single thermodynamic potential will be feasible for a discussion of the continuous or the discontinuous phase transition at the TCP [4].

The presence of an external magnetic field implies that in a metamagnet we deal with two parameters: An order parameter connected with the difference in the sub-lattice magnetizations and a parameter connected with the resultant magnetization of the crystal. In a two-sub - lattice metamagnet we define the order parameter by

$$\vec{M}_s(\vec{x}) = \vec{M}_A(\vec{x}) - \vec{M}_B(\vec{x} + \vec{a}) \tag{1}$$

and we also introduce the parameter

$$\vec{M}(\vec{x}) = \vec{M}_A(\vec{x}) + \vec{M}_B(\vec{x} + \vec{a}) \tag{2}$$

where $\vec{M}_A(\vec{x})$ and $\vec{M}_B(\vec{x} + \vec{a})$ are the mean magnetic mo- ments per sublattice site at the points $\vec{x}$ and $\vec{x} + \vec{a}$ of the A and B sublattices, respectively.

The thermodynamic potential density Φ can be written in the form

$$\Phi = \Phi(\vec{M}) + \Phi(\vec{M}_s) + \Phi(\vec{M},\vec{M}_s) \tag{3}$$

where $\Phi(\vec{M},\vec{M}_s)$ represents the coupling terms between $\vec{M}$ and $\vec{M}_s$. We assume that this potential density has the symmetry G_o of the paramagnetic phase P. According to the Curie principle, the symmetry group G_o is the maximal common subgroup of the crystal symmetry group and the magnetic field symmetry group, for a fixed mutual posi- tion of their symmetry elements [5].

The order parameter $\vec{M}_s(\vec{x})$ is expanded in the terms of the basis functions $\vec{\Psi}_i(\vec{x})$ of an active corepresentation

$$\vec{M}_s(\vec{x}) = \sum_i c_i \vec{\Psi}_i(\vec{x}) = \eta(T,H) \sum_i \gamma_i \vec{\Psi}_i(\vec{x}) \qquad (4)$$

with $\sum \gamma_i^2 = 1$. An active corepresentation has to fulfil the following three conditions: (1) The condition of reality or of physical irreducibility. (2) The Landau condition. (3) The Lifshitz condition.

The potential density $\Phi(\vec{M}_s)$ can be written in the form

$$\Phi(\vec{M}_s) = A(T,H)\eta^2 + \eta^4 \sum_p C_p(T,H) f_p^{(4)}(\gamma_i) + \eta^6 \sum_r D_r(T,H) f_r^{(6)}(\gamma_i) + \ldots - \eta \vec{h} \cdot \sum_i \gamma_i \vec{m}_i \qquad (5)$$

where $f_p^{(4)}$ and $f_r^{(6)}$ are invariants constructed from the γ_i's, and where $\vec{h} = \vec{h}_A$ is the staggered field acting on the A sublattice. The last term in eqn.(5) has been obtained from the expression

$$V^{-1} \int \vec{h} \cdot \vec{M}_s(\vec{x}) d^3\vec{x} = V^{-1} \vec{h} \cdot \sum_i \gamma_i \eta \sum_q \int \vec{e}_q \Psi_{iq}(\vec{x}) d^3\vec{x}$$

where V denotes the magnetic unit cell, and $\vec{e}_q$ are axial unit vectors. Since $\vec{h}$ has the symmetry of $\vec{M}_s$, the staggered field term in eqn.(5) has the symmetry G_o.

The term

$$-V^{-1} \int \vec{M}(\vec{x}) \cdot \vec{H} d^3\vec{x}$$

which appears in the potential density $\Phi(\vec{M})$, has the symmetry G_o since $\vec{M}(\vec{x})$ is invariant under the integral translations $\vec{t}_n \in G_o$, and the symmetry group of the external field $\vec{H}$ is a supergroup of G_o.

Since $\vec{M}(\vec{x})$ has the symmetry G_o, the lowest-degree invariant which appears in $\Phi(\vec{M},\vec{M}_s)$ is of the form

$$|\vec{M}|\, A'\, \eta(T,H) f^{(2)}(\gamma_i) = \chi_T H A'\, \eta(T,H) f^{(2)}(\gamma_i) \qquad (6)$$

where χ_T is the isothermal susceptibility and $A' = A'(T,H)$ while $f(\gamma_i)$ is a second-degree invariant. We also could envisage higher-order coupling terms.

We know [1,2,3] that the temperature and field dependence of the parameter η should be

$$\eta \sim (T_t - T)^{1/4}, \quad \text{for} \quad H = H_t$$

$$\eta \sim (H_t - H)^{1/4}, \quad \text{for} \quad T = T_t$$

where T_t and H_t determine the tricritical point. We therefore have to assume that the temperature and field dependence of the coupling term in eqn. (6) is given by

$$\chi_T H A' = a'_T \cdot (T - T_t) + a'_H \cdot (H - H_t) \qquad (7)$$

Consequently, we can incorporate the coupling term in eqn. (6) into the potential density $\Phi(\vec{M}_s)$ in eqn. (5), and we then can rewrite the potential density in eqn. (3) in the form

$$\Phi = \Phi^{(o)} + A(T,H)\eta^2 + \eta^4 \sum_p C_p(T,H) f_p^{(4)}(\gamma_i) + \eta^6 \sum_r D_r(T,H) f_r^{(6)}(\gamma_i) + \ldots - \eta \vec{h} \cdot \sum_i \gamma_i \vec{m}_i \qquad (8)$$

where $\Phi^{(o)}$ depends on $\vec{M}$ and $\vec{H}$, and the remaining terms vanish for $T \geqslant T_t$, and $H \geqslant H_t$, and where we assume that

$$A(T,H) = a_T \cdot (T - T_t) + a_H \cdot (H - H_t) \tag{9}$$

$$C_1(T,H) = c_T \cdot (T - T_t) + c_H \cdot (H - H_t) \tag{10}$$

with $a_T \cdot c_H \neq a_H \cdot c_T$. With the assumption eqn.(10) the type of the phase transition at the TCP depends on the sign of the coefficient $C_1(T,H)$. For a positive C_1 we can have a continuous transition, while for a negative C_1 we can have a discontinuous transition. The minimum conditions for Φ with respect to η and γ_i, restrain the values of the phenomenological coefficients C_p, with $p > 1$, and of the coefficients D_r. For a one-dimensional order parameter the potential density eqn.(8) is identical to that which has been discussed by Boccara [3].

For a continuous transition at the TCP, the crystal symmetry change is determined when the sets of the coefficient values c_i, to which correspond the minima of the thermodynamic potential density, are calculated and the symmetry of the order parameter $\vec{M}_s(\vec{x})$ is determined. The crystal symmetry after the phase transition is equal to that of the order parameter.

For a discontinuous phase transition at the TCP, the minimum conditions for the potential density in eqn.(8) are supplemented with the condition

$$\Phi = \Phi^{(o)}, \quad \text{for } \vec{h} = 0 \tag{11}$$

This condition yields the three minima at $\eta = 0$, and at $\eta \neq 0$, which characterize a discontinuous phase transition.

3. THE CALCULATION OF THE SYMMETRY CHANGE AT THE TRICRITICAL POINT OF $FeCl_2$

Without magnetic field and for temperatures above or at the Néel point the space group of $FeCl_2$ crystal is $R\bar{3}m$ (D^5_{3d}). When the crystal is placed in an external magnetic field along the three-fold rotation axis, then we deal with a composite system: crystal + field. According to the Curie principle, the symmetry group of the composite system is the maximal common subgroup of the crystal symmetry group and of the field symmetry group, i.e. $R\bar{3}m'$. This is the symmetry group G_o of the paramagnetic phase P.

Since we know from experiments that the phase transition at the TCP of $FeCl_2$ crystal occurs for the reciprocal lattice vector $\vec{k}_1 = (\vec{b}_1 + \vec{b}_2 + \vec{b}_3)/2$, where $\vec{b}_j$, $j = 1,2,3$, are the basis vectors of the reciprocal lattice, we will determine the active irreducible corepresentations at that point. The basis vectors of the Bravais lattice are [6]

$$\vec{a}_1 = (0,-a,c), \qquad \vec{a}_2 = (a\sqrt{3}/2,\ a/2,\ c)$$

$$\vec{a}_3 = (-a\sqrt{3}/2,\ a/2,\ c) \tag{12}$$

The unitary subgroup of the group $R\bar{3}m'$ is C^2_{3i}, and therefore, taking $A = \Theta\sigma_{d1}$ as the antiunitary element (where Θ is the time reversal operator) we can write the magnetic group G_o in the form: $R\bar{3}m' = C^2_{3i} + AC^2_{3i}$.

The six one-dimensional representations of the unitary subgroup C^2_{3i} at the point $\vec{k}_1$ can be found in the tables [7]. The respective irreducible corepresentations of the magnetic group G_o, all belong to type (a), and are one-dimensional. Two of them are real and from the four complex corepresentations we can form two physically irreducible corepresentations. In the following we will concentrate on the corepresentation $D\Gamma_1^+$, derived from the identity representation Γ_1^+. This corepresentation is real and it fulfils the Landau condition and the Lifshitz condition, and therefore it is active.

The stability conditions for a continuous or for a discontinuous phase transition are fulfilled by the potential density in eqn.(8), provided that $D > 0$.

The order parameter $\vec{M}_s(\vec{x}) = \eta\vec{\psi}(\vec{x})$, is invariant under the symmetry operations of the quotient group $G_o/\mathcal{T}$, where $\mathcal{T}$ is the group of integral translations $\vec{t}_n$ of the paramagnetic phase P, and under the new integral translations $\vec{t}'_n$, which are determined from the equation

$$T(\{1\!\!1|n_1\vec{a}_1 + n_2\vec{a}_2 + n_3\vec{a}_3\})\vec{\psi}(\vec{x}) = (-1)^{n_1+n_2+n_3}\ \vec{\psi}(\vec{x})$$

$$n_1, n_2, n_3 = 0, \pm 1, \ldots \qquad (13)$$

where $\vec{a}_1$, $\vec{a}_2$, $\vec{a}_3$, are given in eqn.(12). Consequently, the new basic translations are

$$\vec{a}'_1 = \vec{a}_1 + \vec{a}_2, \quad \vec{a}'_2 = \vec{a}_2 + \vec{a}_3, \quad \vec{a}'_3 = \vec{a}_3 + \vec{a}_1 \qquad (14)$$

and the unit cell volume is increased by the factor 2. The order parameter $\vec{M}_s(\vec{x})$ changes its sign under the translations in eqn.(12), however, it is invariant under the translations in eqn.(14). Its symmetry group is therefore equal to the symmetry group of the experimentally determined AF phase. This symmetry group is $R\bar{3}m'$.

We have shown that the symmetry change at the tricritical point of $FeCl_2$ crystal can be explained in the framework of Landau type theory. For a discussion of the consequences which follow from the remaining corepresentations we refer to the original paper [4].

REFERENCES

1) Cohen, E.G.D. and Kincaid, J.M., Phys. Reports 22C, 58 (1975).
2) Stryjewski, E. and Giordano, N., Advances in Phys., 26, 487 (1977).
3) Boccara, N., "Symetries Brisees", Hermann Paris 1976.
4) Kociński, J. and Osuch, K., International Conference on Magnetism 1985.
5) Kociński, J., "Theory of Symmetry Changes at Continuous Phase Transitions", PWN-Polish Scientific Publishers, Warszawa, Elsevier Amsterdam-Oxford-New York, 1983.
6) Bradley, C.J. and Cracknell, A.P., "The Mathematical Theory of Symmetry in Solids", Clarendon Press, Oxford, 1972.
7) Kovalev, O.V., "Irreducible Representations of the Space Groups", Gordon and Breach, New York 1965.

STATES AND REPRESENTATIONS OF PARTIAL *-ALGEBRAS

J-P. Antoine

Institut de Physique Théorique, Université Catholique de Louvain
B-1348 Louvain-la-Neuve
BELGIUM

1. INTRODUCTION

Spontaneous symmetry breakdown may be defined in many different ways, depending on one's favorite approach to physics. The general meaning is the existence of a state which has less symmetry than the whole system. In the algebraic formulation of quantum theory [1], the basic object is the C*-algebra of observables, and symmetries are realized by automorphisms. Each state determines, through the familiar Gel'fand-Naimark-Segal (GNS) construction, a representation of the algebra by bounded operators in a Hilbert space. In that language, spontaneous symmetry breakdown occurs when a state exists which is not invariant under some automorphism group. In the corresponding representation, these automorphisms are not unitarily implementable.

Now many authors consider this framework too narrow, especially for field theory. On one hand, unbounded operators may be more natural (e.g. boson field operators or symmetry generators). On the other hand, there are systems, typically spin systems with long range interaction, where the thermodynamical limit fails to exist in a C*-topology [2]. Therefore structures more general than normed algebras have been proposed. First came algebras of unbounded operators, the field algebra being the prime example [3][4], and by now a full-fledged theory exists, that of the so-called Op*-algebras [5-7]. More recently Lassner [8] has introduced quasi *-algebras, which are obtained by taking the completion of certain topological *-algebras.

It turns out that these two objects are particular instances of a more general structure, that of <u>partial *-algebra</u>, introduced by W. Karwowski and ourselves [9][10]. If we want to describe spontaneous symmetry breakdown in that wider framework, we have to solve two problems : first extend the usual concepts of states and GNS-representation, then study automorphisms and their possible implementation.

In these lectures we will treat the former only, but first we will review the theory of partial *-algebras. All the material presented is based on joint work with Witold Karwowski (Wrocław), Gerd Lassner (Leipzig) and Françoise Mathot (Louvain-la-Neuve), both published [9)10)] and in progress [11)12)].

2. GENERAL DEFINITIONS AND EXAMPLES

We begin with abstract partial *-algebras, using a definition originally due to Borchers [13)].

2.1. Definition.- A partial *-algebra is a (complex) vector space $\mathfrak{A}$, with an antilinear involution $x \mapsto x^+$ and a subset $\Gamma \subset \mathfrak{A} \times \mathfrak{A}$ such that :

(i) $(x,y) \in \Gamma$ iff $(y^+, x^+) \in \Gamma$;

(ii) if $(x,y) \in \Gamma$ and $(x,z) \in \Gamma$, then $(x, \lambda y + \mu z) \in \Gamma$ for all $\lambda, \mu \in \mathbb{C}$;

(iii) whenever $(x,y) \in \Gamma$, there exists an element $x \circ y \in \mathfrak{A}$

with the usual properties of the product :

$$x \circ (y + \lambda z) = (x \circ y) + \lambda (x \circ z), \ \lambda \in \mathbb{C}$$
$$(x \circ y)^+ = y^+ \circ x^+$$

The element $e \in \mathfrak{A}$ is called a unit if $e^+ = e$, and for every $x \in \mathfrak{A}$ one has $(e,x) \in \Gamma$ and $e \circ x = x \circ e = x$. In this work we will consider only partial *-algebras with unit.

2.2. Definition.- A *-subalgebra of a partial *-algebra $\mathfrak{A}$ is a vector subspace $\mathfrak{M}$ of $\mathfrak{A}$ such that :

(i) $e \in \mathfrak{M}$ (if any);

(ii) $\mathfrak{M}^+ = \mathfrak{M}$;

(iii) whenever $x,y \in \mathfrak{M}$ and $(x,y) \in \Gamma$, then $x \circ y \in \mathfrak{M}$.

It follows that the intersection of any family of *-subalgebras of $\mathfrak{A}$ is one again. Thus given any subset $\mathfrak{N} \subset \mathfrak{A}$, there exists a smallest *-subalgebra containing it, denoted $\mathfrak{M}[\mathfrak{N}]$, and called the *-subalgebra generated by $\mathfrak{N}$.

Since not every product is defined in a partial *-algebra, a special role is played by the set of elements that can multiply a given element, from the left or from the right. Similarly, for any subset $\mathfrak{N} \subset \mathfrak{A}$, we define the set of its <u>left</u>, resp. <u>right multipliers</u> :

$$L\mathfrak{N} = \{ x \in \mathfrak{A} \mid (x,y) \in \Gamma \text{ for all } y \in \mathfrak{N}\}$$

$$R\mathfrak{N} = \{ x \in \mathfrak{A} \mid (y,x) \in \Gamma \text{ for all } y \in \mathfrak{N}\} .$$

This suggests to use a simpler notation :

$$(x,y) \in \Gamma \Leftrightarrow x \in L(y) \quad \Leftrightarrow \quad y \in R(x).$$

The sets of multipliers $L\mathfrak{N}$ and $R\mathfrak{N}$ are vector subspaces of $\mathfrak{A}$ and both contain e. Let now $\mathfrak{N}$ run over all subsets of $\mathfrak{A}$. Then the set of all spaces of multipliers exhibits a remarkable lattice structure [9)10)], due to the fact that the maps $L : \mathfrak{N} \mapsto L\mathfrak{N}$ and $R : \mathfrak{N} \mapsto R\mathfrak{N}$ form a Galois connection. The smallest of such spaces are $L\mathfrak{A}$ and $R\mathfrak{A}$, which are interchanged under the involution. Their elements, the so-called <u>universal multipliers</u>, will play a crucial role throughout.

In Def. 2.1, no requirement of associativity was made for the $\circ$ partial multiplication. Several definitions of that concept are possible for partial algebras.

2.3. <u>Definition</u>.- The partial *-algebra $\mathfrak{A}$ is called <u>associative</u> if the following holds for any $x,y,z \in \mathfrak{A}$: whenever $x \in L(y)$, $y \in L(z)$ and $x \circ y \in L(z)$, then $y \circ z \in R(x)$ and one has :

$$(x \circ y) \circ z = x \circ (y \circ z) \tag{2.1}$$

Although it looks natural this condition is too strong and rarely realized in practice, not even for quasi *-algebras (see below). However, for our purposes, a weaker notion is sufficient.

2.4. <u>Definition</u>.- The partial *-algebra $\mathfrak{A}$ is called <u>semi-associative</u> if the conditions of Def. 2.3 are verified for every element $x,y \in \mathfrak{A}$, $z \in R\mathfrak{A}$. In other words, if $y \in R(x)$ implies $y \circ z \in R(x)$ for every $z \in R\mathfrak{A}$ and Eq .(2.1) holds.

Thus, in a semi-associative partial *-algebra, the sets $L\mathfrak{A}$,$R\mathfrak{A}$ are in fact algebras. Notice also that, contrary to associativity, semi-associativity is not automatically inherited by a *-subalgebra, $\mathfrak{M} \subset \mathfrak{A}$, since $\mathfrak{M}$ may have more universal right multipliers than $\mathfrak{A}$.

We give some examples of such situations .

(1) Quasi *-algebras

Let $\mathfrak{A}_o$ be a non-complete topological *-algebra, with separately, but not jointly, continuous multiplication, $\mathfrak{A} \equiv \tilde{\mathfrak{A}}_o$ the completion of $\mathfrak{A}_o$. Then the multiplication can be extended by continuity to $\mathfrak{A}$, if one of the factors belongs to $\mathfrak{A}_o$. In other words, $\mathfrak{A}$ is a partial *-algebra, with $\Gamma = (\mathfrak{A} \times \mathfrak{A}_o) \cup (\mathfrak{A}_o \times \mathfrak{A})$, called by Lassner [8] a quasi *-algebra. A typical example is $\mathfrak{A}_o = C^o(\Delta)$, the space of continuous functions on a finite interval $\Delta \subset \mathbb{R}$, with pointwise multiplication and a L^p norm $(1 \leq p < \infty)$. Then $\mathfrak{A} = L^p(\Delta; dt)$ is a quasi *-algebra, and therefore $L\mathfrak{A} = R\mathfrak{A} = \mathfrak{A}_o$. It is semi-associative like all quasi *-algebras, but not associative. For instance, if $y \in \mathfrak{A}_o$ vanishes in a neighborhood of some interior point $t_o \in \Delta$, x is continuous except for a simple jump at t_o and z is discontinuous, with at least a jump in the support of y, then $x \circ y \in \mathfrak{A}_o$ and $y \circ z \notin \mathfrak{A}_o$, so that associativity breaks down.

(2) Operators on scales or lattices of Hilbert spaces

Let $(H_n)_{n \in \mathbb{Z}}$ be a scale of Hilbert spaces [14] (the argument is the same for a general lattice) :

$$H_\infty \equiv \bigcap_n H_n \subset \ldots \subset H_2 \subset H_1 \subset H_o \subset H_{\bar{1}} \subset H_{\bar{2}} \subset \ldots \subset H_{\bar{\infty}} \equiv \bigcup_n H_n$$

An operator on such a scale is defined by a unique maximal representative, that is, a bounded linear operator $A : H_p \to H_q$ $(p,q \in \mathbb{Z})$ with p minimal and q maximal, which in turn is extended by natural injections to a linear map $A : H_\infty \to H_{\bar{\infty}}$. Clearly the product $A \circ B$ of two such operators is well defined only if it can be factorized continuously through some H_s :

$$H_p \xrightarrow{B} H_s \xrightarrow{A} H_r .$$

Then the set of all operators on the scale is a partial *-algebra [9].

(3) Closed operators on a fixed dense domain

Let H be a Hilbert space, $\mathcal{D} \subset H$ a fixed dense domain, and consider the following set of closed linear operators :

$$\overline{C}\ (\mathcal{D},H) = \{A \text{ closed} \mid \mathcal{D} \subset \mathcal{D}(A) \cap \mathcal{D}(A^*)\},$$

where A^* is the adjoint of A and $\mathcal{D}(A)$ its domain. If we consider the subset of $\overline{C}(\mathcal{D},\mathcal{H})$ of those operators A such that $A\mathcal{D} \subset \mathcal{D}$, $A^*\mathcal{D} \subset \mathcal{D}$ and take their restrictions to $\mathcal{D}$, we get a *-algebra of (in general) unbounded operators, denoted $L^+(\mathcal{D})$ by Lassner [7)8)]. This is the arena for the theory of Op*-algebras. So $\overline{C}(\mathcal{D},\mathcal{H})$ provides a natural generalization, and it turns out that several structures of partial *-algebras may be introduced on subsets of $\overline{C}(\mathcal{D},\mathcal{H})$. This will occupy us for most of the sequel. In particular, abstract partial *-algebras will be represented (Sec.5) by operators in $\overline{C}(\mathcal{D},\mathcal{H})$ for some $\mathcal{D}$ and $\mathcal{H}$.

3. PARTIAL *-ALGEBRAS OF CLOSED OPERATORS

Given $\mathcal{D} \subset \mathcal{H}$ as above and $A,B \in \overline{C}(\mathcal{D},\mathcal{H})$, we may try to define a product $A \circ B$ as the closure of $A(B \upharpoonright \mathcal{D})$, the restriction to $\mathcal{D}$ of the ordinary product. Of course, this will make sense only if A and B verify a compatibility condition.

3.1. Lemma - Let $A,B \in \overline{C}(\mathcal{D},\mathcal{H})$ verify the conditions :

(i) $B\mathcal{D} \subset \mathcal{D}(A)$

(ii) $A^*\mathcal{D} \subset \mathcal{D}(B^*)$

Then the operator $A(B \upharpoonright \mathcal{D})$ is closable and the domain of its adjoint contains $\mathcal{D}$.

When (i), (ii) are satisfied, we have in fact two natural ways of defining a product that belongs to $\overline{C}(\mathcal{D},\mathcal{H})$, namely :

$$A.B = \overline{A(B \upharpoonright \mathcal{D})} \quad \text{and} \quad A*B = [B^*(A^* \upharpoonright \mathcal{D})]^*.$$

The first one, A.B, has $\mathcal{D}$ as a core (such operators are called $\mathcal{D}$-minimal [15)]), whereas A*B is the adjoint of a $\mathcal{D}$-minimal operator, thus called $\mathcal{D}$-maximal. In order to clarify the situation, we first analyse these two types of operators.

Given $A \in \overline{C}(\mathcal{D},\mathcal{H})$, define $A^{\ddagger} \equiv \overline{A^* \upharpoonright \mathcal{D}}$, $A^{\dagger} \equiv [A \upharpoonright \mathcal{D}]^*$.

Then one gets the following picture :

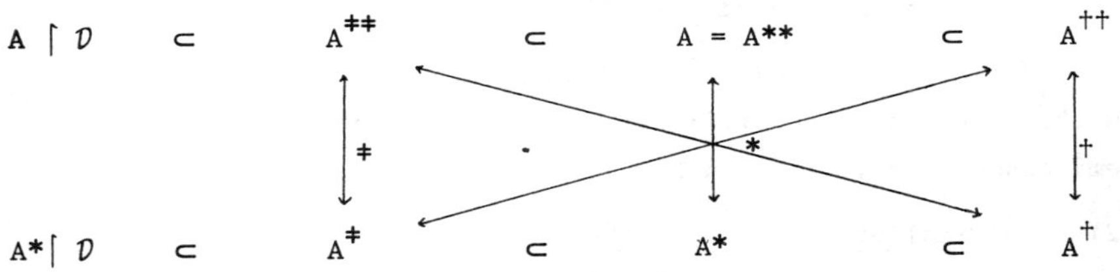

The operators $A^{\ddagger}$, $A^{\ddagger\ddagger}$ are $\mathcal{D}$-minimal and exchanged by $\ddagger$; $A^{\dagger}$, $A^{\dagger\dagger}$ are $\mathcal{D}$-maximal and exchanged by $\dagger$; finally *, the usual adjoint, exchanges the two types. Let us denote by $\mathfrak{C}(\mathcal{D})$, resp. $\mathfrak{C}^*(\mathcal{D})$, the set of all $\mathcal{D}$-minimal, resp. $\mathcal{D}$-maximal, operators in $\overline{C}(\mathcal{D},\mathcal{H})$. Then the map $j : A \mapsto A^{\dagger\dagger}$ is a bijection of $\mathfrak{C}(\mathcal{D})$ onto $\mathfrak{C}^*(\mathcal{D})$, with inverse $j^{-1}: B \mapsto B^{\ddagger\ddagger}$, and moreover the elements of $\mathfrak{C}^*(\mathcal{D})$ are exactly the adjoints of the elements of $\mathfrak{C}(\mathcal{D})$. Now we are in a position to introduce partial *-algebras.

(1) <u>$\mathcal{D}$-minimal operators : $\mathfrak{C}(\mathcal{D})$</u>

On the set $\mathfrak{C}(\mathcal{D})$ of all $\mathcal{D}$-minimal operators, i.e. such that $A = A^{\ddagger\ddagger}$, we consider the following operations :

. vector space structure : $A \hat{+} B = \overline{(A + B) \restriction \mathcal{D}}$, $\lambda A = \overline{(\lambda A) \restriction \mathcal{D}}$

. involution : $A \mapsto A^{\ddagger} = \overline{A^* \restriction \mathcal{D}}$

. partial multiplication : $A.B = \overline{A(B \restriction \mathcal{D})}$, defined whenever $A \in L(B)$, meaning : $B\mathcal{D} \subset \mathcal{D}(A^{\ddagger\ddagger}) = \mathcal{D}(A)$

$$A^{\ddagger}\mathcal{D} \subset \mathcal{D}(B^{\ddagger})$$

. unit : I, the identity operator.

Then, with these operations, $\mathfrak{C}(\mathcal{D})$ verifies all the requirements of Def.2.1 , except (ii); namely it might happen that $C \in R(A) \cap R(B)$ and $C \notin R(A \hat{+} B)$, the reason being that $\mathcal{D}(A \hat{+} B)$ need not contain $\mathcal{D}(A) \cap \mathcal{D}(B)$ [16] (this fact was overlooked in Refs. 9-10). Thus we get a first class of partial *-algebras, by considering, as in Def.2.3, vector subspaces $\mathfrak{M}$ of $\mathfrak{C}(\mathcal{D})$ containing I and stable under the . multiplication, with the additional requirement of distributivity, i.e. that L(C) be a vector subspace for every $C \in \mathfrak{M}$.

The universal multipliers of $\mathfrak{C}(\mathcal{D})$ are easily characterized, in terms of the domain $\mathcal{D}(\mathfrak{C}) \equiv \bigcap_{A \in \mathfrak{C}(\mathcal{D})} \mathcal{D}(A) \supset \mathcal{D}$:

$$L\mathfrak{C}(\mathcal{D}) = \{A \text{ bounded} \mid A^{\ddagger}\,\mathcal{D} \subset \mathcal{D}(\mathfrak{C})\}$$

$$R\mathfrak{C}(\mathcal{D}) = \{A \text{ bounded} \mid A\;\;\mathcal{D} \subset \mathcal{D}(\mathfrak{C})\}$$

Furthermore, the . multiplication on $\mathfrak{C}(\mathcal{D})$ is not associative, but semi-associative, provided $\mathcal{D} = \mathcal{D}(\mathfrak{C})$.

(2) <u>$\mathcal{D}$-maximal operators : $\mathfrak{C}^*(\mathcal{D})$</u>

As defined above, $A \in \mathfrak{C}^*(\mathcal{D})$ iff $A = A^{\ddagger\ddagger}$ or $A^* \in \mathfrak{C}(\mathcal{D})$.

On the set $\mathcal{L}^*(\mathcal{D})$ we consider the following operations :

. vector space structure : $A \check{+} B = [(A^* + B^*) \restriction \mathcal{D}]^*$, $\lambda A = [\bar{\lambda}A^* \restriction \mathcal{D}]^*$

. involution : $A \mapsto A^\dagger = [A \restriction \mathcal{D}]^*$

. partial multiplication : $A*B = [B^\dagger(A^\dagger \restriction \mathcal{D})]^*$, defined whenever $A \in L^*(B)$, meaning : $B\mathcal{D} \subset \mathcal{D}(A^{\dagger\dagger}) = \mathcal{D}(A)$

$$A^\dagger\mathcal{D} \subset \mathcal{D}(B^\dagger).$$

Then $\mathcal{L}^*(\mathcal{D})$ becomes a partial *-algebra with unit I (distributivity is now verified). As above, the universal multipliers consist of bounded operators that map $\mathcal{D}$ into $\mathcal{D}(\mathcal{L}^*) = \bigcap_{A \in \mathcal{L}^*(\mathcal{D})} \mathcal{D}(A) \supset \mathcal{D}(\mathcal{L}) \supset \mathcal{D}$. Finally $\mathcal{L}^*(\mathcal{D})$ is not associative, contrary to the statement made in Ref.10, but it is semi-associative whenever $\mathcal{D} = \mathcal{D}(\mathcal{L}) = \mathcal{D}(\mathcal{L}^*)$. Thus we get a second class of partial *-algebras, namely *-subalgebras of $\mathcal{L}^*(\mathcal{D})$.

(3) <u>$\mathcal{D}$-minimal operators : $\mathcal{L}^w(\mathcal{D})$</u>

The bijection $j : A \mapsto A^{\dagger\dagger}$ of $\mathcal{L}(\mathcal{D})$ onto $\mathcal{L}^*(\mathcal{D})$ respects the product structure, $(A.B)^{\dagger\dagger} = A*B$, but j^{-1} does not, since $L(B) \not\subset L^*(B)$ in general (notice that $A*B = A^{\dagger\dagger}*B^{\dagger\dagger}$ for $A,B \in \mathcal{L}(\mathcal{D})$). This suggests to consider on $\mathcal{L}(\mathcal{D})$ a weaker structure, pulled back from $\mathcal{L}^*(\mathcal{D})$ by the bijection j^{-1} :

. vector space structure : $A \hat{+} \lambda B = (A + \lambda B)^{\ddagger\ddagger}$

. involution : $A \mapsto A^\ddagger = (A^\dagger)^{\ddagger\ddagger}$

. partial multiplication : $A \,\square\, B = (A*B)^{\ddagger\ddagger}$, for $A \in L^*(B)$.

With these operations, the set $\mathcal{L}(D)$ becomes a partial *-algebra with unit I, denoted $\mathcal{L}^w(\mathcal{D})$, which is semi-associative whenever $\mathcal{D} = \mathcal{D}(\mathcal{L}) = \mathcal{D}(\mathcal{L}^*)$. So we get a third class of partial *-algebras, namely *-subalgebras of $\mathcal{L}^w(\mathcal{D})$.

The relation between the three types of partial *-algebras is best understood in terms of the notion of <u>homomorphism</u>, that we will also need in Sec.5 for defining representations.

3.2. <u>Definition</u>. - A <u>homomorphism</u> of a partial *-algebra $\mathfrak{M}$ into another one $\mathfrak{N}$ is a linear map $\sigma : \mathfrak{M} \to \mathfrak{N}$ such that :

(i) $\sigma(x^+) = [\sigma(x)]^+$

(ii) if $x \in L(y)$ in $\mathfrak{M}$, then $\sigma(x) \in L(\sigma(y))$ in $\mathfrak{N}$ and $\sigma(x)\circ\sigma(y) = \sigma(x\circ y)$.

The map σ is an isomorphism if it is a bijection and $\sigma^{-1} : \mathfrak{N} \to \mathfrak{M}$ is also an homomorphism.

Let now $\mathfrak{M} \subset \mathfrak{C}(\mathcal{D})$ be a partial *-algebra, $\mathfrak{M}^* \equiv j(\mathfrak{M}) \subset \mathfrak{C}^*(\mathcal{D})$ and $\mathfrak{M}^w \equiv j^{-1}(\mathfrak{M}^*) \subset \mathfrak{C}^w(\mathcal{D})$. Then $j : \mathfrak{M} \to \mathfrak{M}^*$ is a homomorphism, the identity $i : \mathfrak{M} \to \mathfrak{M}^w$ is a homomorphism, and $j : \mathfrak{M}^w \to \mathfrak{M}^*$ is an isomorphism. The same relationship exists between $\mathfrak{C}(\mathcal{D})$, $\mathfrak{C}^*(\mathcal{D}) = j(\mathfrak{C}(\mathcal{D}))$ and $\mathfrak{C}^w(\mathcal{D}) = j^{-1}(\mathfrak{C}^*(\mathcal{D}))$ (although, strictly speaking, $\mathfrak{C}(\mathcal{D})$ is not a partial *-algebra, the notion of homomorphism still makes sense).

We conclude this section with some examples. The first three types are trivial :

(i) For any dense domain $\mathcal{D}$, the *-algebra of bounded operators $\mathcal{B}(\mathcal{H})$ is a *-subalgebra both of $\mathfrak{C}(\mathcal{D})$ and $\mathfrak{C}^*(\mathcal{D})$, and similarly for any *-subalgebra of $\mathcal{B}(\mathcal{H})$.

(ii) Given $\mathcal{D}$, the *-algebra $\overline{L^+(\mathcal{D})} = \{A \in \mathfrak{C}(\mathcal{D}) \mid A\mathcal{D} \subset \mathcal{D},\ A^\ddagger \mathcal{D} \subset \mathcal{D}\}$ is a *-subalgebra of $\mathfrak{C}(\mathcal{D})$, and so is any *-subalgebra of it, that is, the set $\overline{\mathfrak{A}}$, where $\mathfrak{A}$ is any Op*-algebra on $\mathcal{D}$.

(iii) Similarly, given any Op*-algebra $\mathfrak{A} \subseteq L^+(\mathcal{D})$, $\mathfrak{A}^*$ is a subalgebra of $\mathfrak{C}^*(\mathcal{D})$.

(iv) To get a concrete example, let $\mathcal{H} = L^2(\mathbb{R})$, $\mathcal{D} = C_0^\infty(\mathbb{R})$ and consider the following set of closed operators : $\mathfrak{N} = \{I, \chi, x\frac{d}{dx}, S\}$. Here χ is multiplication by the characteristic function of some interval, $x\frac{d}{dx} = x\frac{d}{dx} \restriction \mathcal{D}$, I is the unit operator and S is the bounded operator defined as

$$(Sf)(x) = \begin{cases} f(x + a), & x \geq 0 \\ f(x - a), & x < 0 \end{cases}$$

where a is fixed positive number. Schematically :

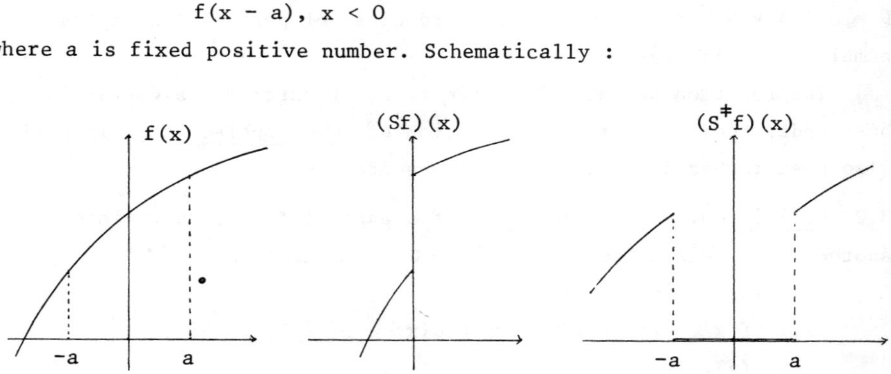

The operators I and χ are self-adjoint, $(x\frac{d}{dx})^{\ddagger} = I \hat{+} (x\frac{d}{dx})$ and $S^{\ddagger}$ is bounded. Hence every element A of $\mathfrak{N}$ verifies $A^{\ddagger\ddagger} = A = A^{\dagger\dagger}$ (such operators are called standard [9)10)]). Let us compute products. First we have $S\mathcal{D} \subset \mathcal{D}(x\frac{d}{dx})$, since the factor x "kills" the discontinuity at 0 created by S, hence $S \in R(x\frac{d}{dx})$. On the other hand, $S^{\ddagger}\mathcal{D}$ is not contained in $\mathcal{D}((x\frac{d}{dx})^{\ddagger})$, so that $S \notin L(x\frac{d}{dx})$. Finally $\chi\mathcal{D}$ is contained neither in $\mathcal{D}(x\frac{d}{dx})$, nor in $\mathcal{D}((x\frac{d}{dx})^{\ddagger})$, so that χ and $x\frac{d}{dx}$ are not multipliers of each other. In conclusion, the allowed products are $\chi . S$, $S . \chi$, $(x\frac{d}{dx}) . S$, I with everything, their adjoints and all powers of each. Thus one may visualize the partial *-algebra $\mathfrak{M}[\mathfrak{N}]$ generated by $\mathfrak{N}$ as the set of all polynomials in the operators in $\mathfrak{N}$ which contain only allowed products.

4. TOPOLOGICAL CONSIDERATIONS

In the theory of Op*-algebras, it is well-known that the so-called closed and self-adjoint algebras have better properties. What is the corresponding situation for partial *-algebras ?

Let $\mathfrak{M} \subset \mathfrak{L}(\mathcal{D})$ be a partial *-algebra on $\mathcal{D}$. It defines on $\mathcal{D}$ the so-called $t_{\mathfrak{M}}$ topology, given by the seminorms $\| \phi \|_A = \| A\phi \|$, $A \in \mathfrak{M}$. For this (projective) topology, the natural domain $\mathcal{D}(\mathfrak{M}) = \bigcap_{A \in \mathfrak{M}} \mathcal{D}(A)$ is complete, but need not be the completion of $\mathcal{D}[t_{\mathfrak{M}}]$. Similarly for $\mathfrak{M}^* = j(\mathfrak{M}) \subset \mathfrak{L}^*(\mathcal{D})$ (notice that $t_{\mathfrak{M}} = t_{\mathfrak{M}^*}$ on $\mathcal{D}$). Hence one has in general :

$$\mathcal{D} \subseteq \tilde{\mathcal{D}}[t_{\mathfrak{M}}] \subseteq \mathcal{D}(\mathfrak{M}) \subseteq \mathcal{D}(\mathfrak{M}^*) . \tag{4.1}$$

We say that $\mathfrak{M}$ is <u>closed</u> if $\mathcal{D} = \tilde{\mathcal{D}}[t_{\mathfrak{M}}]$ and <u>fully closed</u> if $\mathcal{D} = \mathcal{D}(\mathfrak{M})$.

For the case of $\mathfrak{L} \equiv \mathfrak{L}(\mathcal{D})$ itself, one may consider in addition the natural domains associated to $L^{+}(\mathcal{D})$, $\underline{\mathcal{D}} \equiv \mathcal{D}(L^{+}(\mathcal{D}))$ and $\mathcal{D}_{*} \equiv \mathcal{D}([L^{+}(\mathcal{D})]^{*})$. Thus one gets :

$$\begin{array}{ccccccc} \mathcal{D} \subseteq \tilde{\mathcal{D}}[t_{\mathfrak{L}}] & \subseteq & \mathcal{D}(\mathfrak{L}) & \subseteq & \mathcal{D}(\mathfrak{L}^*) \\ & & \cap\!\!| & & \cap\!\!| \\ & & \underline{\mathcal{D}} & \subseteq & \mathcal{D}_{*} \end{array} \tag{4.2}$$

Hence if $L^{+}(\mathcal{D})$ is closed (in the usual sense : $\mathcal{D} = \underline{\mathcal{D}}$), $\mathfrak{L}(\mathcal{D})$ is automatically fully closed. Furthermore, in all known cases, this implies that $L^+(\mathcal{D})$ is self-adjoint, i.e. $\mathcal{D} = \mathcal{D}_*$. Then, both $\mathfrak{C}(\mathcal{D})$ and $\mathfrak{C}^*(\mathcal{D})$ are fully closed.

An interesting consequence of closedness is that, for a fully closed partial *-algebra $\mathfrak{M} \subset \mathfrak{C}(\mathcal{D})$, its universal right multipliers, $R\mathfrak{M}$, map $\mathcal{D}$ into itself. The same is true for $\mathfrak{M}^w$ if $\mathcal{D} = \mathcal{D}(\mathfrak{M}^*)$. This fact will be useful for the GNS construction in Sec.7.

As is well-known, an Op*-algebra $\mathfrak{A}$ may always be extended by continuity to a (fully) closed Op*-algebra $\overline{\mathfrak{A}}$, on the domain $\mathcal{D}(\mathfrak{A})$, isomorphic to $\mathfrak{A}$. Here there are two cases :

(i) If $\mathfrak{M}$ is a *-subalgebra of $\mathfrak{C}(\mathcal{D})$, it can be extended by continuity to a closed partial *-algebra $\overline{\mathfrak{M}}$ on $\tilde{\mathcal{D}}[t_{\mathfrak{M}}]$, isomorphic to $\mathfrak{M}$; $\overline{\mathfrak{M}}$ consists of the same operators as $\mathfrak{M}$, but considered as closures of their restrictions to $\tilde{\mathcal{D}}[t_{\mathfrak{M}}]$.

(ii) If $\mathfrak{M} = \mathfrak{M}^w$ is a *-subalgebra of $\mathfrak{C}^w(\mathcal{D})$, it can be extended by continuity to a closed partial *-algebra $\overline{\mathfrak{M}}$ on $\tilde{\mathcal{D}}[t_{\mathfrak{M}}]$, and to a fully closed partial *-algebra $\hat{\mathfrak{M}}$ on $\mathcal{D}(\mathfrak{M})$, both isomorphic to $\mathfrak{M}$. Again all three sets $\mathfrak{M}$, $\overline{\mathfrak{M}}$ and $\hat{\mathfrak{M}}$ consist of the same operators, only the common core is different.

Notice that in case (i) one can also extend $\mathfrak{M}$ to a fully closed $\hat{\mathfrak{M}}$ on $\mathcal{D}(\mathfrak{M})$, but it may have less multipliers since $B\mathcal{D} \subset \mathcal{D}(A)$ need not imply $B\,\mathcal{D}(\mathfrak{M}) \subset \mathcal{D}(A)$; in other words the identity $i : \hat{\mathfrak{M}} \to \mathfrak{M}$ is a homomorphism, and not necessarily an isomorphism.

Now we come to the question of topologizing partial *-algebras themselves. We will consider only the case $\mathfrak{M} \subset \mathfrak{C}^w(\mathcal{D})$, the other ones are similar. Let $\mathfrak{N}$ be a subset of $\mathfrak{C}^w(\mathcal{D})$, containing I. Then its spaces of multipliers $L^*\mathfrak{N}$ and $R^*\mathfrak{N}$ generate on $\mathfrak{N}$ two natural, quasi-uniform, topologies [7)17)] :

(i) the topology $\tau_*^{\ell}(L^*\mathfrak{N})$, defined by the seminorms :

$$\| B \|_{\ell}^{A,M} = \sup_{\phi \in M} \{\| (A \,\square\, B)\phi\| + \| (B^{\ddagger} \,\square\, A^{\ddagger})\phi \|\}$$

where $A \in L^*\mathfrak{N}$ and M is a bounded set of $\mathcal{D}[t_{\mathfrak{C}}]$.

(ii) the topology $\tau_*^{r}(R^*\mathfrak{N})$, defined by the seminorms :

$$\| B \|_{r}^{C,M} = \| B^{\ddagger} \|_{\ell}^{C^{\ddagger},M} \qquad (C \in R^*\mathfrak{N} \Leftrightarrow C^{\ddagger} \in L^*\mathfrak{N}^{\ddagger}).$$

Similar topologies may be defined on $\mathcal{N}$ from arbitrary subsets of $L^*\mathcal{N}$ or $R^*\mathcal{N}$. On $\mathcal{L}^w(\mathcal{D})$ itself, one considers the topology $\tau_* = \tau_*^{\ell}(I) = \tau_*^{r}(I)$, with the seminorms :

$$\| B \|^M = \sup_{\phi \in M} \{ \| B\phi \| + \| B^{\ddagger}\phi \| \}$$

If we restrict the sets M to finite subsets of $\mathcal{D}$, this is just the strong*-topology [1].

With these topologies, one may prove the following results [10] :

(i) $\mathcal{L}^w(\mathcal{D})$ is complete in τ_*.

(ii) The partial multiplication $L^*\mathcal{N} \times \mathcal{N} \to \mathcal{L}^w(\mathcal{D})$ is separately continuous for the following topologies : $\tau_*^r(\mathcal{N})$ on $L^*\mathcal{N}$, $\tau_*^{\ell}(L^*\mathcal{N})$ on $\mathcal{N}$ and τ_* on $\mathcal{L}^w(\mathcal{D})$.

(iii) All spaces of multipliers are complete in appropriate topologies : $R^*\mathcal{N}$ in $\tau_*^{\ell}(\mathcal{N})$, $L^*\mathcal{N}$ in $\tau_*^r(\mathcal{N})$ and, for $\mathcal{N}^{\ddagger} = \mathcal{N}$, $L^*\mathcal{N} \cap R^*\mathcal{N}$ in $\tau_*^{\ell}(\mathcal{N}) = \tau_*^r(\mathcal{N})$.

Finally we will consider several notions of commutants. Given a subset $\mathcal{N} = \mathcal{N}^{\ddagger}$ of $\mathcal{L}(\mathcal{D})$, we may introduce (among others) the following objects :

(i) the unbounded commutant in $\mathcal{L}^w(\mathcal{D})$:

$$\mathcal{N}'_* = \{B \in \mathcal{L}(\mathcal{D}) \mid B \in L^*\mathcal{N} \cap R^*\mathcal{N},\ B \,\square\, A = A \,\square\, B,\ \forall A \in \mathcal{N}\}$$

(ii) the bounded commutant in $\mathcal{L}^w(\mathcal{D})$:

$$\mathcal{N}'_{*b} = \mathcal{N}'_* \cap \mathcal{B}(\mathcal{H}) .$$

The last one must be compared with the standard bounded commutants [4)6)18)] :

(iii) the weak bounded commutant :

$$\mathcal{N}'_w = \{B \in \mathcal{B}(\mathcal{H}) \mid \langle B^*\phi, A\psi \rangle = \langle A^*\phi, B\psi \rangle\ \forall \phi,\psi \in \mathcal{D},\ A \in \mathcal{N} \}$$

(iv) the strong bounded commutant :

$$\mathcal{N}'_s = \{B \in \mathcal{B}(\mathcal{H}) \cap L^+(\mathcal{D}) \mid BA\phi = AB\phi,\ \forall \phi \in \mathcal{D},\ A \in \mathcal{N} \}.$$

Then the following results are known [11] :

(1) $\mathcal{N}'_s \subseteq \mathcal{N}'_{*b} \subseteq \mathcal{N}'_w$.

(2) If $\mathcal{N}\mathcal{D} \subset \mathcal{D}$, in particular if $\mathcal{N}$ is an Op*-algebra, then $\mathcal{N}'_{*b} = \mathcal{N}'_w$.

(3) $\mathfrak{N}'_*$ and $\mathfrak{N}'_{*b}$ are not *-subalgebras of $\mathcal{L}^w(\mathcal{D})$, in general.

(4) A natural notion of bicommutant may be defined in $\mathcal{L}^w(\mathcal{D})$, namely $\mathfrak{N}''_{**} \equiv (\mathfrak{N}'_*)'_*$. It has the usual algebraic properties : $\mathfrak{N} \subset \mathfrak{N}''_{**}$ and $\mathfrak{N}'''_{***} = \mathfrak{N}'_*$, but it is an open question whether $\mathfrak{N}''_{**}$ is the completion of $\mathfrak{N}$ in a suitable topology. Tha answer is positive in certain cases, namely when $\mathfrak{N}$ contains "enough" bounded operators, as it is the case for Op*-algebras [11)19)].

(5) Irreducibility : a possible definition is to declare $\mathfrak{N}$ irreducible whenever $\mathfrak{N}'_{*b} = \{\lambda I, \lambda \in \mathbb{C}\}$; this choice is suggested by the properties of the GNS representation (Sec.7). The obvious definition $\mathfrak{N}'_* = \{\lambda I\}$ seems too restrictive.

5. REPRESENTATIONS OF PARTIAL *-ALGEBRAS

Although abstract partial *-algebras are interesting mathematical objects, what the physicist needs for applications, especially in quantum theories, are concrete realizations, i.e. representations by operators on a Hilbert space. This is a pattern familiar from the theory of C*-algebras [1)] and similarly for algebras of unbounded operators [6)7)]. In the sequel of this work we will initiate a representation theory for partial *-algebras. Our strategy will be to follow, as closely as possible, the corresponding theory of *-algebras as expounded in the references just quoted.

We start with the basic definition. Of course the key notion is that of homomorphism of partial *-algebras defined in 3.2 above.

5.1. Definition.- A hermitian representation of a partial *-algebra $\mathfrak{M}$ consists of a Hilbert space H, a dense domain $\mathcal{D} \subset H$ and a homomorphism π of $\mathfrak{M}$ into $\mathcal{L}(\mathcal{D})$. Similarly a weakly hermitian, resp. conjugate hermitian, representation is a homomorphism into $\mathcal{L}^w(\mathcal{D})$, resp. $\mathcal{L}^*(\mathcal{D})$.

5.2. Definition.- A representation π_1 on $\mathcal{D}_1$ is an extension of π_2 on $\mathcal{D}_2$, and π_2 a restriction of π_1, if $\mathcal{D}_2 \subseteq \mathcal{D}_1$ and $\pi_2(x) \subset \pi_1(x)$ for all $x \in \mathfrak{M}$.

Here, as in many other instances, the difference with the corresponding situation for *-algebras is that we consider the closed operators $\pi(x)$ themselves, not their restriction to $D(\pi) = \bigcap_{x \in \mathfrak{M}} \mathcal{D}(\pi(x))$, as is done customarily. This will show up also in the definition of adjoint representation. We have to distinguish two cases.

(i) Let first π be hermitian or weakly hermitian; then the adjoint representation π^* is defined by the relation :

$$\pi^*(x) = [\pi(x^+)]^* = \pi(x)^{\dagger\dagger} = j(\pi(x)) \qquad (x \in \mathfrak{M}) \tag{5.1}$$

so that π^* is conjugate hermitian;

(ii) Let σ be a conjugate hermitian representation; its adjoint σ^* is

$$\sigma^*(x) = [\sigma(x^+)]^* = \sigma(x)^{\ddagger\ddagger} = j^{-1}(\sigma(x)), \tag{5.2}$$

and it is weakly hermitian.

Furthermore, each representation coincides with its second adjoint, $\pi^{**} = \pi$, $\sigma^{**} = \sigma$. But in the case of a conjugate hermitian representation σ, the partial *-algebras $\sigma(\mathfrak{M})$ and $\sigma^*(\mathfrak{M})$ are isomorphic, so that we may replace σ by its adjoint σ^* without losing any information. Thus from now on we will consider two kinds of representations only : hermitian ones, in $\mathfrak{C}(\mathcal{D})$, and weakly hermitian ones, in $\mathfrak{C}^w(\mathcal{D})$. Of course every hermitian representation is a fortiori weakly hermitian.

Next we consider closed representations. As in Sec.4, the topology t_π defined by π and the topology t_{π^*} defined by π^* coincide on $\mathcal{D}$, where they are given by the seminorms $\phi \mapsto \|\pi(x)\phi\|$, $x \in \mathfrak{M}$. Again $\mathcal{D}(\pi) = \bigcap_{x \in \mathfrak{M}} \mathcal{D}(\pi(x))$ is complete in t_π, and $\mathcal{D}(\pi^*) = \bigcap_{x \in \mathfrak{M}} \mathcal{D}(\pi(x)^{\dagger\dagger})$ is complete in t_{π^*}, but $\mathcal{D}$ need not be dense in either of them, so that we get :

$$\mathcal{D} \subseteq \tilde{\mathcal{D}}[t_\pi] \subseteq \mathcal{D}(\pi) \subseteq \mathcal{D}(\pi^*). \tag{5.3}$$

Accordingly we say that π is closed if $\mathcal{D} = \tilde{\mathcal{D}}[t_\pi]$ and fully closed if $\mathcal{D} = \tilde{\mathcal{D}}[t_\pi] = \mathcal{D}(\pi)$. The extension theorem discussed in Sec.4 yields the following result :

(i) If π is hermitian, it admits a unique minimal closed hermitian extension $\tilde{\pi}$, with domain $\tilde{\mathcal{D}}[t_\pi]$.

(ii) In any case, π admits a unique minimal fully closed extension $\hat{\pi}$, which is a weakly hermitian representation on $\mathcal{D}(\pi)$.

The last concept to generalize is that of cyclic vector. Let again π be a representation on $\mathcal{D}$, hermitian or weakly hermitian. Then, as usual [6)7)], we say that a vector $\Omega \in \mathcal{D}$ is <u>cyclic</u> for π if the set $\pi(\mathfrak{M})\Omega$ is dense in $\mathcal{H}$. However, this condition is too weak, even for representations of Op*-algebras. As we will see in Sec.7 below, the domain to consider is $\mathcal{D}^\circ = \pi(R\mathfrak{M})\Omega$. For every $z \in R\mathfrak{M}$, the vector $\pi(z)\Omega$ belongs to $\mathcal{D}(\pi)$ if π is hermitian, and to $\mathcal{D}(\pi^*)$ if π is weakly hermitian, since π is a homomorphism; but this is not enough, in view of Eq.(5.3). We say that the vector Ω is <u>strongly cyclic</u> for π if the domain $\mathcal{D}^\circ$ is actually contained in $\tilde{\mathcal{D}}[t_\pi]$ and is dense in it for the topology t_π, i.e. $\tilde{\mathcal{D}}^\circ[t_\pi] = \tilde{\mathcal{D}}[t_\pi]$. Clearly, if Ω is strongly cyclic for π, it is also strongly cyclic for the closed extension $\tilde{\pi}$, but <u>not</u> for the fully closed extension $\hat{\pi}$, if $\tilde{\mathcal{D}}[t_\pi] \neq \mathcal{D}(\pi)$.

The main result of this section is the following uniqueness theorem, according to which a (fully) closed strongly cyclic representation is uniquely determined by its two-point functions, up to unitary equivalence.

5.3. <u>Theorem</u> - For i = 1,2, let π_i be a closed representation of the partial *-algebra $\mathfrak{M}$ on the domain $\mathcal{D}_i \subset \mathcal{H}_i$, with a strongly cyclic vector $\Omega_i \in \mathcal{D}_i$. Assume that, for every $x,y \in \mathfrak{M}$, one has

$$< \pi_1(x)\,\Omega_1,\ \pi_1(y)\,\Omega_1 > \;=\; < \pi_2(x)\,\Omega_2,\ \pi_2(y)\,\Omega_2 >. \tag{5.4}$$

Then there exists a unitary operator $U : \mathcal{H}_2 \to \mathcal{H}_1$ such that $U\mathcal{D}_2 = \mathcal{D}_1$, $U\Omega_2 = \Omega_1$ and $\pi_2(x) = U^*\,\pi_1(x)\,U$ for every $x \in \mathfrak{M}$, i.e. the representations π_1 and π_2 are unitarily equivalent. The same is true also for their fully closed extensions $\hat{\pi}_1, \hat{\pi}_2$: $U\,\mathcal{D}(\pi_2) = \mathcal{D}(\pi_1)$ and $\hat{\pi}_2(x) = U^*\,\hat{\pi}_1(x)\,U$.

We sketch the main steps of the proof. Define U on $\pi_2(\mathfrak{M})\Omega_2$ by the relation $U\pi_2(x)\Omega_2 = \pi_1(x)\Omega_1$. By Eq.(5.4), $U : \pi_2(\mathfrak{M})\Omega_2 \to \pi_1(\mathfrak{M})\Omega_1$ is an isometry, which extends to a unitary operator $U : \mathcal{H}_2 \to \mathcal{H}_1$. Let $\mathcal{D}_i^\circ = \pi_i(R\mathfrak{M})\Omega_i$. Then Eq.(5.4) implies that $U : \mathcal{D}_2^\circ \to \mathcal{D}_1^\circ$ extends to a unitary operator from $\mathcal{D}(\pi_2(x))$ onto $\mathcal{D}(\pi_1(x))$, for each $x \in \mathfrak{M}$, and thus $U\mathcal{D}(\pi_2) = \mathcal{D}(\pi_1)$. This also means that U is a homeomorphism of $\mathcal{D}_2^\circ\,[t_{\pi_2}]$ onto $\mathcal{D}_1^\circ\,[t_{\pi_1}]$, and therefore it extends to the respective completions, i.e. $U\mathcal{D}_2 = \mathcal{D}_1$. The rest is easy.

6. STATES ON PARTIAL *-ALGEBRAS

So far we have discussed the first steps of an abstract theory of representations of partial *-algebras, but we did not provide a method of construction. In the case of *-algebras, the canonical method is the so-called Gel'fand-Naimark-Segal (GNS) construction [1]. Given a *-algebra $\mathfrak{A}$ and a state ϕ on $\mathfrak{A}$, i.e. a normalized positive linear functional, a positive semidefinite scalar product on $\mathfrak{A}$ is defined as

$$< A,B > = \phi(A^+B) \tag{6.1}$$

and the representation space is the completion of $\mathfrak{A}/\mathrm{Ker}\ \phi$. Clearly this definition is not applicable as its stands to a partial *-algebra, since Eq. (6.1) defines the scalar product $< A,B >$ only when $A^+ \in L(B)$. For that reason, we shall replace linear forms by <u>sesquilinear</u> forms, defined on every pair. A similar procedure was used by Inoue [20] in a different context.

6.1. <u>Definition</u>.- A <u>h-form</u> on a partial *-algebra $\mathfrak{M}$ is a sesquilinear form ω on $\mathfrak{M} \times \mathfrak{M}$ which is

(i) positive : $\omega(x,x) \geq 0$ for all $x \in \mathfrak{M}$;

(ii) multiplication-invariant : if $x \in L(z)$ and $x^+ \in L(y)$, then $\omega(x^+ \circ y,z) = \omega(y,x \circ z)$.

The h-form is called a <u>h-state</u> if $\omega(e,e) = 1$.

As for *-algebras, every positive sesquilinear form ω is hermitian and satisfies the Cauchy-Schwarz inequality :

$$\omega(x,y) = \overline{\omega(y,x)} \quad , \quad |\omega(x,y)|^2 \leq \omega(x,x)\ \omega(y,y) .$$

It follows that its kernel Ker ω may be defined in two equivalent ways :

$$\begin{aligned} \mathrm{Ker}\ \omega &= \{x \in \mathfrak{M} \mid \omega(x,y) = 0,\ \forall\ y \in \mathfrak{M} \} \\ &= \{x \in \mathfrak{M} \mid \omega(x,x) = 0 \} . \end{aligned}$$

These h-states are a natural generalization of states; indeed, when $\mathfrak{M}$ is a *-algebra with unit, the two concepts are equivalent : every state ϕ defines a h-state ω by $\omega(x,y) = \phi(x^+y)$, and conversely every h-state ω defines a state ϕ by $\phi(x) = \omega(e,x)$.

We give some examples of states.

(1) Consider the quasi *-algebra $\mathfrak{A} = L^2(\Delta;dt) \supset \mathfrak{A}_o = C^o(\Delta)$ described in Sec.2. Let ρ be a non-negative, essentially bounded function on Δ, $0 \leq \rho \in L^\infty(\Delta;dt)$, with $\int_\Delta \rho(t)\, dt = 1$.
Then ρ defines a h-state ω_ρ over $\mathfrak{A}$ by the relation :

$$\omega_\rho\ (f,g) = \int_\Delta \overline{f(t)}\ g(t)\ \rho(t)\ dt\ .$$

This h-state is jointly continuous in the L^2 norm of $\mathfrak{A}$, and it is the most general one with that property. One gets in the same fashion h-states over the quasi *-algebra $L^p(\Delta;dt)$, for $p \geq 2$, using functions $\rho \in L^{p/p-2}$.

(2) Let $\mathfrak{M}$ be any *-subalgebra of $\mathcal{L}^\dagger(\mathcal{D})$. Then every pair $\phi,\psi \in \mathcal{D}$ defines a sesquilinear form $\omega_{\phi\psi}$ over $\mathfrak{M}$ by the relation $\omega_{\phi\psi}(A,B) = < A\phi, B\psi >$. The form $\omega_{\phi\psi}$ is a h-form if $\phi = \psi$, and a h-state if in addition $< \phi, \phi > = 1$. Such h-states are called vector states. The GNS construction given in Sec.7 below shows that every h-state over $\mathfrak{M}$ is a vector state for some representation of $\mathfrak{M}$.

(3) Given a Hilbert space $\mathcal{H}$, let H be a self-adjoint operator in $\mathcal{H}$ such that $\exp(-\beta H)$ is nuclear for every $\beta > 0$. Let $\{\phi_n, \lambda_n\}$ be the eigenvectors and the corresponding eigenvalues of H. Define

$$\mathcal{D} = \mathcal{D}^\infty(e^H) = \bigcap_{\beta > 0} \mathcal{D}(e^{\beta H}).$$

Then the following relation defines a h-state ω_β over $\mathcal{L}^\dagger(\mathcal{D})$:

$$\omega_\beta(A,B) = Z^{-1} \sum_{n=1}^{\infty} < A\phi_n, B\phi_n > e^{-\beta\lambda_n}\ ,$$

where $Z = \mathrm{Tr}(e^{-\beta H})$. The restriction of ω_β to $\mathcal{B}(\mathcal{H})$ is the familiar state $Z^{-1}\ \mathrm{Tr}(A^*B\ e^{-\beta H})$. This h-state ω_β was used by Bouziane and Martin [2)] in their proof of the Bogoliubov inequality for unbounded operators.

7. THE GNS REPRESENTATION

Let $\mathfrak{M}$ be a partial *-algebra, ω a h-state over $\mathfrak{M}$, with kernel Ker ω. Then the vector space $\mathfrak{M}_\omega = \mathfrak{M}/\mathrm{Ker}\ \omega$ becomes a prehilbert space with the scalar product :

$$< \phi_x, \phi_y >_\omega = \omega(x,y) \tag{7.1}$$

where $\phi_x = x + \mathrm{Ker}\ \omega$. Denote by $\mathcal{H}_\omega$ the completion of $\mathfrak{M}_\omega$ in the corresponding norm. In the usual GNS construction, one defines next a representation of $\mathfrak{M}$ on $\mathfrak{M}_\omega$ by the multiplication map $\sigma(x) : \phi_y \to \phi_{x \circ y}$. Here this is not possible, unless $y \in R\mathfrak{M}$. But then we need an additional restriction on ω. We shall say that the h-state ω is <u>weakly GNS</u> if the set $\mathcal{R}_\omega = \{\phi_u | u \in R\mathfrak{M}\}$ is dense in $\mathcal{H}_\omega$.

So we assume from now on that $\mathfrak{M}$ is semi-associative, and ω is a weakly GNS h-state. As a first step in the GNS construction, we define, for $x \in \mathfrak{M}$ and $u \in R\mathfrak{M}$, the map :

$$\pi^\circ_\omega(x)\, \phi_u = \phi_{x \circ u} .$$

Then one can show that $\pi^\circ_\omega(x)$ is a well-defined closable operator in $\mathcal{H}_\omega$, with domain $\mathcal{D}^\circ_\omega = \mathcal{R}_\omega$, and it verifies the following relations :

(i) $\pi^\circ_\omega(\lambda x + \mu y) = \lambda\, \pi^\circ_\omega(x) + \mu\, \pi^\circ_\omega(y), \ \forall\, x,y \in \mathfrak{M}\ , \ \lambda,\mu \in \mathbb{C}$

(ii) $\pi^\circ_\omega(x^+) = [\pi^\circ_\omega(x)]^* \restriction \mathcal{D}^\circ_\omega \quad , \ \forall\, x \in \mathfrak{M} \ ;$

(iii) $\pi^\circ_\omega(x \circ v) = \pi^\circ_\omega(x)\, \pi^\circ_\omega(v) \quad , \ \forall\, x \in \mathfrak{M}\ , \ v \in R\mathfrak{M}$.

To get a representation on $\mathcal{D}^\circ_\omega$, we have to extend the relation (iii) above to arbitrary elements $x,v \in \mathfrak{M}$. This can be done, but it yields only a weakly hermitian representation.

Next we define, for each $x \in \mathfrak{M}$, $\pi_\omega(x)$ as the closure of the operator $\pi^\circ_\omega(x)$ and consider the domain

$$\mathcal{D}_\omega \equiv \mathcal{D}(\pi_\omega(\mathfrak{M})) = \bigcap_{x \in \mathfrak{M}} \mathcal{D}(\pi_\omega(x)).$$

As usual the domain $\mathcal{D}_\omega$ is complete in the topology t_{π_ω}, but need not be the completion of $\mathcal{D}^\circ_\omega$. Clearly, for each $x \in \mathfrak{M}$, $\pi_\omega(x)$ belongs to $\mathfrak{C}^w(\mathcal{D}_\omega)$ and $\pi_\omega(x^+) = \pi_\omega(x)^\ddagger$. Then we get our main theorem, which gives the full GNS construction for partial *-algebras with unit.

7.1. <u>Theorem</u> - Let $\mathfrak{M}$ be a semi-associative partial *-algebra with unit e, and ω <u>a</u> weakly GNS h-state over $\mathfrak{M}$. Then the map $\pi_\omega : x \mapsto \pi_\omega(x) \equiv \overline{\pi^\circ_\omega(x)}$ is a weakly hermitian, fully closed, representation of $\mathfrak{M}$ on $\mathcal{D}_\omega$, cyclic with respect to the vector $\Omega_\omega \equiv \phi_e$, and one has, for every $x,y \in \mathfrak{M}$: $\omega(x,y) = \langle \pi_\omega(x)\, \Omega_\omega, \pi_\omega(y)\, \Omega_\omega \rangle_\omega$.

The restriction of π_ω to $\tilde{\mathcal{D}}^\circ_\omega[t_{\pi_\omega}]$, the completion of $\mathcal{D}^\circ_\omega$, is closed and strongly cyclic with respect to Ω_ω.

Finally, the GNS representation π_ω is uniquely determined by the h-state ω, up to unitary equivalence.

Again we sketch the main steps of the proof. Take $u,v \in R\mathfrak{M}$ and a pair $x,y \in \mathfrak{M}$ such that $x \in L(y)$.
Then we may write successively :

$$\begin{aligned} < \phi_u, \pi^\circ_\omega(x\circ y)\, \phi_v >_\omega &= \omega(u,(x\circ y)\circ v) \\ &= \omega(u,x\circ(y\circ v)), \text{ by semi-associativity of } \mathfrak{M}, \\ &= \omega(x^+\circ u,\ y\circ v), \text{ by multiplication-invariance of } \omega, \\ &= < \phi_{x^+\circ u}\ ,\ \phi_{y\circ v} >_\omega \\ &= < \pi^\circ_\omega(x^+)\ \phi_u\ ,\ \pi^\circ_\omega(y)\ \phi_v >_\omega . \end{aligned} \tag{7.2}$$

By a similar argument, this quantity is also equal to :

$$\ldots = < \pi^\circ_\omega\ (y^+\circ x^+)\ \phi_u, \phi_v >_\omega \quad . \tag{7.3}$$

Thus Eqs.(7.2),(7.3) and the relations (i) - (iii) above show that π°_ω is a weakly hermitian representation of $\mathfrak{M}$ on $\mathcal{D}^\circ_\omega$. Then, by the extension theorem discussed in Sec.5, it extends to a fully closed, weakly hermitian representation on $\mathcal{D}_\omega$, namely π_ω. More precisely, Eqs.(7.2), (7.3) extend to arbitrary vectors $\phi,\psi \in \mathcal{D}_\omega$, which shows that $\pi_\omega(x) \in L^*(\pi_\omega(y))$ and $\pi_\omega(x\circ y) = \pi_\omega(x) \,\Box\, \pi_\omega(y)$. This means that π_ω is a homomorphism from $\mathfrak{M}$ into $\mathfrak{C}^w(\mathcal{D}_\omega)$. Finally, the uniqueness statement results from Theorem 5.3.

Thus the general result is that,whenever ω is a weakly GNS h-state, the corresponding representation π_ω is only weakly hermitian. Of course it might happen that π_ω is actually hermitian. When this is the case, we say that ω is a <u>GNS h-state</u>. An example is given below.

As a consequence of Theorem 7.1 , we get the usual result that any automorphism of $\mathfrak{M}$ that leaves the h-state ω invariant is unitarily implemented in the GNS representation π_ω. It follows that spontaneous symmetry breaking requires non-invariant h-states, or better, h-states invariant under a proper subgroup of the symmetry group considered.

7.2. <u>Corollary</u>.- Let τ be an automorphism of $\mathfrak{M}$ that leaves the h-state ω invariant : $\omega(\tau x,\tau y) = \omega(x,y)$ for all $x,y \in \mathfrak{M}$. Then τ is represented in the Hilbert space H_ω by a unitary operator $U_\omega(\tau)$ with verifies the following relations :

(i) $U_\omega(\tau)\,\Omega_\omega = \Omega_\omega$

(ii) $\pi_\omega(\tau x) = U_\omega(\tau)\,\pi_\omega(x)\,U_\omega(\tau)^*,\ \forall\, x \in \mathfrak{M}$.

Our last question concerns the irreducibility of the GNS representation π_ω. As for *-algebras, we say that the h-state ω is pure if it cannot be decomposed into a convex combination of two h-states : $\omega = \lambda\omega_1 + (1-\lambda)\omega_2$, $0 < \lambda < 1$. Then the analysis leads to the following result, entirely similar to the corresponding statement for Op*-algebras [6] .

7.3. Theorem.- Let ω be a weakly GNS state over $\mathfrak{M}$, π_ω the corresponding GNS representation. Then ω is pure iff the bounded commutant $[\pi_\omega(\mathfrak{M})]'_{*b}$ contains only multiples of the identity. If ω is a GNS h-state, the statement is, in addition, equivalent to the triviality of the weak bounded commutant $[\pi_\omega(\mathfrak{M})]'_w$.

We conclude this section with an example Let $\mathfrak{A} = L^p(\Delta;dt)$ be the quasi *-algebra described in Sec.2, ω_ρ a h-state of the type discussed in Sec.6, with $\rho \in L^{p/p-2}(\Delta;dt)$, $\rho > 0$. Then Ker $\omega_\rho = \{0\}$ and $H_{\omega_\rho} = L^2(\Delta;\rho dt)$. In this space, the GNS representation takes the following form :

(1) $\pi^\circ_{\omega_\rho}(f)$ is the operator of multiplication by $f \in L^p$ on the domain $\mathcal{D}^\circ_\omega = \mathfrak{A}_o = C^o(\Delta)$.

(2) $\pi_{\omega_\rho}(f)$ is the operator of multiplication by $f \in L^p$ on the domain $\mathcal{D}_{\omega_\rho} = L^\infty(\Delta;\rho dt)$.

This representation is of course highly reducible. Its commutants consist also of multiplication operators and we get :

$$[\pi_{\omega_\rho}(\mathfrak{A})]'_* = [\pi_{\omega_\rho}(\mathfrak{A})]'_{*b} = [\pi_{\omega_\rho}(\mathfrak{A})]'_w = L^\infty(\Delta;\rho dt) .$$

8. FINAL REMARKS

The results presented so far may be extended in several directions and they suggest a number of questions.

8.1. Partial *-Algebras Without Unit

When a *-algebra $\mathfrak{A}$ has no unit, it can be embedded into a

*-algebra with unit $\mathfrak{A}_e = \mathfrak{A} + \mathbb{C}e$ and the whole GNS machinery may be extended [1][20]. As will be shown in Ref. 12, the same construction works for partial *-algebras, extendable h-states may be characterized and the GNS construction goes through, extending previous results by Inoue [20] and Bhatt [22].

8.2. Bounded vs. Unbounded Commutants

Theorem 7.3 suggests that the appropriate commutant in $\mathcal{L}^w(\mathcal{D})$ is the bounded one, $[\pi_\omega(\mathfrak{M})]'_{*b}$. In fact, one can prove more [12]. If the h-state ω is pure, the unbounded commutant $[\pi_\omega(\mathfrak{M})]'_*$ contains no nontrivial positive element B, where B is either bounded or of the form $B = \sum_i C_i^{\#} \ast C_i$. In general, and already in the case of polynomial algebras, there exists positive elements which are not representable as a sum of squares : when this is the case, we cannot conclude that the unbounded commutant is trivial.

8.3. Open Problems

The work described here is only a first step towards a genuine theory of partial *-algebras, and many questions remain open. We would like to single out three of them.

(1) Abelian partial *-algebras : Several definitions are possible, and it is not clear which one is the most appropriate. This is of course linked to a choice of commutant. Preliminary results of W. Karwowski suggest, here also, that the unbounded commutant $\mathfrak{N}'_*$ might be too large.

(2) Continuity properties of h-states and representations : The definitions given in Secs.6-7. cover only the algebraic aspect. In the case of C*-algebras, continuity of positive states and corresponding representations is automatic, but this is not so in the case of unbounded operators. More precisely one should study the continuity properties of h-states and GNS representations with respect to the various topologies described in Sec.4.

(3) Examples : explicit examples should be studied systematically, for instance, Example (3) of Sec.6, which may yield interesting applications in statistical mechanics. In particular quasi *-algebras

can probably be worked out completely, and possibly classified.

9. REFERENCES

1) Bratteli O., Robinson D.W., "Operator Algebras and Quantum Statistical Mechanics I", Springer, Berlin et al. 1979

2) See for instance, Emch G.G., Knops H.J.F., J. Math. Phys. 11 3008 (1970) or Narnhofer H., Acta Phys. Austriaca 49, 207 (1978)

3) Borchers H.J., Nuovo Cim. 24, 214 (1962)

4) Vasil'ev A.N., Theor. Math. Phys. 2, 113 (1970)

5) Ascoli R., Epifanio G., Restivo A., Commun. Math. Phys. 18, 291 (1970); Rivista Mat. Univ. Parma 3, 21 (1974)

6) Powers R.T., Commun. Math. Phys. 21, 85 (1971); Trans. Amer. Math. Soc. 187, 261 (1974)

7) Lassner G., Rep. Math. Phys. 3, 279 (1972); Wiss. Z. Karl-Marx-Univ. Leipzig, Math.-Naturwiss. R. 24, 465 (1975); and subsequent papers

8) Lassner G., Physica A 124, 471 (1984)

9) Antoine J-P., Karwowski W., in "Quantum Theory of Particles and Fields", pp. 13-30; Lukierski J. and Jancewicz B., (eds), World Scientific Publ. Co, Singapore 1983

10) Antoine J-P., Karwowski W., Publ. RIMS, Kyoto 21 (1985), in press

11) Antoine J-P., Mathot F., preprint in preparation

12) Antoine J-P., Lassner G., preprint in preparation

13) Borchers H.J., in RCP 25 (Strasbourg) 22, 26 (1975); and also in "Quantum Dynamics : Models and Mathematics", L. Streit (ed.), Acta Phys. Austriaca Suppl. XVI, 15 (1976)

14) Antoine J-P., J. Math. Phys. 21, 2067 (1980)

15) Epifanio G., Trapani C., J. Math. Phys. 20, 148 (1979)

16) Kürsten K-D., Lassner G., private communication

17) Lassner G., Wiss. Z. Karl-Marx-Univ. Leipzig, Math. Naturwiss. R. 30, 572 (1981); also in "Algèbres d'opérateurs et leurs applications en physique mathématique" (Proc. Marseille 1977) pp. 249-260; Connes A., Kastler D. and Robinson D.W. (eds), Editions du CNRS, Paris 1979

18) Borchers H.J., Yngvason J., Commun. Math. Phys. 42, 231 (1975)

19) Mathot F., J. Math. Phys. 26 (1985) in press

20) Inoue A., Japan. J. Math. 9, 247 (1983)

21) Bouziane M., Martin Ph. A., J. Math. Phys. 17, 1848 (1976)

22) Bhatt S.J., Yokohama Math. J. 29, 7 (1981)

SPONTANEOUS BREAKING IN SUPERSYMMETRY

L. O'Raifeartaigh

Dublin Institute for Advanced Studies
10 Burlington Road
Dublin 4, Ireland

ABSTRACT

Spontaneous symmetry breaking is considered for supersymmetric quantum mechanics and supersymmetric field theory.

1. INTRODUCTION

For our purposes a supersymmetric system (e.g. a quantum theory, a group representation) is one for which[1)]

(i) The underlying function space (usually a Hilbert space) $\mathcal{H}$ splits into two parts $\mathcal{H}_+ + \mathcal{H}_-$ (which may be thought of as bosonic and fermionic respectively) characterized by an operator σ such that $\sigma\mathcal{H}_\pm = \pm\mathcal{H}_\pm$.

(ii) There exist a finite number of (self-adjoint) operators Q_i (supersymmetric generators) which anti-commute with σ,

$$\{Q_i , \sigma\} = 0 , \tag{1.1}$$

and

(iii) The operator of interest for the system (e.g. Hamiltonian or Laplacian) is a sum of squares of the Q_i, and commutes with each Q_i i.e.

$$H = \sum_i Q_i^2 , \qquad [H,Q_i] = 0. \tag{1.2}$$

It is easy to see from (1.1) and (1.2) that the operator H commutes with σ. From this and (1.1) one sees that the eigenspaces of H with zero and non-zero eigenvalues behave quite differently. Since

$$H\psi_0 = 0 \quad \Leftrightarrow \quad Q_i\psi_0 = 0 \qquad (i=1...n), \tag{1.3}$$

one sees that the eigenstates with eigenvalue zero, and only these eigenstates, are symmetric with respect to the Q_i. Further, since the Q_i and σ anti-commute, any eigenspace $\mathcal{V}_\varepsilon$ of H with non-zero eigenvalue ε must consist of two non-zero components $\mathcal{V}_\varepsilon^\pm$ corresponding to $\sigma = \pm 1$ and thus the eigenspaces of H with non-zero eigenvalues must be (at least) doubly degenerate.

Some examples of supersymmetric systems are

(i) $Q_1 = \partial$, $Q_2 = \partial^*$, $H = \partial\partial^* + \partial^*\partial$ where ∂ is the outer derivative for a differentiable manifold and ∂^* is its dual. Examples of this kind have been used by Witten[1] and Gaume[2] to obtain new insights into Morse theory and the Atiyah-Singer index theorem.

(ii) $a_i = Q_i + iQ_{i+p}$ where $\{a_i, a_j^*\} = \delta_{ij}$ are the usual creation and destruction operators. These Q_i are often used to construct the finite-dimensional representations of the compact simple Lie groups[3], and then $H = \sum_i Q_i^2$ plays the role of a Casimir operator.

(iii) $Q = Q_1 \pm i\, Q_2$, where $\{Q_+, Q_-\} = \{Q_1^2 + Q_2^2\} = H$ and H is the Hamiltonian for a one-dimensional quantum-mechanical system[4] (This is the case that will be considered in the present lecture)

(iv) Q_α, $\alpha = 1, \ldots, 4$ where the Q are hermitian operators which transform according to the Majorana representation of the Lorentz group, and satisfy the relations $\{Q_\alpha, Q_\beta\} = (C\gamma_\mu)_{\alpha\beta} P^\mu$ $[P_\mu, Q_\alpha] = 0$ where C is the charge conjugation matrix and P_μ is the four-momentum. (This is the case of ordinary quantum field theory. and will be considered in the following three lectures).

(v) Case (iv), but with the restriction $[P_\mu, Q_\alpha] = 0$, or constancy of the Q_α, relaxed to $[P_\mu[P_\lambda, Q_\alpha]] = 0$, or linearity of the Q_α in x. This is the case of conformally invariant QFT.

(vi) Case (iv), but with the x-dependence of the Q_α left arbitrary (supergravity theory).

The literature on all these aspects of supersymmetry is enormous, but as we shall be interested only in its spontaneous breakdown we shall confine our attention to the quantum mechanical and the (non-gravitational) field theoretical cases (iii) and (ii) respectively, and concentrate only on the breaking in these cases. The quantum mechanical case will be considered in this first lecture and the field theoretical case in the following two lectures.

LECTURE 1. SUPERSYMMETRIC QUANTUM MECHANICS

2. Construction of Supersymmetric Quantum Mechanical Hamiltonians

Let us now consider case (iii) above — supersymmetric quantum mechanics in one dimension. Given any ordinary, one-dimensional, quantum-mechanical system

$$H_0 = \frac{p^2}{2} + V(x) \quad , \tag{1.4}$$

(where $V(x)$, which has to be bounded below for stability in any case, is normalized to be positive) a supersymmetric counterpart can be constructed as follows: Let U(X) denote a square root of 2V(X)

$$U^2(X) = 2V(X) \, , \tag{1.5}$$

and let

$$Q^{\pm}(x) = \left(U(x) \pm \partial/\partial x\right)\sigma^{\pm} \quad , \tag{1.6}$$

where $\sigma^{\pm}$ are the usual Pauli matrices $(\sigma_1 \pm i\sigma_2)/2$, and $\partial/\partial x = ip$. Then the Hamiltonian

$$H = \frac{1}{2}\{Q^+, Q^-\} = \frac{1}{2}U(x)^2 - \frac{1}{2}\frac{\partial^2}{\partial x^2} + \frac{\sigma_3}{2}\frac{\partial U(x)}{\partial x} = H_0 + \frac{\sigma_3}{2}\frac{\partial}{\partial x}(V(x))^{1/2} \, , \tag{1.7}$$

is the supersymmetric counterpart of (1.4). Note that

$$[H, Q^{\pm}] = 0 \quad , \quad [H, \sigma_3] = 0 \quad , \quad \{Q^{\pm}, \sigma_3\} = 0 \quad , \tag{1.8}$$

so that σ_3 plays the role of σ. Note also that for $\sigma_3 = \pm 1$ H reduces to S^+S and SS^+ respectively where $S = U - \partial/\partial x$. It follows that H is non-negative and that if a state with zero eigenvalue of H exists, then it is unique and satisfies either

$$\sigma_3 \psi_0 = \psi_0 \;, \quad S\psi_0 = 0 \qquad \text{or} \qquad \sigma_3 \psi_0 = -\psi_0 \;, \quad S^+ \psi_0 = 0 \tag{1.9}$$

All other energy eigenstates are doubly degenerate[5)6)]. From (1.7) it is clear that the rule for supersymmetrizing a given one dimensional quantum mechanical Hamiltonian $p^2/2 + V(x)$ is simply to add a term $\frac{\sigma_3}{2}\frac{\partial U(x)}{\partial x}$ where $U^2(X) = 2V(X)$. The only ambiguity that arises in this procedure is in the sign of the square root. If V(X) is real analytic the ambiguity can be reduced to an overall sign of U(X) by requiring that U(X) also be real analytic. However, this is not a necessary condition, because there is no need for a potential to be analytic, differentiable or even continuous (piece-wise continuity is all that is required, as in the case of square-well potentials for example). In particular, if $V(x)$ has zeros, natural choices of $U(x)$ are obtained by choosing different signs on either side of the zeros. For example for the monomial potentials $2V(x) = \omega^2 x^{2n}$ the analytic square-root is ωx^n but other admissible square-roots are $U(x) = \pm\omega x^n \theta(x)$ where $\theta(x) = \pm 1$ for $x \gtrless 0$. Similarly, for $2V(x) = \prod_i (x-a_i)^2$ admissible square roots are $U(x) = \pm\prod_i (x-a_i)$ where the $\pm$ may be different in the different sections $x < a_1$, $a_1 < x < a_2$, $a_2 < x < a_3 \ldots$. We shall return to the question of choosing U(X) later as it plays an important role in symmetry breaking.

3. Supersymmetric Harmonic Oscillators

In order to obtain some intuitive feeling for the supersymmetric Hamiltonian (1.7) let us consider some examples. First, for the conventional harmonic oscillator potential $2V(x) = \omega^2 x^2$ with analytic square root $U(x) = \omega x$ the quantum mechanical counterpart is

$$H = \frac{p^2}{2} + \frac{\omega^2 x^2}{2} + \frac{\sigma_3}{2}\omega \,, \tag{1.10}$$

and since the additional term $\omega\sigma_3/2$ is constant it is easy to see that it converts the conventional harmonic oscillator spectrum

$$E = \frac{\omega}{2}, \frac{3\omega}{2}, \frac{5\omega}{2}, \ldots \,, \tag{1.11}$$

into the spectrum

$$E = \quad , \omega , 2\omega , 3\omega , \dots \qquad (\sigma_3 = 1)$$
$$= 0 , \omega , 2\omega , 3\omega , \dots \qquad (\sigma_3 = -1) . \qquad (1.12)$$

Thus, as expected, the ground state (described by the wave function $\exp(-\omega^2 x^2/2)$) is unique and has zero energy, and all the other energy levels are doubly degenerate.

Consider next the (anharmonic) oscillator with potential $\omega^2 x^4/2$. An analytic square root is $U(x) = \omega x^2$ and then the supersymmetric counterpart is

$$H = \frac{p^2}{2} + \frac{\omega^2 x^4}{2} + \sigma_3 \omega x . \qquad (1.13)$$

It is easy to see that if $\binom{f(x)}{g(x)}$ is an eigenstate of this Hamiltonian then $\binom{g(-x)}{f(-x)}$ is an eigenstate with the same eigenvalue, and that there is no state with eigenvalue zero, since the solutions $\exp(\pm\omega x^3/3)$ of the equations $S\Psi_0 = 0$ and $S^{+}\Psi_0 = 0$ are not square-integrable. Thus the energy spectrum must consist of a doubly degenerate set

$$E = \varepsilon_0, \varepsilon_1, \varepsilon_2, \dots \ (\sigma_3 = \pm 1) \qquad \text{where} \qquad \varepsilon_0 > 0 . \qquad (1.14)$$

4. Symmetry Breaking.

The Hamiltonian (1.7) is formally supersymmetric in the sense that it commutes with the generators $Q^{\pm}$ of supersymmetry. However, following convention, the system will be said to be supersymmetric if, and only if, in addition, the ground state Ψ_{ε_0} is supersymmetric i.e.

$$Q^{\pm}\Psi_{\varepsilon_0}(x) = 0 , \qquad (1.15)$$

and, as we have already seen, this will be true if, and only if, the ground state energy ε_0 is zero (and Ψ_{ε_0} is unique). Thus the basic condition for supersymmetry breaking is $\varepsilon_0 > 0$. In analogy with the situation in field theory the supersymmetry in this case $(\varepsilon_0 > 0)$ is often said to be spontaneously broken. However, it should be recalled that because quantum mechanics has only a finite number of degrees of freedom there is no spontaneous breakdown in the usual field-theoretic sense of the word. (The ground states $\psi^{\pm}_{\varepsilon_0}$ where $\sigma_3 \psi^{\pm}_{\varepsilon_0} = \psi^{\pm}_{\varepsilon_0}$ can be transformed into one another by unitary transformations).

A simple practical criterium for supersymmetry breaking $(\varepsilon_0 > 0)$ can be obtained from the condition, $S\psi_0 = 0$ or $S^{+}\psi_0 = 0$ where $S = U - \partial/\partial x$,of Section 2, because, since $U - \partial/\partial x$ is a first-order differential operator these equations can be solved at once[5] to yield

$$\psi_{\varepsilon_0} = \text{const.} \times \exp\left(\pm \int_0^x U(y)\,dy\right) , \tag{1.16}$$

and so the question reduces to the square-integrability of the expression in (1.16). This question in turn reduces to the question as to whether the leading term in U(x) as $x \to \pm\infty$ is even or odd, since if U(X) is odd/even, then $\int^x U(y)dy$ is even/odd and (1.16) is/is not square-integrable. Thus the criterium for symmetry breaking is simply that $U(x)$ be even for large x. Examples of this have already been seen in Section 3, where for the harmonic oscillator with $U(x) = \omega x$ the symmetry is unbroken and for the anharmonic oscillator with $U(x) = \omega x^2$ it is broken. More generally, for monomial potentials $2V(x) = \omega^2 x^{2n}$, $U(x) = \omega x^n$ it will be unbroken for odd n and broken for even n.

5. Non-Analytic Square Roots

The result just obtained for monomial potentials holds only for the analytic square-roots. If, for example, one chooses the alternative square-roots

$$U(x) = \omega x^n \theta(x) \quad \text{where} \quad \theta(x) = \pm 1 \text{ for } x \gtrless 0 , \tag{1.17}$$

then it is easy to see that just the opposite situation obtains, since in this case $U(x)$ is odd/even according as n is even/odd. Note that the Hamiltonian (1.7) constructed with $U(x)$ in (1.17) is well-defined for $x \gtrless 0$ so long as $n \geqslant 1$, and is continuous and well-defined even at $x=0$ for $n \geqslant 2$. Thus the simplest case, namely the case of the non-analytic harmonic oscillator

$$H = \frac{p^2}{2} + \frac{\omega^2 x^2}{2} + \frac{\sigma_3}{2} \omega \theta(x) , \tag{1.18}$$

is the only case in which the potential for (1.17) is discontinuous, and even in this case it is no more discontinuous than a potential well. It is easy to verify that for (1.18) the lowest energy is strictly positive, and the lowest energy state is doubly degenerate.

In contrast, if one takes the non-analytic square-root $U(x) = \omega x^2 \theta(x)$ for the anharmonic oscillator $2V(x) = \omega^2 x^4$ then

$$H = \frac{p^2}{2} + \frac{\omega^2 x^4}{2} + \sigma_3 \omega \theta(x) x , \tag{1.19}$$

and it is easy to verify that

$$\psi_0(x) = \begin{pmatrix} 0 \\ 1 \end{pmatrix} \exp\left(-\omega |x|^3/3\right) , \tag{1.20}$$

is a (unique) ground state.

It is perhaps amusing to consider in the context of the supersymmetric harmonic oscillator (1.18) the (non-supersymmetric) modified harmonic oscillator

$$H = \frac{p^2}{2} + \frac{\omega^2 x^2}{2} + \lambda\theta(x) , \qquad (1.21)$$

where λ is an arbitrary parameter. It is easily seen that if $\lambda = 2m$ where m is an integer, the eigenstates of this Hamiltonian are linear combinations of two ordinary harmonic oscillator states $|n)$ and $|n + 2m)$, but if $\lambda \neq 2m$ (as in (1.18)) then the eigenstates are not given by any finite combination of the $|n)$.

6. The Higgs Potential.

A final interesting example is provided by the standard Higgs potential

$$2V(x) = \lambda^2(x^2 - a^2)^2 . \qquad (1.22)$$

The analytic square root of (1.22) is evidently

$$U(x) = \pm\lambda(x^2 - a^2) , \qquad (1.23)$$

and, according to the criterium of Section 4, the supersymmetric counterpart of (1.22) constructed with (1.23) does not have a supersymmetric ground state since the leading term in (1.17) for large x is even. Thus the supersymmetry of the analytic supersymmetric counterpart of (1.22) is broken.

On the other hand, there are actually six natural square-roots of (1.22), namely (1.23) with the choices $\pm$ not universal, but chosen independently in the three sections

$$x \leq -a \qquad -a \leq x \leq a \qquad x \geq a \quad , \qquad (1.24)$$

From (1.16), (1.23) and (1.24) it is easy to see that the supersymmetry will be unbroken if, and only if, one makes one of the two choices $(-,\pm,+)$ for the three regions in question. In that case

$$\frac{\partial u(x)}{\partial x} = \pm 2\lambda x \qquad \text{for } x \gtrless a \text{ and } -a \quad , \tag{1.25}$$

and the ground-state wave-function is

$$\begin{pmatrix}1\\0\end{pmatrix} \exp\left(\pm \frac{\lambda x}{3}(x^2-3a^2)\right) \qquad \text{for } \quad x \gtrless -a, a \quad , \tag{1.26}$$

respectively. Note that in these cases the supersymmetry is preserved but the reflexion-symmetry $a \leftrightarrow -a$ is broken.

A Remark on the Partition Function. Since supersymmetry is characterized by the degeneracy of the energy levels, and its breaking by the non-existence of a ground state with zero energy, a useful function for describing it is the so-called partition function $\chi(T)$ defined as

$$\chi(T) = \sum_n m(n)\, e^{-E_n T} \quad , \tag{1.27}$$

where E_n are the energy levels and m(n) their multiplicities. In particular, the limit $T \to \infty$ of $\chi(T)$ is sufficient to determine whether the supersymmetry is broken or not, since $\chi(T) \to 0,1$ in the respective cases. Note that $\chi(T)$ can be expressed as the analytic continuation from t to T = it of the Feynman-Kac integral $\int K(a,a,t)da$ where

$$K(a,b,t) = \langle b, e^{iHt} a\rangle = \int_a^b d[x] \exp i \int_0^t dt\, \mathcal{L}(x,\dot{x}) \quad . \tag{1.28}$$

Let us consider for example the case of the harmonic oscillator. It is well-known that in the non-supersymmetric case

$$K(a,b,t) = \left(\frac{\omega}{2\pi\hbar \sin\omega t}\right)^{1/2} \exp\left[\frac{\omega}{2\hbar\sin\omega t}\{(a^2+b^2)\cos\omega t - 2ab\}\right] \tag{1.29}$$

so the partition function is

$$\chi(T) = \frac{1}{\sinh \omega T} = e^{-\frac{\omega T}{2}}\left(1 + e^{-\omega T} + e^{-2\omega T} + \ldots\right), \tag{1.30}$$

reflecting the fact that the energy levels are $E_n = (n+\frac{1}{2})\omega$ and that they all have multiplicity one. As already seen in Section 1.2 the (analytic) supersymmetric analogue is obtained by simply adding a term $\sigma_3\omega/2$ to the Hamiltonian, and from (1.10) one sees that this makes the simple change

$$\chi(T) \rightarrow \chi_{ss}(T) = \frac{\mathrm{tr}_R\, e^{\sigma_3\omega T/2}}{\sinh \omega T} = \frac{\cosh \omega T}{\sinh \omega T} = \left(1 + 2e^{-\omega T} + 2e^{-2\omega T} + \ldots\right), \tag{1.31}$$

in the partition function. As expected, $\chi_{ss}(\infty) = 1$ since the supersymmetry is unbroken, and the non-zero eigenvalues of H are multiples of ω and are doubly degenerate.

It might be worth remarking that if one changes the trace in (1.31) to the supersymmetric trace

$$\mathrm{str}_R\, e^{\sigma_3\omega T/2} \rightarrow \mathrm{str}\, e^{\sigma_3\omega T/2} = \mathrm{tr}(-1)^F e^{\sigma_3\omega T/2} = \mathrm{tr}\,\sigma_3 e^{\sigma_3\omega T/2}, \tag{1.32}$$

where F is the 'fermion' number, which in this case is just ± 1 for $\sigma_3 = \pm 1$ one finds that

$$s\chi_{ss} \equiv \frac{\mathrm{str}\, e^{\sigma_3\omega T/2}}{\sinh \omega T} = 1, \tag{1.33}$$

for all T. This is actually the quantum-mechanical-harmonic-oscillator analogue of the Nicolai[7] result

$$\frac{1}{N}\int d(A,\chi)\, e^{-\int \mathcal{L}_{ss}(A,\chi)\,dx} = 1, \qquad (1.34)$$

in unbroken supersymmetric field theory. This result states that the Schwinger functional for zero external current (J=0) is independent of the parameters in the supersymmetric Lagrangian $\mathcal{L}_{ss}(A,\chi)$. It implies that in such a theory the vacuum graphs must vanish, a result that had already been observed in perturbation theory[8]

References

1) Witten, E., Journ. Diff. Geom. 17 661 (1982).
2) Alvarez-Gaumé, L., Comm. Math. Phys. 90, 161 (1983).
3) Lipkin, H., Lie Groups for Pedestrians (North-Holland, Amsterdam, 1965).
4) Witten, E., Nucl. Phys. B185, 513 (1981). For a review see Lancaster, D., Nuovo Cim. 79A, 28 (1984).
5) For a general discussion of ground state functions see Claudson, M. and Halpern, M., Nucl. Phys. B250, 689 (1985).
6) For a determination of non-zero energy levels see Giler, S., Kosińsky, P., Rembielinski, J. and Maslanka, P., Phys. Lett. 152B, 185 (1985).
7) Nicolai, H., Phys. Lett. 89B (1979) 341.
8) Iliopoulos, J. and Zumino, B., Nucl. Phys. B76 (1974) 310.

LECTURE 2. SUPERSYMMETRY IN QUANTUM FIELD THEORY.

2.1. Superfields and Supertranslations

In this and the following two lectures we shall consider the spontaneous breakdown of supersymmetry in quantum field theory (QFT). But to do this it is necessary to describe supersymmetry in QFT, and that will be the purpose of the present lecture[1]. As already mentioned in the first lecture the basic commutation relations to be satisfied in the QFT case are

$$\{ Q_\alpha , Q_\beta \} = i(C\gamma^\mu)_{\alpha\beta} P_\mu \quad , \tag{2.1}$$

where the Q_α are the supersymmetric generators, C is the charge conjugation matrix, P_μ is the four-momentum, and it is understood that

$$[P_\mu, Q_\alpha] = 0 , \qquad [L_{\mu\nu}, Q_\alpha] = M^{\alpha\beta}_{\mu\nu} Q_\beta \quad , \tag{2.2}$$

where $M^{\alpha\beta}_{\mu\nu}$ are the generators of the Lorentz group in the Majorana (real Dirac) répresentation. The problem is how to implement the Q_α on quantized fields.

The solution to this problem is most neatly expressed by introducing the ideas of superspace and superfields[2]. Superspace is obtained by adding to the usual coordinates x_μ for a four-dimensional space with Minkowski metric, four anti-commuting coordinates θ_α

$$\{\theta_\alpha, \theta_\beta\} = 0 \quad , \tag{2.3}$$

for a 4-dimensional space with a symplectic metric. Superfields $\Phi(x,\theta)$ are then fields over the resulting eight dimensional space. They satisfy the usual Lorentz transformation law

$$\left(U(\Lambda,a)\Phi\right)(x,\theta) = D(\Lambda)\,\Phi\left(\Lambda^{-1}(x-a), M^{-1}\theta\right) , \tag{2.4}$$

where $D(\Lambda)$ is some finite-dimensional representation of Λ (usually chosen to be trivial, $D(\Lambda) = 1$)) and $M(\Lambda)$ is the Majorana representation of Λ . The new feature is that they also satisfy the 'supertranslation' law

$$\left(U(\varepsilon)\Phi\right)(x_\mu,\theta) = \Phi\left(x_\mu + i\,\bar{\theta}\gamma_\mu\varepsilon\ ,\ \theta+\varepsilon\right), \tag{2.5}$$

where ε_α are a second set of anti-commuting coordinates (analogues of a_μ for the Poincaré group) which also anti-commute with the θ's. Thus (2.5) is essentially a translation in the θ-part of the superspace, but it induces also a formal translation $x_\mu \to x_\mu + i\bar{\theta}\gamma_\mu\varepsilon$ in the Minkowski part. (Of course, one may take more than one set of θ's, i.e. θ^i_α , i=1...n, in which case the supersymmetry is called n-supersymmetry, but we shall consider only n=1 in these lectures).

By taking the infinitesimal (linear in ε) version of the supertranslations (2.5) one sees that

$$\delta\Phi = U(\varepsilon)\Phi - \Phi = \varepsilon_\alpha Q_\alpha \Phi \ , \tag{2.6}$$

where

$$Q_\alpha = i\frac{\partial}{\partial\theta_\alpha} + (\bar{\theta}\gamma)_\alpha \ . \tag{2.7}$$

It is then easy to verify that the Q_α satisfy the relations (2.1) and (2.2), so these relations have been implemented at the level of the superfields.

To implement them at the level of ordinary fields, one notes that since the θ_α anti-commute, the superfields have a finite expansion in terms of ordinary fields

$$\Phi(x,\theta) = A + \bar{\theta}\cdot\eta + \bar{\theta}\,(\gamma^\mu A_\mu + F_1 + \gamma_5 F_2)\theta + (\bar{\theta}\cdot\theta)(\bar{\theta}\cdot\lambda) + (\bar{\theta}\cdot\theta)^2 D, \tag{2.8}$$

where the ordinary component fields not only have the usual Lorentz properties as indicated (Φ scalar $\rightarrow$ A,F_1,D scalars, F_2 pseudo-scalar, A_μ vector, η,λ Majorana spinors) but also have the conventional commutation and anti-commutation relations when the superfields have them (e.g. the commutation of $\Phi(x,\theta)$ and $\Phi(y,\theta)$ implies the commutation of A(x) and A(y) and the anti-commutation of $\eta(x)$ and $\eta(y)$). By applying (2.7) to (2.8) one finally obtains the operation of the Q_α on the ordinary fields as

$$(\varepsilon\cdot Q)A(x) = \bar{\varepsilon}\gamma_5\eta(x)$$

$$(\varepsilon\cdot Q)\eta(x) = \left\{\not{A}(x) + \underset{\sim}{\gamma}\,\underset{\sim}{F}(x) + \gamma_5\not{\partial}C(x)\right\}\varepsilon$$

$$(\varepsilon\cdot Q)\underset{\sim}{F}(x) = \bar{\varepsilon}\,\underset{\sim}{\gamma}\left(\lambda(x) + \not{\partial}\eta(x)\right) \qquad (2.9)$$

$$(\varepsilon\cdot Q)A_\mu(x) = \bar{\varepsilon}\left(\gamma_\mu\lambda(x) + \partial_\mu\eta(x)\right)$$

$$(\varepsilon\cdot Q)\lambda(x) = \left\{\gamma_5 D(x) + \vec{\sigma}\cdot\vec{B}(x)\right\}\varepsilon$$

$$(\varepsilon\cdot Q)D(x) = \bar{\varepsilon}\gamma_5\not{\partial}\lambda(x)$$

$$\underset{\sim}{F}\cdot\underset{\sim}{\gamma} = F_1 + \gamma_5 F_2 \qquad \vec{\sigma}\cdot\vec{B} = \sigma_{\mu\nu}(\partial_\mu A_\nu - \partial_\nu A_\mu)\,.$$

Thus the superfield and supertranslations may be thought of as just a convenient way of summarizing (2.9). Note that the super multiplit ($A,F, D,A_\mu, \eta,\lambda$) contains component fields of different spin, including both fermions and bosons, but that (modulo spontaneous supersymmetry breaking) all the component fields must have the same mass since $P_\mu P^\mu$ commutes with the Q_α . Note also that the two parts, $\partial/\partial\theta_\alpha$ and $(\bar{\theta}\not{\partial})_\alpha$ of the Q_α may be thought of as step-up and step-down operators respectively, which are similar to those in the representations of SU(3) or indeed of any compact simple group, except that they are both accompanied by a spin change $\Delta s = \pm\frac{1}{2}$ and that the step-down operation is accompanied by a Euclidean-space derivative.

2.2 Reduction of Superfields

As it stands the field $\Phi(x,\theta)$ does not carry an irreducible representation of the supertranslations. Indeed there are two ways that it can be reduced, corresponding roughly to the reduction of a Dirac spinor to two Majorana spinors and to two chiral (Weyl) spinors respectively.

The first (Majorana) reduction is simply to demand that $\Phi(x,\theta)$ be real, and this puts the following reality conditions on the component fields

$$\Phi^*(x,\theta) = \Phi(x,\theta) \iff (A, F_1, F_2, D \text{ real: } \eta, \lambda \text{ Majorana}). \qquad (2.10)$$

Such a superfield is called a real vector superfield because the highest spin it contains is a vector.

The second (chiral) reduction is more complicated. First, one notices that

$$\{Q_\alpha, D_\beta\} = 0 \qquad \text{where} \qquad D_\beta = i\frac{\partial}{\partial\theta_\beta} - (\bar\theta\gamma)_\beta , \qquad (2.11)$$

that is, if one introduces the operators D_β which are the differences rather than the sums of the step-up step-down operators then they commute with the Q_β. It follows that the conditions

$$D_\alpha \Phi = 0 \qquad \text{or} \qquad D_\alpha^* \Phi = 0 , \qquad (2.12)$$

on a superfield are supersymmetric invariant. Applied to a real superfield the conditions (2.12) give nothing (they kill it) but applied to a complex superfield, they reduce the independence of the component fields as follows:

$$D \to \Box A \qquad \lambda \to \partial\!\!\!/ \eta \qquad A_\mu \to \partial_\mu A , \qquad (2.13)$$

Superfields satisfying (2.12) are denoted $\Phi^\pm$ and are called chiral scalar superfields (because the vector has been reduced to a scalar and the fermion η to its chiral components).

The conditions (2.12) can also be expressed in superfield language in the following way. Let $\theta, \theta^* = (\theta_a, \theta_{\dot a})$, $a, \dot a = 1,2$ be the chiral (Weyl) components of θ_α, $\alpha = 1 \ldots 4$ and

$$\Psi\,(x,\theta) = A + \bar{\theta}\cdot\chi + F(\bar{\theta}\cdot\theta)\,, \tag{2.14}$$

and its complex conjugate $\Psi^*(x,\theta^*)$ be superfields constructed with the two-component θ (and θ^*) alone. Then

$$\Phi^+(x,\theta) = \Psi(\,x_\mu + i\bar{\theta}\gamma_\mu\theta\,,\,\theta), \tag{2.15}$$

and similarly for the complex conjugate. In other words the dependent multiplets $(A, \eta, F, \partial_\mu A, \not\partial\eta, \Box A)$ are just the simpler multiplets (A, χ, F) boosted by letting $x \to x + i\bar{\theta}\gamma\theta$. Note that for the chiral multiplets $\Psi(x,\theta)$ one can introduce the chiral transformations

$$\Psi(x,\theta) \;\rightarrow\; e^{in\alpha}\,\bar{\Psi}(x, e^{i\alpha}\theta), \tag{2.16}$$

where n is an arbitrary integer. Symmetry with respect to this transformation is called R-symmetry, and the integers n are called R-charges.

We shall see later that the chiral superfields play the role of matter fields (spin 0 and ½ fields) and we should like the real vector fields Φ to play the role of gauge-fields. For this purpose, however, one has to generalize to superfields the concept of gauge transformations

$$A_\mu \to G^{-1} A_\mu G + G^{-1}\partial_\mu G \qquad \text{where} \qquad G = \exp i\,\Lambda(x)\,, \tag{2.17}$$

where A_μ and Λ lie in the Lie algebra of a compact Lie group. To find the generalization is not so easy but, once found, it turns out to be even simpler than (2.17), as follows. Consider r real vector fields Φ_a assigned to the adjoint representation of a simple Lie group or the 1-dimensional representation of an abelian one and let $\Phi_A^{\pm}$ be n chiral scalar superfields assigned to any representation of the Lie group (and its conjugate). Then the required generalization of the gauge-transformation (2.17) is

$$\exp\Phi \to \exp i\Phi^{+} \exp\Phi \exp(-i\Phi^{-}) \qquad \text{where} \quad \Phi = \Phi_a t_a \tag{2.18}$$

and t_a are the generators of the Φ representation.

Since $\Phi = (A, \eta, F, A_\mu, \lambda, D)$ and $\Phi^{+} = (B, \chi, F)$ it is intuitively evident that a gauge can be chosen so that the fields (A, η, F) of are cancelled by the corresponding fields in $\Phi^{\pm}$. Such a gauge is called the Wess-Zumino gauge[3], and since in this gauge Φ consists only of (A_μ, λ, D) it is obviously the most convenient gauge for the component field formulation. The only residual gauge freedom in the Wess-Zumino gauge is the usual one (2.16) where $\Lambda = B + B^*$, and where λ and D are gauge-invariant.

2.3 Construction of Lagrangians

The construction of supersymmetric Lagrangians is based on the following two simple principles

(i) Products of superfields, such as $\Phi^{+}\Phi^{-}$, Φ^{2}, $(\Phi^{+})^{3}$ are again superfields.

(ii) The variation of the "highest" components $(\Phi^{+})_h = F$, $(\Phi)_h = D$, $(\Phi^{2})_h$, $(\Phi^{+3})_h$ etc. of superfields are total derivatives.

It follows that the supersymmetric variation of a quantity such as

$$\mathcal{L} = a(\Phi^{+}\Phi^{-})_h + b(\Phi)_h + c(\Phi^{2})_h + d(\Phi^{+4} + \Phi^{-4})_h , \tag{2.19}$$

consists only of a derivative. Hence, for conventional boundary conditions the variation of $\int dx \mathcal{L}_{ss}$ is zero.

Before going on to discuss the particular Lagrangians used for the matter and gauge superfields of Section 2, it might be worthwhile to illustrate first the importance of the boundary conditions by noting that for finite temperature the argument fails and the supersymmetry[4] becomes broken. The point is that for finite temperature the time-integral in $\int \mathcal{L}\, d^4x$ is finite, $t_0 \leq t = t_0 + 2\pi/T$, where T is the termperature). At the same time, because the Lagrangian is bosonic its variation must be fermionic,

$$\delta \mathcal{L}(x) = \partial_\mu \mathcal{J}_\mu(x) \,, \tag{2.20}$$

where $\mathcal{J}_\mu(x)$ is fermionic, and since for fermions the boundary conditions are anti-periodic, one obtains

$$\delta \int_{t_0}^{t_0+\frac{2\pi}{T}} dt \int d^3x\, \mathcal{L}(x) = Q(t_0+\tfrac{2\pi}{T}) + Q(t_0) = 2Q(t_0) \text{ where } Q(t_0) = \int d^3x\, \mathcal{J}_0(x,t_0), \tag{2.21}$$

and $Q(t_0)$ does not vanish since otherwise the charge would be trivial. Thus, in contrast to the usual variation $Q(t_1) - Q(t_2)$ which vanishes either because $t_1,\ t_2 \to \pm\infty$ and $Q(\pm\infty)$ vanishes, or because of periodic boundary conditions $Q(t_1) = Q(t_2) (\neq 0)$, the variation in (2.21) does not vanish and there is a spontaneous breakdown of the supersymmetry. Physically this may be understood from the fact that, in spite of the supersymmetry, the 'filling-up' of boson and fermion states is not the same on account of the different statistics, so that at any temperature $n_B \neq n_F$. All that the supersymmetry implies is that the change in these numbers in a single process be equally probable, $\Delta n_B = \Delta n_F$.

2.4 Supersymmetric Lagrangians for Matter Superfields

For the matter superfields it turns out that the most general renormalizable Lagrangian for the chiral scalar superfields $\Phi^{\pm}$ is

$$\mathcal{L}_{SS} = \int d^4x \left\{ (\Phi^+\Phi^-)_{h} + f(\Phi^+ + \Phi^-)_{h} + m(\Phi^{+2} + \Phi^{-2})_{h} + g(\Phi^{+3} + \Phi^{-3})_{h} \right\}. \tag{2.22}$$

On expansion in terms of the conventional fields (A, χ, F) in the supermultiplets this turns out to be

$$\mathcal{L}_{SS} = \int d^4x \left\{ \tfrac{1}{2}(\bar{\chi}\partial\!\!\!/\chi + (\partial A)^2 + F^2) + fF + m(\bar{\psi}\psi + AF + \bar{A}\bar{F}) \right. \\ \left. + g(\bar{\psi}\psi A + FA^2 + h.c.) \right. . \tag{2.23}$$

Since the field f has no kinetic term it can be eliminated using the Euler-Lagrangian equations

$$\frac{\delta \mathcal{L}}{\delta F} = F + f + mA + gA^2 = 0, \tag{2.24}$$

and then one obtains

$$\mathcal{L}_{ss} = \int d^4x \left\{ \bar{\Psi}(\not{\partial} + m + gA)\Psi + \tfrac{1}{2}|F|^2 \right\} \tag{2.25}$$

where

$$\bar{\Psi} A \Psi = \bar{\Psi}(A_1 + iA_2\gamma_5)\Psi, \qquad (-F) = f + mA + gA^2. \tag{2.26}$$

In other words one obtains the usual renormalizable Lagrangian for a Majorana field coupled to a scalar and pseudo-scalar field, but with the masses and coupling constants strongly correlated, indeed reduced to the three parameters f, m, g. The generalization to any number of superfields Φ_a is evidently

$$\mathcal{L}_{ss} = \int d^4x \left\{ \bar{\Psi}_a \not{\partial} \Psi_a + \bar{\Psi}_a m_{ab} \Psi_b + g_{abc} \bar{\Psi}_a A_b \Psi_c + \tfrac{1}{2} F_a \bar{F}_a \right\}, \tag{2.27}$$

where

$$- F_a = f_a + m_{ab} A_b + g_{abc} A_b A_c \ , \tag{2.28}$$

and the couplings m_{ab} and g_{abc} are totally symmetric.

5. Lagrangians for Gauge-Superfields.

In order to construct the Lagrangian for gauge superfields we first recall the construction for ordinary gauge-fields, namely, if $\mathcal{L}(\phi, \partial_\mu \phi)$ is a non-gauge Lagrangian then the gauge-equivalent is

$$\mathcal{L}(\phi, D_\mu \phi) + \frac{1}{4} \mathrm{tr}\, F_{\mu\nu} F^{\mu\nu} , \qquad (2.29)$$

where D_μ is the covariant derivative $\partial_\mu + e A_\mu$, where $A_\mu = A_\mu^k \sigma_k$ are the gauge-potentials, and $F_{\mu\nu}$ are the gauge-fields

$$F_{\mu\nu} = \partial_\mu A_\nu - \partial_\nu A_\mu + e [A_\mu, A_\nu] . \qquad (2.30)$$

The procedure for superfields is quite analogous, but in order to describe it, one needs to have the supersymmetric analogues of the gauge-fields $F_{\mu\nu}$. The form of these is a little unexpected, namely,

$$W_\alpha = D^*_\beta D^{*\beta} (e^{-e\Phi} D_\alpha e^{e\Phi}) , \qquad (2.31)$$

where D_α and D^*_α are the differential operators defined in Section 2 and which anti-commute with the supersymmetric generators Q_α. On expansion, the W_α take the form

$$W_\alpha = (\lambda_\alpha, F_{\mu\nu}, D, \ldots) , \qquad (2.32)$$

where the fields not written vanish in the Wess-Zumino (WZ) gauge. The procedure for gauging the supersymmetric matter-Lagrangian (2.27) is then quite analogous to (2.29), namely,

$$\mathcal{L}_{ss}^{matter} \Rightarrow \mathcal{L}_{ss} = \int d^4x \{ \Phi^+ e^{e\Phi} \Phi^- + \frac{1}{4} \mathrm{tr}\, W_\alpha W^\alpha + d\, \mathrm{tr}\, D \} , \qquad (2.32)$$

where the second term, like $\mathrm{tr}\, F_{\mu\nu} F^{\mu\nu}$ in (2.29), is the supersymmetric Lagrangian for the super-gauge-field alone. On expansion in terms of the conventional fields (in the WZ - gauge) this term yields

$$\mathrm{tr}_R\, W_\alpha W^\alpha = \mathrm{tr}_R\left(\bar{\lambda}\not{D}\lambda + F_{\mu\nu}F^{\mu\nu} + \tfrac{1}{2}D^2\right), \tag{2.33}$$

where D^2 plays no role (the field equations for (2.33) imply D = 0) and the first two terms are just the conventional terms for a gauge-field interacting with a Majorana field in the adjoint representation. Thus, such a conventional (F, λ) system is automatically supersymmetric!

The matter-gauge-field interaction that results from (2.32) (in the WZ - gauge) is just

$$\begin{aligned} &\bar{\chi}(\not{D} + m + gA)\chi + \tfrac{1}{2}|D_\mu A|^2 + \tfrac{1}{2}|F|^2 \\ &+ e\bar{\chi}A\lambda + eA^\dagger DA + d\,\mathrm{tr}_R D \quad . \end{aligned} \tag{2.34}$$

Thus it is just the usual term that one obtains from the substitution $\partial \to D$, together with a term $e\bar{\chi}A\lambda$ which links the two fermion fields, and two D-field terms. Since there are no kinetic terms for the D fields the latter can be eliminated by using the Euler-Lagrange equations, and on eliminating them, and introducing the Dirac fermion fields

$$\psi = \chi + i\lambda, \tag{2.35}$$

the expression (2.34) reduces to

$$\bar{\chi}(\not{D} + m + gA)\chi + \tfrac{1}{2}|DA|^2 + \tfrac{1}{2}|F|^2 + \tfrac{1}{2}|G|^2, \tag{2.36}$$

where

$$(-G) = (d + eA^\dagger A)^2 + e_0^2(A^\dagger \sigma_k A)^2, \tag{2.37}$$

where σ_k are the generators and e_0 is/(are) the coupling constant (s) for the semi-simple part of the group.

References

1) This is naturally only a skeleton review. Extensive reviews are given in:
Fayet, P. and Ferrara, S., Phys. Reports 32 (1977) 249.
Bagger, J. and Wess, J., Supersymmetry and Supergravity (Princeton Univ. Press 1983).
and some recent reviews with references are
Supergravity, Supersymmetry 1984 (Proc. Trieste Spring School, World Scientific Singapore 1984).
Supersymmetry and Supergravity (ed. M. Jacob, World Scientific Singapore 1984).
Grand Unification with and without Supersymmetry (Kounnas et al. World Scientific Singapore 1984).
2) Salam, A. and Strathdee, J., Nucl. Phys. B76 (1974) 477.
3) Wess, J. and Zumino, B., Nucl. Phys. B78 (1974) 1.
4) Girardello, L., Grisaru, Salomenson, P., Nucl. Phys. B178 (1981) 513. (For a counter-interpretation see van Hove, L., Nucl. Phys. B207 (1982) 15).

LECTURE 3. SPONTANEOUS SYMMETRY BREAKING IN SUPERSYMMETRIC FIELD THEORY

1. Variety of Symmetry Breakdown

Having set up the standard supersymmetric Lagrangians for matter and gauge fields we can now turn to the real purpose of these lectures, namely the spontaneous breakdown of these Lagrangians. There are actually two distinct kinds of spontaneous symmetry breaking namely (a) the spontaneous breakdown of an internal symmetry in the presence of supersymmetry and (b) the spontaneous breakdown of supersymmetry itself. Let us consider the general features of each kind in turn.

(a) Spontaneous breakdown of internal symmetry. Such a spontaneous breakdown is characterized by

$$\langle A \rangle \neq 0 \qquad\qquad \langle F \rangle = \langle D \rangle = 0 , \tag{3.1}$$

where A is the lowest and F,D the highest scalar fields in a supermultiplet. The reason that the supersymmetry is not broken in this case is that the fields A are themselves trivial superfields $\langle A \rangle = \Phi(\;) = \Phi(x,\theta,\bar{\theta})$ By $\langle A \rangle$ in (3.1) is meant, of course, the value of A at the potential minimum in a classical theory or the vacuum value in a quantized theory. A breakdown of the internal symmetry is very easy to achieve because the matter-field potential is $|F|^2/2$, and so any non-zero solution of the equation

$$F = f + mA + gA^2 = 0 , \tag{3.2}$$

will break the internal symmetry. Two novel features of internal symmetry breaking in the supersymmetric case are:

(i) The existence of Goldstone multiplets (including Goldstone fermions). This is because the supersymmetry is unbroken and therefore each Goldstone boson must be accompanied by a whole multiplet.

(ii) The existence of a Higgs mechanism for fermions. That is, the merging of scalar and gauge-fields in the usual Higgs mechanism is accompanied by a merging of matter-field and gauge field fermions.

The way in which this can happen has been foreshadowed by the merging of χ and λ into the Dirac field ψ, in (2.35).

(b) Spontaneous breakdown of supersymmetry itself. The spontaneous breakdown of supersymmetry itself is characterized by the opposite situation to (3.1), namely

$$\langle A \rangle = 0 \qquad\qquad \langle F \rangle \text{ or } \langle D \rangle \neq 0 . \tag{3.3}$$

One sees that (3.3) implies supersymmetry breaking because

$$F = \{ Q_\alpha , \chi \} \qquad\qquad D = \{ Q_\alpha , \lambda \} \tag{3.4}$$

from (2 .9) and hence (3.3) implies that $Q_\alpha |0\rangle$ cannot be zero. In contrast to the breakdown of internal symmetry, (3.3) is very difficult to achieve because if there is both a supersymmetric vacuum $|s\rangle$ and a non-supersymmetric vacuum $|ns\rangle$, then

$$H = \sum_\alpha Q_\alpha^2 \quad \Rightarrow \quad \langle s|H|s\rangle = 0 \ , \quad \langle ns|H|ns\rangle > 0 \tag{3.5}$$

so the supersymmetric vacuum is always the one with lower energy. Hence the only way in which supersymmetry can be broken is when no supersymmetric vacuum exists. In the case that a breakdown of supersymmetry does occur, there are again some novel features, namely,

(i) The existence of Goldstone fermions. However, in this case the Goldstone fermions exist in their own right because Goldstone fields are those in the directions of the variations of the vacuum state, and since

$$\delta F = \varepsilon\cdot\chi \qquad\qquad \delta D = \varepsilon\cdot\lambda \ , \tag{3.6}$$

one sees that in the supersymmetric case these are just the directions of χ and λ for $\langle F\rangle \neq 0$ and $\langle D\rangle \neq 0$ respectively.

(ii) The existence of anomalies similar to the Adler anomalies for axial vector currents[1]. More precisely, if Γ is the effective potential, and Q_α the supersymmetric generators, then, in contrast to what one would expect for a spontaneously broken symmetry, $Q_\alpha\Gamma$ is not zero.

2. Spontaneous Breaking of Internal Symmetry for Matter-Superfields.

Let us consider in this section the case where there are no gauge-fields i.e. the Lagrangian is that of (2.27) for matter fields alone, with potential

$$V = \tfrac{1}{2}|F|^2, \qquad \text{where} \qquad F_\alpha = f_\alpha + m_{\alpha\beta}A_\beta + g_{\alpha\beta\gamma}A_\beta A_\gamma \ . \tag{3.7}$$

Let us first consider the spontaneous breakdown of internal symmetry, and in particular consider two examples, in one of which the breaking is optional, in the other mandatory.

Example (1): Φ_a belongs to the adjoint representation of SU(n). Then

$$f_a = 0 \qquad m_{ab} = m\,\delta_{ab} \qquad \text{and} \qquad g_{abc} = g\, d_{abc} \ , \tag{3.8}$$

where d_{abc} is the totally symmetric invariant tensor used to construct the cubic Casimir invariant, equation (3.7) reduces to

$$F_a = m A_a + g\, d_{abc} A_b A_c \ , \tag{3.9}$$

and there are two solutions of $F_a = 0$ namely

$$A_a = 0 \qquad \text{and} \qquad A_a = \frac{m}{g}\,\mathrm{diag}(1,1,1,\ldots 1,\ 1-n) \tag{3.10}$$

The first solution leaves the internal symmetry unbroken, and the second solution breaks it down to U(n-1). Neither solution is preferred at this level since V=0 for each one and thus the breakdown is optional. As a matter of fact, radiative corrections do not change this situation[2], but gravitational corrections may do so[3].

<u>Example</u> (2): To remove the degeneracy found in Example (1), and actually present for all irreducible representations, one uses reducible representations. The simplest example is obtained by using two representations of a group, one of which (S) is a singlet, and the other (A) any non-trivial (real) representation. Then, on choosing

$$V_{SS} = f(\Phi_S^+ + \Phi_S^-) + m(\Phi_A^{+2} + \Phi_A^{-2}) + g(\Phi_S^+\Phi_A^{+2} + \Phi_S^-\Phi_A^{-2}), \tag{3.11}$$

one has

$$2V = |F_S|^2 + |F|^2, \qquad F_S = f + gA^2, \quad F_A = mA + gSA, \tag{3.12}$$

and for $f/g < 0$ the potential minimum is at

$$V = F = 0 \quad \Rightarrow \quad S = -\frac{m}{g} \qquad |A|^2 = -\frac{f}{g}\ (>0), \tag{3.13}$$

and since $|A| \neq 0$ the internal symmetry is necessarily broken.

At first sight the choice (3.11) would seem to be an ad hoc one, which would change after radiative corrections, because although a term of the form of A^3 in V could be avoided by using a representation with no cubic invariant, there is no way in which terms such as aS and bS^2 could be avoided by means of arguments based on internal symmetry, and in the presence of such terms there is an unbroken solution of (3.12), namely,

$$S = -\frac{f}{a} \qquad \text{or} \qquad S^2 = -\frac{b}{f} \quad , \quad A = 0 \; . \tag{3.14}$$

However, it turns out that terms such as aS, bS^2 (and A^3) can be eliminated by means of another symmetry, namely, the R-symmetry introduced in Section 2.2. By inspection of the Lagrangian (2.22) one sees that if Φ_a are different multiplets, with R-charges then one has the following selection rules:

$$\begin{aligned} f_a &= 0 \quad \text{unless} \quad N_a = 2 \\ m_{ab} &= 0 \quad \text{"} \quad N_a + N_b = 2 \\ g_{abc} &= 0 \quad \text{"} \quad N_a + N_b + N_c = 2 \; . \end{aligned} \tag{3.15}$$

Hence if one chooses $N_S=2$ and $N_A=-1$ as the R-charges of the fields S and A of the model the Lagrangian can only take the form shown in (3.11). Thus the form (3.11) follows from symmetry principles and is stable with respect to radiative corrections.

3. Spontaneous Breakdown of Matter Supersymmetry

We have seen that the condition for a spontaneous breakdown of the supersymmetry itself is $F_a \neq 0$ where

$$F_a = \lambda_a + m_{ab} A_b + g_{abc} A_b A_c \; . \tag{3.16}$$

However, since every quadratic equation (for complex variables) has a root it is evident that for a single matter-superfield Φ there always exists an A such that F=0. Thus for a single matter-superfield there can be no breakdown of supersymmetry. The question, therefore, is whether F_a in (3.17) can be strictly non-zero for $a=1\ldots n$, $n \geq 2$. It seems that for n=2 the answer is again negative (although I know of no formal proof). But for $n \geq 3$ the answer is positive. Thus for $n \geq 3$ there can be a spontaneous breakdown of supersymmetry. In this lecture we shall describe the simplest example, which is for n=3. The general conditions for which F_a in (3.16) cannot be zero (for arbitrary $n \geq 3$ and complex

A_α) are not known and it might be an interesting problem to try to find them.

In any case, the example for n=3 is constructed as follows[4]: Let Φ_0, Φ_1, Φ_2 be three superfields, with R-charges

$$N_0 = N_2 = 1,\ N_1 = 0\ , \tag{3.17}$$

respectively, and let the Lagrangian be R-invariant. Let us also suppose that the Lagrangian is invariant with respect to the following reflexion symmetry:

$$\Phi_0 \leftrightarrow \Phi_0\ , \qquad \Phi_i \leftrightarrow -\Phi_i\ , \qquad i=1,2\ . \tag{3.18}$$

It is easy to see that the most general renormalizable potential which is invariant with respect to these symmetries is unique and that it takes the form

$$V_{SS} = f(\Phi_0^+ + \Phi_0^-) + m(\Phi_1^+\Phi_2^+ + \Phi_1^-\Phi_2^-) + g(\Phi_0^+\Phi_1^{+2} + \Phi_0^-\Phi_1^{-2})\ . \tag{3.19}$$

One then has

$$F_0 = f + gA_1^2 \qquad F_1 = mA_2 + gA_0A_1 \qquad F_2 = mA_2\ . \tag{3.20}$$

From (3.20) one easily sees that $|F|^2 \geq f^2$ and that the lower bound is attained for

$$A_1 = A_2 = 0\ , \quad A_0 \text{ arbitrary.} \tag{3.21}$$

The arbitrariness of at least one of the fields at the potential minimum seems to be a general feature of the spontaneous breakdown of supersymmetry for matter-superfields, (but again I do not know of any general proof, and it is found that the arbitrariness does not survive the radiative corrections).

It is easy to see that for $F \neq 0$ the fields χ_0 and A_0 are massless i.e. are the Goldstone fields. On computing the masses of the other fields at the potential minimum one finds that

$$m^2(A_2) = m^2(\chi_2) = m^2 \qquad \text{but} \quad \begin{aligned} m^2(\mathrm{Re}\,A_1) &= m^2+2fg \\ m^2(\mathrm{Im}\,A_1) &= m^2-2fg\,, \\ m^2(\chi_1) &= m^2 \end{aligned} \tag{3.22}$$

Thus the masses of the Φ_1 supermultiplet are split. It is, perhaps, worth noting that the mass-splitting of a supermultiplet is the best guarantee that the supersymmetry really is broken. Note, that even after breaking, the sum-rule

$$m^2(\mathrm{Re}\,A_1)+m^2(\mathrm{Im}\,A_1)+m^2(\chi_1^{\uparrow})+m^2(\chi_1^{\downarrow}) = 4m^2, \tag{3.23}$$

is preserved (a result that is also true for the soft explicit breaking that is obtained[5] simply by adding to the Lagrangian a term of the form $\int A dx$. As mentioned earlier, the Ward identities i.e.

$$Q_\alpha \int d^4x\, \mathcal{L} = 0, \tag{3.24}$$

are not preserved by the radiative corrections. In fact, for the one-loop effective action $\Gamma^{(1)}$ one obtains[1])

$$Q_\alpha \Gamma^{(1)} = f^2 \int d^4x \left(A_0(x)\chi_0(x)\right), \tag{3.25}$$

where A_0, χ_0 are the Goldstone fields. Note that the anomaly vanishes when $f = 0$.

4. Spontaneous Symmetry Breaking for Gauge Superfields

In order to consider spontaneous breakdown in the presence of gauge superfields let us recall that the Lagrangian that describes the interaction of matter with gauge-fields takes the form

$$\begin{aligned}\mathcal{L}_{SS}^{gauge}(x) = {} & \Phi^+(\exp e\Phi)\Phi^- + \tfrac{1}{4}\,\mathrm{tr}\left(W_a W^a + dD\right) + \\ & + \left\{ f\Phi + m\Phi^2 + g\Phi^3 + \bar{\chi}(\not{\partial}+m+gA)\chi \right\} ?\end{aligned} \tag{3.26}$$

where the last bracket is a short-hand notation for the self-interaction of the matter-fields, and that when it is ex panded in terms of conventional fields $\mathcal{L}$ becomes

$$\mathcal{L}^{gauge}_{ss}(x) = tr\left(\bar{\lambda}\not{\partial}\lambda + F_{\mu\nu}F^{\mu\nu} + dD\right) + e\bar{\lambda}A\chi + \bar{\chi}(\not{\partial} + m + gA)\chi + \tfrac{1}{2}\left(|F|^2 + |D|^2\right), \quad (3.27)$$

where F,λ are in the adjoint representation of the gauge-group, the matter-fields χ,A are in an arbitrary representation and

$$|F(A)|^2 = |f + mA + gA^2|^2 \qquad |D(A)|^2 = \left(d + e|A|^2\right)^2 + e_0\left(A^{+}\sigma A\right)^2. \quad (3.28)$$

The spontaneous breakdown is governed by $|F(A)|^2$ and $|D(A)|^2$ since these are the potentials, but since the case of F(A) alone has previously been discussed we shall assume for simplicity that F(A) plays no role (in particular that F(0)=0) and concentrate on D(A)[6]. Once again there are two possibilities, namely,

(1) The internal symmetry is broken but the supersymmetry is not. This happens when $ed<0$ because then the potential minimum occurs for D(A)=0, $|A|^2 = |d/e|$.

(2) The supersymmetry is broken but the internal symmetry is not. This happens when $ed>0$ because then the potential minimum occurs for

$$A = 0 \,, \qquad |D(A)| = |D(0)| = |d| \neq 0 \,. \quad (3.29)$$

Of course, in more complicated models there could be a spontaneous breakdown of both internal and super symmetry. But for simplicity we shall consider only the cases (1) and (2). We shall also assume for simplicity that the gauge-group is U(1) and that there is only one matter-field.

5. Spontaneous Breakdown of Internal Group Symmetry.

Let us first consider the case $ed<0$ when the internal symmetry is broken. For simplicity, and because it is the most interesting case, we shall assume that the gauge - and matter-field fermions λ and χ have opposite chirality. Note that when χ has definite chirality the mass-terms and Yukawa couplings vanish, so the vanishing of F(A) is automatic and does not have to be assumed. In that case the Lagrangian (3.27) reduces to

$$\mathcal{L}_{ss}^{gauge} = \tfrac{1}{4} F_{\mu\nu} F^{\mu\nu} + \bar{\lambda}\not{\partial}\lambda + \bar{\chi}\not{\partial}\chi + e\bar{\chi}A\lambda + \tfrac{1}{2}|D_\mu A|^2 + \tfrac{1}{2}|d+e|A|^2|^2 \,. \quad (3.30)$$

Since the potential $|d+e|A|^2|^2$ in (3.30) is just the standard Higgs potential, with minimum at $\langle A\rangle^2 = |d/e|$ one sees that the non-supersymmetric part of the Lagrangian undergoes the usual Higgs mechanism

$$\mathcal{L}_{ss}^{gauge} \Rightarrow \tfrac{1}{4} F_{\mu\nu} F^{\mu\nu} + \tfrac{1}{2}|D_\mu A|^2 + \frac{m^2}{2} A_\mu^2 + \text{interaction terms}, \quad (3.31)$$

in which the gauge-field acquires a mass $m = e|d/e|^{1/2} = |ed|^{1/2}$. However, in addition to this conventional Higgs mechanism one sees from (3.30) that the fermion terms also undergo a Higgs mechanism, namely,

$$e\bar{\chi}A\lambda \rightarrow e\bar{\chi}\lambda|\mathring{A}| + \dots \quad = m\bar{\chi}\lambda + \dots \,, \quad (3.32)$$

where .... denotes interaction terms. Furthermore, if one introduces the Dirac field $\psi = \chi + i\lambda$ of Section and recalls that $\gamma_0\gamma_\mu$ and γ_0 connect the same and opposite chiralities respectively, one sees that the mechanism (3.32) can be written more fully as

$$\bar{\lambda}\not{\partial}\lambda + \bar{\chi}\not{\partial}\chi + e\bar{\chi}A\lambda = \bar{\psi}(\not{\partial} + eA)\psi = \bar{\psi}(\not{\partial} + m + eB)\psi, \quad (3.33)$$

where $B = A - \langle A\rangle$. Thus the Higgs mechanism for the fermions combines the gauge-field and matter-field fermions into a Dirac fermion with the same mass as the gauge-field.

Finally one notes that since

$$\tfrac{1}{2}|d+e|A|^2|^2 = \tfrac{1}{2}|2e|\mathring{A}|B+e|B|^2|^2 = \tfrac{m}{2}|B|^2 + O(B^3)\,, \tag{3.34}$$

the scalar field B also acquires the mass m = $|ed|^{\frac{1}{2}}$. So the gauge-superfield and the matter-superfield have merged to form a single superfield

$$\tilde{\Phi} = \{B, \psi, A_\mu\}\,, \tag{3.35}$$

consisting of a real scalar B , a real vector A_μ, and a Dirac field χ , all with the same mass $m = |ed|^{1/2}$.

6. Spontaneous Breakdown of Gauge Supersymmetry (D-breaking)

Let us now consider Case (2) in which the supersymmetry is broken by D(A) but the internal symmetry is not. We shall not assume here that the fermion fields have definite chirality or that F(A) is zero (although we shall assume that F(A) does not cause any symmetry breakdown). Then the U(1) supersymmetric gauge Lagrangian is

$$\mathcal{L}^{gauge}_{ss}(x) = \frac{F^2}{4} + \bar{\lambda}\ \lambda + \bar{\chi}(\not{D}+m+gA)\chi + \tfrac{1}{2}|D_\mu A|^2 + \tfrac{1}{2}|F(A)|^2 + \tfrac{1}{2}|d+e|A|^2|^2\,. \tag{3.36}$$

The nice thing about this case is that, since the potential minimum occurs at A=0, there is no need to shift the scalar field, so the mass-spectrum can be read directly from (3.36), and is

$$m_\chi^2 = m^2 \qquad m_A^2 = m^2 + 2ed \qquad\qquad m^2(\lambda) = m^2(A_\mu) = 0\,. \tag{3.37}$$

Hence the masses of the gauge supermultiplet remain degenerate (and zero) but the masses of the matter-multiplet split. Thus the gauge-supermultiplet acts as a catalyst for the breaking of the matter-superfield, but is not itself broken. Note that in this case there is no sum-rule such as (3.23) which is preserved after the supersymmetry breakdown.

References

1) Clark, J. Piguet, O. and Sibold, K., Nucl. Phys. B119 (1977) 292.
Nemeschansky, D. and Rohm, R., Nucl. Phys. B249 (1985) 157.
Feruglio, F., Helayut-Neto, J. and Legovini, F., Nucl. Phys. B249 (1985) 533.

2) Capper, D. Ramon-Medrano, M., J. Phys. G2 (1976) 269.
O'Raifeartaigh, L. and Parravicini, G., Nucl. Phys. B111 (1976) 516.
Weinberg, S., Phys. Rev. Lett. 62B (1976) 111.
Lang, W., Nucl. Phys. B114 (1976) 123.

3) Ross, G., in Supersymmetry, Supergravity (Proc. XVth Gift Seminar, World Scientific Singapore 1984).

4) O'Raifeartaigh, L., Nucl. Phys. B96 (1975) 331. For generalizations and applications see
Nilles, H.P., Phys. Reports 110 (1984) 1.

5) Iliopoulos, J. and Zumino, B., Nucl. Phys. B76 (1974) 310.

6) Fayet, P. and Iliopoulos, J., Phys. Lett. 51B (1974) 461.
Fayet, P., Nuovo Cim. 31A (1976) 626.
Mainland, G.B. and Tanaka, K., Phys. Rev. D12 (1975) 2394.
Likhtman, E., JETP Lett. 21 (1975) 109.

RECONCILIATION OF THE UNIFIED FIELD THEORY WITH PHENOMENOLOGY

Jerzy Rayski
Institute of Physics
Jagellonian University
Cracow, POLAND

Abstract

A reconciliation of super-symmetric field theory with the phenomenology of quark and lepton gauge interactions may be achieved by performing a suitable compactification of seven subsidiary dimensions to form a product of spheres $S^2 x S^5$.

1. Introduction

The usual gauge theories proved to be useful for a formulation of a satisfactory theory of weak interactions and their fusion with electrodynamics as well as for a formulation of a chromodynamics. Also their fusion called Grand Unification (GUT) possesses a certain predictive power. However, the standard gauge theories are not quite satisfactory in their roles of unifying theories of all fields and their interactions. First of all, the GUT based on SU(5) does not predict correctly the life-time of proton and does not unify all types of forces leaving out the most common force in nature: the gravitational force. Moreover, it is unable to explain the nature of the "imaginary" spaces like the isospace (more generally the spaces of colour and flavour). The mere fact that the gauge transformations involve functions of the usual coordinates x^k and the analogy among the roles of the indices a,b,...

of the gauge groups and the tensor indices from the ordinary space time is not superficial and cannot be accidental.

What is also not satisfactory in the standard approach to gauge theories is the fact that several features have been introduced as one uses to say "by hand", i.e. without any deeper justification. All free fields constituting the sources of the gauge fields have been introduced "by hand".

There are at least two distinct possibilities to remedy, at least partly, the above mentioned drawbacks. One of them consists in assuming a generalized d-dimensional metrical space-time with four ordinary dimensions and d-4 compact dimensions of a space-like character. The compact subspace is to be identified with the isospace. It must be closed very tightly, otherwise it would be directly accessible in experiment. The gauge group reflects the internal symmetry of this compact subspace. This version may be called a "metrical gauge theory" in as much as the vector fields A_k^a may be shown to be (up to a factor of proportionality) nothing else but the mixed metric tensor components g_{ak} where k = 0,1,2,3 is the index enumerating the ordinary coordinates of the open subspace whereas a = 1,...,d-4 is the shifted index running originally from 4 to d-1 and enumerating the subsidiary coordinates forming the compact subspace.

Such metric gauge theories provide a geometrical interpretation of the gauge fields and gauge symmetries, but the source fields are still to be introduced "by hand". This drawback may be remedied by assuming the point of view of the so called superspace and supersymmetries.

As is already well known, the fields of tensor and spinor character may be combined into supermultiplets. The theories involving supersymmetries may be put into a geometrical scheme involving, besides the ordinary coordinates x^k also a number N of Grassmann coordinates θ^a_α where a = 1,...,N. The case N = 1 is called supersymmetry, whereas in the cases N > 1 one uses to speak about extended supersymmetries. The advantages of the idea of supersymmetries are undeniable inasmuch as it enables for a geometrical interpretation of physical properties in terms of a generalized geometry whereby the particles with half-integral spin fit naturally and are no more introduced "ad hoc".

In view of the above it seems that a truly satisfactory unified theory may be obtained by unifying both above mentioned extensions: the idea of an extended number of ordinary dimensions with the idea of superspace. The usual gauge procedure is applicable within the supersymmetric multiplets. In this way it should be possible to achieve a truly grand unified theory (TGUT).

2. Unification of gravity with electromagnetism

Let us begin with Kaluza-Klein assumption that the world is five-dimensional with a space-like fifth dimension. Denote the usual coordinates by x^k (k = 0,...,3) and the fifth coordinate by x^5. The indices j, k,... are reserved for the four-dimensional world, the Greek indices μ , ν ,... for the five-dimensional world. The index 4 is missing in order not to mix x^4 with ict. Assume also that g_{55} is either equal unity as in the original papers of Kaluza, or that $g_{55} = 1 + \phi$ where ϕ is an arbitrary function, as in the Brans-Dickie version of the theory. Both versions are a priori equally well possible though Kaluza's version cannot be regarded as a limiting case $\phi \rightarrow 0$. Assume also that all physically interesting quantities are independent of the fifth coordinate. This assumption will be justified later. It has been shown that the mixed metric tensor components g_{5k} are interpretable as the electromagnetic potentials

$$eA_k = \frac{1}{\ell} g_{5k} \tag{1}$$

where ℓ is a constant with dimension of a length. It must be closely related to the Planck's length.

The identification (1) has been confirmed by computing the scalar curvature of the five-dimensional universe. Assuming the metric tensor of the five-dimensional world to be

$$\left(g^{(5)}_{\mu\nu}\right) = \left(\begin{array}{c|c} \overset{(4)}{g}_{k\ell} + e^2\ell^2 A_k A_\ell & e\ell A_k \\ \hline e\ell A_\ell & 1 \end{array}\right) \tag{2}$$

the five-dimensional scalar curvature becomes

$$R_{(5)} = R_{(4)} - \frac{e^2\ell^2}{4} F_{k\ell} F^{k\ell} \tag{3}$$

with

$$F_{k\ell} = \partial_k A_\ell - \partial_\ell A_k . \tag{4}$$

In this way it is seen that the five-dimensional theory of gravity involves automatically the usual four-dimensional gravity together with the electromagnetic field (in the Brans-Dickie theory there appears also a scalar field ϕ). Gauge transformations appear to be nothing else but x^k-dependent translations of the fifth coordinate. Inasmuch as the electric charge is a generator of the infinitesimal gauge transformations, one gets a nice geometrical interpretation of the electric charge as a generator of infinitesimal translations of the fifth coordinate, or in other words, as the fifth component of momentum; charge conjugation is inversion of x^5. Inasmuch as two inversions mean rotation, the CP-transformation may be achieved by a rotation in the five-dimensional universe. All this was published first by the present author[1]) as early as in 1965.

Kaluza-Klein theory was developed further in the sixties by the present author, but it evoked little interest until recently "rediscovered" by Scherk and Crammer.[2]) It has been criticised for obvious reasons: the fifth coordinate plays a manifestly distinguished role whereas the assumption that the field quantities do not depend upon x^5 seemed to be quite ad hoc and arbitrary. However, the apparent contradiction between the requirement of general covariance in the five-dimensional manifold and the manifestly privileged role of the fifth dimension was explained by the present author[1]).

Assume that the fifth dimension is closed tightly (compact) so that the universe forms a five-dimensional "world tube" with a circumference of the order of Planck's length. Introducing a suitable coordinate system the dependence of the field quantities upon x^5 must be periodic whence any quantity may be developed into a Fourier series

$$\psi(x^\nu) = \sum_n \psi^{(n)}(x^k)\, e^{in x^5/l} \tag{5}$$

whence the quantity ψ splits into an infinite set of functions of the usual four coordinates x^k. Let us assume e.g. that ψ satisfies a Klein-Gordon equation in five dimensions. Neglecting gravity the equation

$$(\pentagon - \kappa^2)\,\psi(x^\nu) = 0 \tag{6}$$

is seen to be equivalent to a set of equation

$$(\Box - n^2/l^2 - \kappa^2)\,\psi^{(n)}(x^k) = 0 \tag{7}$$

in the usual space-time describing particles with the masses $n^2/l^2 + k^2$ instead of k^2. If $1/l$ is very large (of a similar order of magnitude to Planck's length) the separate fields $\psi^{(n)}$ represent particles with gigantic masses being multiples of 10^{15} GeV. Such particles cannot play any role unless one were able to penetrate distances about one or two orders of magnitude larger than Planck's length. Thus, in practice, we are left only with the first term of the expansion (5) n = 0 and any dependence upon x^5 disappears, or looks spurious.

The formalism remains generally covariant locally, but the covariance is broken globally due to the peculiarities of a topological origin. This explains why the concepts of scalars, vectors, and tensors in five dimensions are indispensable notwithstanding of the fact that in other respects the role of x^5 is quite different from the role of the remaining four coordinates.

3. The supermultiplet N = 8 and its gauge interactions*)

After a period of enthusiasm as regards a possibility of formulating a Truly Grand Unified Theory (TGUT) combining the three ideas: gauge theory (i.e. Yang-Mills fields) with Kaluza-Klein theory, and with supersymmetries, we wittnessed recently a tide of disappointments. In spite of the fact that the tasks of constructing supersymmetric Lagrangians of the type N = 8 in d = 4 as well as N = 1 in d = 11 dimensions have been accomplished, there remain several open problems as regards spontaneous symmetry breaking (the problem of masses of the elementary constituents of matter), the problems of uniqueness of the procedure of compactification, and the problems of quantization (see the so called "problem of N = 3 barrier")[3]. Last not least, neither the extensively studied compactification of seven subsidiary dimensions into a seven-torus T^7 nor the discussion of a "squashed" seven-sphere S^7 led to a formulation satisfactory from a phenomenological viewpoint.

---

*) The contents of this section was prepared in collaboration with J.M. Rayski jnr.

The aim of this lecture is to show that the pessimism was unjustified and there are good chances for formulating a unified theory, based on a compactification S^2xS^5 and satisfying all the requirements of phenomenology without necessity of going over from the level of quarks and leptons to a still more elementary level of "preons".

Let us begin with the following two fundamental assumptions:

I. The Universe is eleven-dimensional (d = 11) with seven compact dimensions.

II. The fundamental constituents of matter form a supermultiplet N = 8 (see the table 1)

Table 1

s	2	3/2	1	1/2	0
f	1	8	28	56	70

where s = spin, and f = number of two-component fields for $s > 0$. The table 1 applies to the massless case. In the case of massive fields some of the spin 1/2 fields have to be "swallowed" by the spin 3/2 fields in order to yield four spin orientations (3/2, 1/2, -1/2, -3/2). This leaves us with 48 instead of 56 spin 1/2 fields. Moreover, the whole set N = 8 possibly is to be supplemented by some Higgs fields and s-Higgs fields with spins 0 and 1/2.

The above assumptions I and II fit well together inasmuch as the total number of fermionic degrees of freedom from the table 1 is equal $2(8 + 56) = 2(16 + 48) = 128$ whereas, according to the formula

$$n_{3/2} = \frac{1}{2} 2^{E(d/2)} (d-3) \qquad (8)$$

it is also equal 128 if d = 11. Thus, from the 11-dimensional point of view we have to do with a supermultiplet involving only a single spin 3/2 spinor field that splits into 8 spin 3/2 and 56 (possibly 48) spin 1/2 fields if viewed from Minkowski's space.

In order to interpret properly the spin 1/2 tail and its gauge interactions let us introduce the following assumption.

III. There exist also right-handed neutrinos, though they are not involved into weak interactions.

If it is so, all spinors appearing in the table 1 may be fused into pairs, i.e. into Dirac's bispinors. Recalling once more that 8 from 56 spin 1/2 fields should be swallowed by the 8 spin 3/2 fields to endow them with masses we are left with 48 pure spin 1/2 two-component fields or 24 Dirac fields. According to the formula

$$n_{1/2} = 2^{E(d/2)} \tag{9}$$

for the number of Dirac field components in d dimensions we need 32-component spinors and 32x32 Dirac matrices in d = 11.

The Dirac spinor fields ψ_α where $\alpha = 1,\dots,32$ may be reinterpreted from the four-dimensional viewpoint as $\psi_{\alpha\varrho}$ where $\alpha = 1,\dots,4$ is the usual spinor index in M_4 and $\varrho = 1,\dots,8$ is the index of "internal spin". Inasmuch as we need not 8 but 3 x 8 = 24 Dirac fields involved in the supermultiplet N = 8, these fields may be distinguished one from the others by means of another index g assuming three values g = 1,2,3. Hence $\psi = \psi_{\alpha\varrho g}$ mean the totality of spin 1/2 components. Of course, from the point of view of the original spin 3/2 field in eleven dimensions g is another spinorial index, but from the point of view of the Minkowski world it is reinterpretable as some SU(3)-internal-symmetry index. Also $\varrho = 1,\dots,8$ could be interpreted as another internal symmetry (gauge) index if viewed from the four-dimensional perspective.

If the seven additional dimensions were on equal

footing, i.e. if the compact subspace were of the form S^7 the gauge group generated by the spinor fields would be either SU(8)xSU(3) or SO(8)xSU(3) which necessitated 63 + 8 or 28 + 8 Yang-Mills fields. Since the total number of vector fields in the supermultiplet N = 8 is 28 we have to look for another solution in the Y-M sector. Let us introduce the following assumption.

IV. The compactified subspace is $S^2 x S^5$. The corresponding symmetry is SO(3)xSO(6) isomorphic with SU(2)xSU(4). At the first sight it seems to be not satisfactory since it yields only 3 + 15 = 18 Yang-Mills fields of metrical origin involved into the mixed metric tensor components

$$g_{\mu\xi} = \sum_{a=1}^{3} A_{\mu}^{a} K_{\xi}^{a} + \sum_{b=1}^{15} A_{\mu}^{b} K_{\xi}^{b} \qquad (10)$$

where K_{ξ}^{a} and K_{ξ}^{b} are the Killing vectors of the spheres S^2 and S^5 respectively. However, not all Yang-Mills fields have to be of a metrical origin, some may be genuine vector fields. It should be recalled that besides the spinor indices $\alpha = 1,...,4$ and $\varrho = 1,...,8$ there appears a subsidiary index g = 1,2,3 labelling the spinor components. Altogether we have to do with four indices $\psi_{\alpha\varrho\omega g}$, where α is the usual spinor index from Minkowski's space, $\varrho = 1,2$ is connected with S^2 and $\omega = 1,...,4$ with S^5 whereas g = 1,2,3. This means that the internal symmetry may be SU(2)xSU(3)xSU(4). However, since introduction of any further gauge symmetries of the type U(1) does not necessitate an introduction of any additional indices, we may supplement the above gauge group by an arbitrary number of U(1) factors. In order to get the total number of vector fields 28 two and only two such factors U(1) are necessary and sufficient. In this way we get the assumption

V. The Yang-Mills sector is characterized by the gauge group

$$G = U(1) \times U(1) \times SU(2) \times SU(3) \times SU(4). \qquad (11)$$

It fits well with the assumption IV since one U(1) is a maximal subgroup of SU(2) while U(1)xSU(3) is a maximal subgroup of SU(4).

Comparing the assumptions IV and V it is seen that the 28 vector fields are not of the same origin, but split into two different classes: (i) these involved directly into the metric A^a_μ and B^b_ν appearing in (10) with a = 1,2,3, and b = 1,...,15, and deserving the name of metric constituent fields, and (ii) the remaining Yang-Mills fields to be called U_μ, U'_μ, and V^a_μ (where a = 1,...,8) connected with the gauge groups U(1), another U(1), and SU(3) respectively. These last fields should be, however, not four-vector fields but eleven-vectors, U_M, U'_M, and V^a_M for M = 0,1,...,10. Their "tails" U_ξ, U'_ξ, V^a_ξ (for ξ = 4,...,10) are reinterpretable as scalar fields whose number is (1 + 1 + 8) x 7 = 70 exactly as demanded by the table 1. This cannot be a pure coincidence!

Finally we see that the supermultiplet N = 8 involves, besides one spin 3/2 fermion, also one metric field and a set of 1 + 1 + 8 = 10 multivectors if viewed from the eleven-dimensional point of view. If it is so, the sector $(g_{\xi\eta})$ of the metric field (ξ, η = 4,...,10) should not represent any further scalar fields, i.e. it should be frozen. This assumption seems to be a natural generalization of the original Kaluza's assumption $g_{55} = 1$. However, it seems to be also possible to extend our supermultiplet by adding (according to the version of Brans and Dickie) further 28 scalar fields into the sector $(g_{\xi\eta})$. These scalar fields could possibly play the role of Higgs fields endowing 27 from 28 vector fields with masses while the only massless vector field would remain the electromagnetic field. However, in order not to distroy the equality of the numbers of bosonic and fermionic degrees of freedom we should also introduce 14 s-Higgs fields with spin 1/2 of a Weyl or Majorana type. In this way, however, the supermultiplet would undergo a proliferation and be no more a simple supermultiplet.

Let us go over to an interpretation of the spin 1/2 fields involved into the gauge interactions intermediated by the Yang-Mills fields. The group SU(3) involved into the gauge group G cannot be a colour group, but the group of symmetry of interactions coupling the three generations of particles among themselves, i.e. intermediating some interactions between particles belonging to different generations. On the other hand, the colour group may be a broken SU(4) group according to $SU(4) \to SU(3) \oplus U(1)$. In other words, we had originally four colours, red, green, blue, and lilac, this last denoting leptons. For reasons of economy the same set of fields V_M^a might be involved in both generation and colour groups, although the gluons cannot be just identical with the V_M^a but may differ by a unitary transformation $V^a_{(c)} = T^{ab}V^b_{(g)}$ where the suffix (c) denotes colour, and (g) generation. This is analogous with the appearance of the Weinberg's angle in electro-weak interactions.

It may be objected that the theory with a gauge group G (see (11)) does not deserve to be called Grand Unified Theory since this group is not simple and involves several independent coupling constants. It should be noticed, however, that U(1)xSU(2)xSU(3) as well as U(1)xSU(4) are two maximal subgroups of the group SU(5), and SU(5)xSU(5) is a maximal subgroup of the exceptional group E_8 so that

$$G \subset SU(5) \times SU(5) \subset E_8 \qquad (12)$$

which shows that the different coupling constants appearing

in G are not quite independent from one another. Besides, since it is known that in quantized supersymmetric theories the most harmful divergences cancel, we may try to adjust the coupling constants so as to achieve the best possible cancellation of divergent terms.

Our approach differs principally from the standard procedure where one used to start with a supersymmetric formalism in eleven (open) dimensions ($N = 1$, $d = 11$) and then to compactify seven dimensions by simply folding them into a seven-torus

$$M_{11} \longrightarrow M_4 \times T_7 . \tag{13}$$

Hereby the number of degrees of freedom involved into the metric field is

$$n_2 = \frac{1}{2}(d-1)(d-2) - 1 = 44 \tag{14}$$

before as well as after the compactification. On the other hand, we perform first something that may be called an intertwining of the subspaces

$$M_{11} \rightarrow (adS) \times S^2 \times S^5 \tag{15}$$

whereby the number of degrees of freedom involved into the metric undergoes a change (from 44 to $2 + 2x(3 + 15) + 28 = 66$). It is only after such compactification that we try to satisfy the supersymmetry requirement $N = 8$ in $d = 4$.

At any rate, it is easily seen that compactification is not and cannot be a unique procedure since we may compactify either into $M_4 x T_7$, or $(adS) x S^7$, or $(adS) x S^2 x S^5$, [4] all three of them being solutions of Einstein's equations in vacuo. Most probably all of them are possible, but denote three different phases of the Universe. We happen to live in the phase $S^2 x S^5$.

References.

1) J. Rayski, Acta Phys. Polonica 27, 89 (1965) and 28 (1965)

2) E. Cremmer, B. Julia, Nucl. Phys. B159 (1979) 141

3) V.O. Rivelles, J.G. Taylor, Phys. Letters 121B (1983) 37

4) H. Nicolai, Proc. Trieste 1982 School, ed S. Ferrara, J.G. Taylor, P. Nieuwenhuizen, Singapore

INVARIANT TENSOR FIELDS AND LINEAR CONNECTIONS ON EXTENDED SPACETIME

Lora Nikolova and V.A.Rizov

Bulgarian Academy of Sciences
Institute of Nuclear Research and Nuclear Energy
72, boul.Lenin, Sofia 1184, Bulgaria

ABSTRACT

Let the compact Lie group G act smoothly on the smooth manifold S (dimS>4) with a single orbit type G/H where H is a closed subgroup of G, and let N(H)/H be a Lie group (N(H) denotes the normalizer of H in G). Assuming a given connection in the fibre bundle S→M where M=S/G is the orbit space (to be identified with the 4-dimensional spacetime), and N(H)/H is a gauge group, we establish a kind of "dimensional reduction" for the G-invariant tensor fields and linear connections on S: every field corresponds uniquely to a field on M, the linear connection defines uniquely a linear connection and a set of fields on M, the gauge group becomes N(H)/HxGL(4,ℝ).

INTRODUCTION

The study of field theories on an extended spacetime (i.e. with dimension greater than four) is motivated by several reasons emerging from the picture of the observed basic interactions. Among the others we mention the atempts 1)to give a unified description of gravity and Yang-Mills interactions; 2)to understand why the left- and right-handed fermions transform under different representations of the gauge group; 3)to find a "natural" explanation for the terms with Higgs fields in the Lagrangean of a unified theory.

In these lectures we shall describe tensor fields and linear connections defined on a multidimensional space. Following the point of view suggested by Coquereaux and Jadczyk in ref.[1] we assume that the model of the space is a smooth manifold S equipped with a smooth action of a compact Lie group G such that its orbits are of the same type G/H, H being a closed subgroup of G.

It can be shown (see e.g.[2],Chap.2 and Chap.6) that in this case the projection π which assigns to every point $s\in S$ the G-orbit passing through s: $\pi: S \to S/G$ is a differentiable map and the orbit space M=S/G is a smooth manifold. Moreover, the space S has a structure of a locally trivial fibre bundle over M with G/H as a typical fibre and K=N(H)/H as a gauge group (N(H) is the normalizer of H in G). In this picture the orbit space M is identified with the 4-dimensional spacetime.

We shall make the hypothesis that the gravitational and Yang-Mills interactions in 4-dimensional spacetime result from a dynamics of G-invariant tensor fields and a linear connection on S. It will be shown how the latter can be described in terms of fields and a linear connection defined on the orbit space M=S/G provided a connection is given in the fibre bundle $\pi: S \to M$. This description agrees with the concept of dimensional reduction from the papers [1],[3] and [4]. A similar approach to theories on extended spacetime is proposed in [5] (cf. also [6]).

The material of these lectures is based partly on the paper [7]. The first lecture recalls some standard facts about fibre bundles and the description of fields in terms of cross-sections of fibre bundles. In the second lecture we review the structure of the fibre bundle $\pi: S \to S/G$, the bundle $\mathcal{F}(S)$ of linear frames over S, and define a G-invariant field of tensor type on S. The third lecture demonstrates that the invariant fields admit dimensional reduction in the following sense: every G-invariant field on the multidimensional space S corresponds uniquely to a field on the 4-dimensional space M with gauge group $K \times GL(4,\mathbb{R})$. We also briefly expose there the analogous statement for an invariant linear connection on S: it determines uniquely a linear connection on M and a set of scalar and vector fields on M with gauge group $K \times GL(4,\mathbb{R})$.

We note that a nice and clear introduction to the theory of fibre bundles and its applications to physical problems is contained in the articles [8] and [9]. For a detailed exposition see, for example, the book [10].

In the text we use the notations GL(q) and $\mathfrak{gl}(q)$ for the general linear group $GL(q,\mathbb{R})$ and its Lie algebra, respectively. For two vector spaces V_1 and V_2 the space of linear maps from V_1 into V_2 will be denoted by $L(V_1;V_2)$.

1. FIBRE BUNDLES AND FIELDS

1.1 Let P be a differentiable manifold endowed with a right differentiable and free action of the Lie group K. If a) Q=P/K is the quotient space with respect to the action of K on P and the natural projection $\pi: P \to Q$ is a differentiable map;

b) for every $x \in Q$ there exist a neighbourhood U of x and a diffeomorphism $\Psi: \pi^{-1}(U) \to U \times K$ such that

$\Psi(p)=(\pi(p),\theta(p))$, where the map $\theta:\pi^{-1}(U)\to K$ satisfies $\theta(pa)=\theta(p).a$, $\forall p\in\pi^{-1}(U)$, $a\in K$,
we say that P is the total space of a principal fibre bundle $P(Q,K,\pi)$ with base Q, structure group K and projection π. We shall also use the notation P(Q,K) or simply P.

Let F be a K-space and $\rho: v\mapsto\rho(a)v$, $\forall a\in K$, $v\in F$, denote the left action of K on F. We introduce an equivalence relation $\sim$ in PxF with respect to the action $(p,v)\mapsto(pa,\rho(a^{-1})v)$, $a\in K$, of the group K. The quotient space

$$E=(P\times F)/K \tag{1.1}$$

obtained by factorizing PxF with respect to $\sim$ is called a fibre bundle with base Q and typical fibre F associated to $P(Q,K,\pi)$ by the action ρ of the structure group on F. We shall also use the notation E(Q,F) or simply E. The element from E determined by the pair $(p,v)\in P\times F$ will be often denoted by $[p,v]$. For the projection $\pi_E: E\to Q$ we have $\pi_E([p,v])=\pi(p)$.

Every $p\in P$ determines an isomorphism $i_p: F\to E_{\pi(p)}$ between the typical fibre F and the fibre $E_{\pi(p)}$ at the point $\pi(p)\in Q$ by the formula $i_p(v)=[p,v]$. One checks that

$$i_{pa}(v)=i_p(\rho(a)v) \tag{1.2}$$

for each $a\in K$.

1.2 A field on Q of type $(\rho,F;P)$ is a cross-section φ of the fibre bundle E, i.e. a map $\varphi: Q\to E$ satisfying

$$\pi_E(\varphi(x))=x,\ \forall x\in Q.$$

In the following we shall use the one-to-one correspondence between the set of fields of type $(\rho,F;P)$ on Q and the set of mappings $\tilde\varphi: P\to F$ such that

$$\tilde\varphi(pa)=\rho(a^{-1}).\tilde\varphi(p). \tag{1.3}$$

Indeed, given the mapping $\tilde\varphi: P\to F$ with the property (1.3) one defines a field $\varphi: Q\to E$ on Q by

$$\varphi(x)=[p,\tilde\varphi(p)] \tag{1.4}$$

where $p\in\pi^{-1}(x)$. Conversely, to a given field φ: $Q\rightarrow E$ there corresponds a mapping $\tilde{\varphi}$: $P\rightarrow F$ by

$$\tilde{\varphi}(p)=i_p^{-1}(\varphi(x)), \quad x=\pi(p). \tag{1.5}$$

Due to (1.2) $\tilde{\varphi}$ obeys (1.3):

$$\tilde{\varphi}(pa)=i_{pa}^{-1}(\varphi(x))=i_{pa}^{-1}(i_p(\rho(a)\rho(a^{-1})\tilde{\varphi}(p)))=\rho(a^{-1})\tilde{\varphi}(p).$$

The mappings $\tilde{\varphi}$ with the property (1.3) are called K-equivariant or K-covariant.

2. STRUCTURE OF S AND OF ITS LINEAR FRAME BUNDLE

2.A The Fibre Bundle S

2.1 Let the compact Lie group G act smoothly (on the left) on the smooth manifold S with a single orbit type G/H, where H is a closed subgroup of G. We assume that G acts effectively on G/H. One can prove (see [2]), Chap.2 and Chap.6) that the set S_o of H-invariant points from S,

$$S_o=\{s_o\in S: hs_o=s_o, \forall h\in H\}$$

is a smooth submanifold of S. The normalizer N(H) of H in G, $\quad N(H)=\{n\in G: nhn^{-1}\in H, \forall h\in H\}$

acts transitively on the restriction to S_o of every G-orbit.

Denote by $\pi_o=\pi|S_o$ the projection $S_o\rightarrow M$. One verifies that S_o is the total space of a principal fibre bundle with base M, structure group K=N(H)/H and projection π_o (in comparison with Subsect.1.1 in this case the structure group acts on the left on S_o: $s_o\mapsto k.s_o=ns_o$ if $k=nH\in K$, $n\in N(H)$).

One can identify S with the quotient space $(G/H\times S_o)/K$:

$$S=(G/H\times S_o)/K \tag{2.1}$$

where the action of $k=nH$ ($n\in N(H)$) on $G/H\times S_o$ reads $(gH,s_o)\mapsto(gn^{-1}H,ks_o)$. Clearly, the equivalence class $[gH,s_o]\in(G/H\times S_o)/K$ equals $s=gs_o\in S$. Another way to look at eq.(2.1) is to say that S is a fibre bundle associated to

$S_o(M,K)$ by the action $gH \mapsto gnH$ for every $k=nH \in K$ on the typical fibre G/H.

In the principal fibre bundle $S_o(M,K)$ we fix a connection which induces a connection Γ in the associated bundle $S(M,G/H)$ (see e.g.[10], Chap.2,§7): for every $s \in S$ $T_s(S)$ can be split uniquely into a vertical part V_s (tangent to $\pi^{-1}(\pi(s))$) and a horizontal part H_s:

$$T_s(S)=V_s \oplus H_s \,. \qquad (2.2)$$

It is not difficult to derive that: $H_s=gH_{s_o}$ if $s=gs_o$. Note that every $h \in H$ acts trivially on the horizontal spaces.

2.2 A linear frame r_s at the point s is a basis $(\underline{e}_1,\dots,\underline{e}_p)$ in $T_s(S)$, where $p=\dim S$. There is a right free action of the group $GL(p)$ on the set of linear frames at s: if $a \in GL(p)$ is the matrix with elements $(a^i_{j'})$, then the transform of r_s by a is the frame $r_s a \equiv R_a r_s$ at s consisting of the vectors $\underline{e}_{j'}=\underline{e}_i a^i_{j'}$, $j'=1,\dots,p$. The space $\mathcal{F}(S)$ of all linear frames on S is a principal fibre bundle with base S, structure group $GL(p)$ and projection Π: $\mathcal{F}(S)\rightarrow S$ which maps every linear frame r_s at s into s.

The tangent space $T(S)$ of S is the total space of the fibre bundle $(\mathcal{F}(S)\times\mathbb{R}^p)/GL(p)$ with typical fibre $\mathbb{R}^p$ and base S associated to $\mathcal{F}(S)$ by the natural action of $GL(p)$ on $\mathbb{R}^p$: $v \mapsto av$, $a \in GL(p)$. Indeed, every vector $\underline{t}$ tangent to S at the point s can be represented by the equivalence class (with respect to the action $(r_s,t)\mapsto (r_s a, a^{-1}t)$ of $GL(p)$ on $\mathcal{F}(S)\times\mathbb{R}^p$) $\underline{t}=[r_s,t]$ where $t=(\xi^1,\dots,\xi^p)\in\mathbb{R}^p$ are the coordinates of $\underline{t}$ in the frame r_s.

In general, let V be a space equipped with the action ρ of $GL(p)$: $v\mapsto\rho(a)v$, $\forall a\in GL(p)$. We form the fibre bundle (see(1.1)):

$$E=(\mathcal{F}(S)\times V)/GL(p) \qquad (2.3)$$

associated to $\mathcal{F}(S)$ by the action ρ of the linear group on V; denote by π_E the projection $E\to S$. A <u>tensor field of type</u> (ρ,V) is a cross-section φ: $S\to E$ of the fibre bundle (2.3).

2.3 The action of G on S induces a natural action on the tangent space of S: every vector $\underline{t}\in T_s(S)$ is transformed into a vector $g\underline{t}\in T_{gs}(S)$ for $g\in G$. The G-action on the tangent vectors further induces a natural action on the linear frames: any linear frame r_s at s given by $(\underline{e}_1,\dots,\underline{e}_p)$ is transformed by $g\in G$ into a linear frame $r_{gs}\equiv L_g r_s$ at gs consisting of the vectors $(g\underline{e}_1,\dots,g\underline{e}_p)$. Notice that this (left) action of G on $\mathcal{F}(S)$ commutes with the (right) action of GL(p). So we can introduce a left action of the group $\tilde{G}=G\times GL(p)$ on $\mathcal{F}(S)$ by the formula

$$r_s\mapsto\tilde{g}r_s=L_gR_{a^{-1}}r_s \quad \text{for } \tilde{g}=(g,a)\in\tilde{G}. \tag{2.4}$$

Now we define a left G-action on the fibre bundle (2.3) in the following way: taking each element from the fibre E_s over $s\in S$ as the equivalence class $[r_s,v]$ where $(r_s,v)\in\mathcal{F}(S)\times V$, we write $(\forall g\in G)$

$$g:\ [r_s,v]\in E_s\mapsto g[r_s,v]=[L_gr_s,v]\in E_{gs} \tag{2.5}$$

(the same notation is used for the action of G on S and for the induced G-action on E). By means of the action (2.5) every field φ: $S\to E$ is mapped into the field $\tau_g\varphi$: $S\to E$ where $(\tau_g\varphi)(s)=g\varphi(g^{-1}s)$.

We say that a tensor field φ of type (ρ,V) is G-<u>invariant</u> if $\tau_g\varphi=\varphi$, i.e.

$$\varphi(gs)=g\varphi(s). \tag{2.6}$$

In terms of the GL(p)-equivariant map $\tilde{\varphi}$: $\mathcal{F}(S)\to V$ with

$$\tilde{\varphi}(r_sa)=\rho(a^{-1})\tilde{\varphi}(r_s), \tag{2.7}$$

corresponding to the field φ (see Subsect.1.2), the property (2.6) translates into

$$\tilde{\varphi}(L_gr_s)=\tilde{\varphi}(r_s),\quad \forall g\in G, \tag{2.8}$$

i.e. $\tilde{\varphi}$ is constant on the G-orbits in $\mathcal{F}(S)$. The last two equations can be combined into

$$\tilde{\varphi}(\tilde{g}r_s)=\rho(a^{-1})\tilde{\varphi}(r_s),\ \forall\tilde{g}=(g,a)\in\tilde{G}.$$

2.B $\tilde{G}$-Orbits In $\mathcal{F}(S)$

In order to exhibit the structure of the $\tilde{G}$-orbits in $\mathcal{F}(S)$ we shall start considering the linear frames at the points from S_o.

In the following we denote $m=\dim G/H$, so $p=m+4$. We also suppose that a connection Γ is fixed in the fibre bundle $S(M,G/H)$ with gauge group K (see Subsect.2.1).

2.4 At any point $s_o\in S_o$ the vectors tangent to the G-orbit passing through s_o are the values at s_o of the fundamental vector fields generated by the G-action on S. The m-dimensional linear space spanned by these vectors is V_{s_o} (see eq.(2.2)). We shall choose m fundamental vector fields on S_o in the following way.

The Lie algebra $\mathfrak{g}$ of G admits a direct sum decomposition (as a vector space)

$$\mathfrak{g}=\mathfrak{h}\oplus\mathfrak{m}$$

where $\mathfrak{h}$ is the Lie algebra of H and the subspace $\mathfrak{m}$ is Ad(H)-invariant: $Ad(h)\mathfrak{m}\subseteq\mathfrak{m},\ \forall h\in H$. It can be verified (see e.g.[1]) that $\mathfrak{m}$ is also Ad(N(H))-invariant and admits, in turn, a direct sum decomposition

$$\mathfrak{m}=\mathcal{K}\oplus\mathcal{L}\ . \tag{2.9}$$

Here $\mathcal{K}$ is a subalgebra of $\mathfrak{g}$ isomorphic to the Lie algebra of K and $hlh^{-1}=l$ for each $l\in\mathcal{K}, h\in H$; $\mathcal{L}$ is an Ad(N(H))-invariant subspace. So one obtains [1]:

$$\mathfrak{g}=\underbrace{\mathfrak{h}\oplus\overbrace{\mathcal{K}}^{}}_{\mathfrak{n}}\overbrace{\oplus\mathcal{L}}^{\mathfrak{m}}\ .$$

where $\mathfrak{n}$ denotes the Lie algebra of N(H). In $\mathfrak{m}$ we choose a basis $\{e_\alpha\}$ ($\alpha=1,\dots,m$) adapted to the decomposition (2.9). This allows one to identify $\mathfrak{m}$ with $\mathbb{R}^m$, $\mathcal{K}$ with $\mathbb{R}^q$ ($q=\dim\mathcal{K}$), and $\mathcal{L}$ with $\mathbb{R}^l$ ($l=\dim\mathcal{L}$), respectively, where $\mathbb{R}^m=\mathbb{R}^q\oplus\mathbb{R}^l$. Consequently, the corresponding invertible maps $\mathcal{M}\to\mathcal{M}$, $\mathcal{K}\to\mathcal{K}$, and $\mathcal{L}\to\mathcal{L}$, are identified with the groups GL(m), GL(q) and GL(1), respectively. In particular the representation $n\mapsto Ad(n)|\mathcal{M}$ of N(H) induces a homomorphism

$$\lambda:\ N(H)\to GL(m), \tag{2.10}$$

where $\lambda(n)$ for $n\in N(H)$ is the matrix of the linear map $Ad(n):\mathcal{M}\to\mathcal{M}$ in the basis $\{e_\alpha\}$:

$$ne_\alpha n^{-1}=e_\beta\lambda(n)^\beta_\alpha\ . \tag{2.11}$$

In a similar way, the Ad-action of K on $\mathcal{K}$ induces a homomorphism $\lambda':\ K\to GL(q)$. We note that $\lambda(N)|\mathbb{R}^q=\lambda'(K)$. Denoting $\lambda(N)|\mathbb{R}^1$ by $\lambda^1(N)$, we have

$$\lambda(n)=(\lambda^1(n),\lambda'(k))\in GL(1)xGL(q) \tag{2.12}$$

for every $n\in N(H)$ with $k=nH$.

Now m independent vectors tangent to the G-orbit through a point $s_o\in S_o$ are defined by

$$\underline{e}_\alpha(s_o)=\frac{d}{dt}\exp(te_\alpha)s_o\Big|_{t=0}\qquad(\alpha=1,\ldots,m) \tag{2.13}$$

These vectors form a basis in the space V_{s_o} denoted by $r^V(s_o)$. One finds from (2.11) that the action of every $n\in N(H)$ on a vector from the set (2.13) reads

$$n\underline{e}_\alpha(s_o)=\underline{e}_\beta(ns_o)\lambda(n)^\beta_\alpha\ . \tag{2.14}$$

In order to form a linear frame at each $s_o\in S_o$ we have to select 4 more vectors in $T_{s_o}(S)$. To this end we use the horizontal subspace $H_{s_o}\subset T_{s_o}(S)$ corresponding to the fixed connection Γ in the fibre bundle S(M,G/H) (see (2.2)). Let r(x) be a linear grame at the point $x=\pi(s_o)\in M$ consisting of the vectors $\{\underline{e}_\mu\}$ ($\mu=1,\ldots,4$) and let $\underline{e}^h_\mu(s_o)$ be the horizontal lift of $\underline{e}_\mu$ to the point s_o (with respect to Γ). The set $\underline{e}^h_\mu(s_o)$, $\mu=1,\ldots,4$, forms a basis in H_{s_o} which we denote by hor $r(x)|_{s_o}$ and call the horizontal lift to s_o of the linear frame r(x).

2.5 In the space $\mathcal{F}(S)|S_o$ of linear frames at the points from S_o we define a subspace $\mathcal{F}_o$ by

$$\mathcal{F}_o=\{(r^V(s_o),\ \mathrm{hor}\ r(x)|_{s_o}):\ s_o\in S_o,\ r(x)\in\mathcal{F}(M)\}.$$

Here $\mathcal{F}(M)$ is the bundle of linear frames over M (with structure group GL(4)). Denoting the elements from $\mathcal{F}_o$ by $r(s_o)$, we have

$$r(s_o)=(\underline{e}_1(s_o),\ldots,\underline{e}_m(s_o),\underline{e}_1^h(s_o),\ldots,\underline{e}_4^h(s_o)).\quad(2.15)$$

In the sequel we consider the groups GL(m) and GL(4) canonically imbedded in GL(p).

It follows from (2.4) and (2.14) that the subgroup $\widetilde{H}$ of $\widetilde{G}$ defined by

$$\widetilde{H}=\{(h,\ \lambda(h)a)\in\widetilde{G}:\ h\in H,\ a\in GL(d)\}$$

is the isotropy group of every element from $\mathcal{F}_o$. Denote by $\widetilde{N}$ the following subgroup of $\widetilde{G}$

$$\widetilde{N}=\{(n,\ \lambda(n)a)\in\widetilde{G}:\ n\in N(H),\ a\in GL(4)\}$$

contained in the normalizer of $\widetilde{H}$ in $\widetilde{G}$. According to (2.14) the action of each $\widetilde{n}=(n,\ \lambda(n)a)\in\widetilde{N}$ on the frame

$$r(s_o)=(r^V(s_o),\ \mathrm{hor}\ r(x)|_{s_o})\in\mathcal{F}_o$$

is

$$\widetilde{n}r(s_o)=L_nR_{\lambda(n^{-1})a^{-1}}r(s_o)=(r^V(ks_o),\ \mathrm{hor}(r(x)a)|_{ks_o}),\quad(2.16)$$

where $k=nH\in K$. It can be seen from here that the quotient group $\widetilde{K}=\widetilde{N}/\widetilde{H}$ acts freely on $\mathcal{F}_o$. Note that the groups $\widetilde{K}$ and KxGL(4) are isomorphic, the isomorphism being given by the formula

$$\widetilde{k}=(n,\ \lambda(n)a)\widetilde{H}\in\widetilde{K}\leftrightarrow(nH,a)\in K\mathrm{x}GL(4).\quad(2.17)$$

Moreover, we have [7)]

<u>Proposition 2.1</u> The space $\mathcal{F}_o\subset\mathcal{F}(S)|S_o$ has a structure of a principal fibre bundle with base M, structure group $\widetilde{K}$ and projection $\Pi_o:\ \mathcal{F}_o\to M$ given by $\Pi_o(r(s_o))=\pi(s_o)$, $\forall\, s_o\in S_o$.

The group $\widetilde{G}$ acts transitively on the linear frames lying on a G-orbit in S: for every $r_s\in\mathcal{F}(S)$, using (2.4) we obtain

$$r_s=L_g R_{a^{-1}} r(s_o), \quad s=gs_o, \quad r(s_o)\in\mathcal{F}_o. \tag{2.18}$$

An argument analogous to the derivation of (2.1) leads [7]:

Proposition 2.2 The manifold $\mathcal{F}(S)$ has a structure of a fibre bundle over M associated to the principal bundle $\mathcal{F}_o(M,\tilde{K})$ with typical fibre $\tilde{G}/\tilde{H}$:

$$\mathcal{F}(S)=(\tilde{G}/\tilde{H}\times\mathcal{F}_o)/\tilde{K}. \tag{2.19}$$

Here the action of $\tilde{K}$ on $\tilde{G}/\tilde{H}$ is $(\tilde{g}\tilde{H})\tilde{k}=\tilde{g}\tilde{n}\tilde{H}$ for $\tilde{k}=\tilde{n}\tilde{H}\in\tilde{K}$.

3. DESCRIPTION OF INVARIANT TENSOR FIELDS AND LINEAR CONNECTIONS ON S

3.A Invariant Fields

3.1 From proposition 2.2 we see that every G-invariant tensor field φ: $S\to E$ of type (ρ,V) is completely determined by the values of $\tilde{\varphi}$ for the elements from $\mathcal{F}_o$ since by (2.18)

$$\tilde{\varphi}(r_s)=\rho(a)\tilde{\varphi}(r(s_o)). \tag{3.1}$$

From the invariance of each $r(s_o)$ under the action of $(h,\lambda(h))\in\tilde{H}$:

$$r(s_o)=L_h R_{\lambda(h^{-1})} r(s_o)$$

we derive

$$(r(s_o))=\rho(\lambda(h))\tilde{\varphi}(r(s_o)), \quad \forall h\in H.$$

Therefore the restriction $\tilde{\varphi}|\mathcal{F}_o$ of $\tilde{\varphi}$ to $\mathcal{F}_o$ takes values in the subspace V_o of V containing the $\rho(\lambda(H))$-invariant points:

$$V_o=\{v_o\in V: \rho(\lambda(h))v_o=v_o, \forall h\in H\}. \tag{3.2}$$

V_o carries a natural action $\tilde{\rho}$ of the group $\tilde{K}=\tilde{N}/\tilde{H}$:

$$v_o\mapsto\tilde{\rho}(\tilde{k})v_o=\rho(\lambda(n)a)v_o \tag{3.3}$$

for $\tilde{k}=(n,\lambda(n)a)\tilde{H}\in\tilde{K}$. Thus it follows from (2.16) that

$$\tilde{\varphi}(\tilde{k}r(s_o))=\rho(\lambda(n)a)\tilde{\varphi}(r(s_o))=\tilde{\rho}(\tilde{k})\tilde{\varphi}(r(s_o)), \tag{3.4}$$

i.e. $\tilde{\varphi}|\mathcal{F}_o$ is a $\tilde{K}$-equivariant map $\mathcal{F}_o\to V_o$. Therefore, using the 1:1 correspondence between covariant mappings and cross-sections (see Subsect.1.2), we see that $\tilde{\varphi}|\mathcal{F}_o$ determines a field φ_o of type $(\tilde{\rho},V_o;\mathcal{F}_o)$ on M:

$$\varphi_o: M\to E_o=(\mathcal{F}_o\times V_o)/\tilde{K}.$$

Conversely, given the field φ_o one determines $\tilde{\varphi}|\mathcal{F}_0$ and, by (3.1) the extension $\tilde{\varphi}$ over $\mathcal{F}(S)$. So we obtain [7]

Proposition 3.1 Given a connection in the fibre bundle S(M,G/H), the G-invariant fields on S,

$$\varphi: S \rightarrow E=(\mathcal{F}(S)xV)/GL(p), \quad \varphi(s)=[r_s,\tilde{\varphi}(r_s)],$$

are in one-to-one correspondence with the fields

$$\varphi_o: M \rightarrow E_o=(\mathcal{F}_o xV_o)/\tilde{K}$$

on M defined by

$$\varphi_o(\pi(s_o))=[r(s_o),\tilde{\varphi}(r(s_o))]',$$

where $[,]'$ denotes the equivalence class with respect to the action of $\tilde{K}$ in $\mathcal{F}_o xV_o$.

We say that the G-invariant field φ admits dimensional reduction to the field φ_o; the latter is called dimensionally reduced field, corresponding to the invariant field φ.

3.B Invariant Connections

3.2 Finally in this lecture we shall exhibit the results obtained in [7] concerning the dimensional reduction of a G-invariant linear connection on S.

A connection in the principal fibre bundle $\mathcal{F}(S)(S,GL(p))$ with connection form ω (by definition with values in $\mathfrak{gl}(p)$) is G-invariant (see e.g.[10], Chap.2) if

$$\omega_{L_g r}(L_g\tilde{X})=\omega_r(\tilde{X}) \qquad (3.5)$$

for every $r\in\mathcal{F}(S)$ and $\tilde{X}\in T_r(\mathcal{F}(S))$. From the invariance of ω and its GL(p)-equivariance,

$$\omega_{R_a r}(R_a\tilde{X})=Ad(a^{-1})\,\omega_r(\tilde{X}), \quad \forall a\in GL(p), \qquad (3.6)$$

we are lead to consider the restriction $\omega|\mathcal{F}_0$.

It is convenient to introduce the direct sum decomposition of R^p of the form

$$R^p \rightarrow \begin{pmatrix} R^1 \\ R^q \\ R^4 \end{pmatrix} \qquad (3.7)$$

Correspondingly, we have the following block-matrix forms for the matrices from $\lambda(H)$ and $\lambda(N)xGL(4)$:

$$\lambda(h)\in\lambda(H) \rightarrow \left(\begin{array}{c|c|c} \lambda(h) & & \\ \hline & id_{GL(q)} & \\ \hline & & id_{GL(4)} \end{array}\right) \qquad (3.8)$$

$$\lambda(n)a\in\lambda(N)xGL(4) \rightarrow \left(\begin{array}{c|c|c} \lambda^1(n) & & \\ \hline & \lambda'(k) & \\ \hline & & a \end{array}\right) \qquad (3.9)$$

Here $k=nH\in K$ and we have used the notations from eq.(2.12). We also use the corresponding direct sum decomposition of the Lie algebra $gl(p)$:

$$gl(p) = \left(\begin{array}{c|c|c} gl(1) & A'' & B'' \\ \hline A' & gl(q) & C'' \\ \hline B' & C' & gl(4) \end{array}\right) \qquad (3.10)$$

Defining the commutant $\mathfrak{z}_H$ of $\lambda(H)$ in $gl(p)$:

$$\mathfrak{z}_H=\{f \in gl(p):\ Ad(\lambda(h))f=f,\ \ \forall h\in H\}$$

one obtains

$$\mathfrak{z}_H= \left(\begin{array}{c|c} C_H & \\ \hline & gl(q+4) \end{array}\right) = \left(\begin{array}{c|cc} C_H & & \\ \hline & gl(q) & C'' \\ \cline{2-3} & C' & gl(4) \end{array}\right) \qquad (3.11)$$

where C_H is the commutant of $\lambda(H)$ in $gl(1)$. We also introduce an $Ad(\lambda(N))$-invariant complement J of C_H in $gl(1)$:

$$gl(1)=C_H \oplus J \qquad (3.12)$$

with respect to some $Ad(\lambda(N))$-invariant metric in $gl(1)$. Then for the $Ad(\lambda(N))$-invariant complement I of $\mathfrak{z}_H$ in $gl(p)$ we have

$$I = J \oplus A' \oplus A'' \oplus B' \oplus B''. \qquad (3.13)$$

Let $L_o(\mathbb{R}^p; \mathfrak{gl}(p))$ be the subspace of $L(\mathbb{R}^p; \mathfrak{gl}(p))$ consisting of the linear maps l which satisfy

$$\mathrm{Ad}\,(\lambda(h))\circ l\circ\lambda(h^{-1})=l, \quad \forall h\in H.$$

There is an action of the group $\widetilde{K}(\approx K\times GL(4))$ on this subspace given by $\widetilde{k}: l\mapsto \mathrm{Ad}(\lambda(n)a)\circ l\circ\lambda(n^{-1})$, $\forall\ \widetilde{k}=(n,\lambda(n)a)\widetilde{H}\in\widetilde{K}$.

It follows from (3.5) and (3.6) that $\omega|\mathcal{F}_o$ reduces to a set of $\widetilde{K}$ (resp. K or GL(4))-equivariant maps from $\mathcal{F}_o$ (resp. $S_o=\mathcal{F}_o/GL(4)$ or $\mathcal{F}(M)=\mathcal{F}_o/K$) into $\widetilde{K}$-invariant subspaces of $L_o(\mathbb{R}^p;\mathfrak{gl}(p))$, which give rise to a linear connection on M and a set of cross-sections of the following fibre bundles:

$$E_1=(S_o\times L_o(\mathbb{R}^1;\ J\oplus A'\oplus A''))/K,$$
$$E_2=(\mathcal{F}_o\times L_o(\mathbb{R}^1;\ B'\oplus B''))/\widetilde{K},$$
$$E_3=(S_o\times L_o(\mathbb{R}^q;\ C_H\oplus\mathfrak{gl}(q)))/K,$$
$$E_4=(\mathcal{F}(M)\times L_o(\mathbb{R}^q;\mathfrak{gl}(4)))/GL(4),$$
$$E_5=(\mathcal{F}_o\times L_o(\mathbb{R}^q;\ C'\oplus C'')))/\widetilde{K},$$
$$E_6=(S_o\times\widetilde{C})/\widetilde{K},$$

where

$$\widetilde{C}=C_H\oplus\mathfrak{gl}(q)\oplus C'\oplus C''$$

(see (3.8) - (3.13)).

Thus one can show [7]

<u>Proposition 3.2</u> Given a connection in the fibre bundle S(M,G/H) the G-invariant linear connections on S are in one-to-one correspondence with the set $\{\omega^M,\nu_i\,(i=1,\dots,6)\}$ where ω^M is a linear connection on M, ν_i (i=1,...,5) are fields on M with values in the vector bundles $E_1,\dots,E_5$, respectively, and ν_6 is a vector field on M with values in the fibre bundle E_6 (in other terms, ν_6 is a cross-section of the tensor product bundle $T(M)^*\otimes E_6$).

REFERENCES

1). Coquereaux, R. and Jadczyk, A.Z., Commun. Math.Phys. 90, 79 (1983).
2). Bredon, G., Introduction to Compact Transformation Groups, Academic Press, New-York-London, 1972.
3). Jadczyk, A.Z. and Pilch, K., Lett. Math. Phys. 8, 97 (1984).
4). Coquereaux, R. and Jadczyk, A.Z., Symmetries of Einstein-Yang-Mills fields and dimensional reduction, Marseille preprint CPT-84/P1611, 1984.
5). Hudson, L.B. and Kantowski, R., Higgs fields from symmetric connections - The bundle picture, Univ. of Oklahoma preprint, 1983.
6). Mansouri, F. and Witten, L., J. Math. Phys. 25, 1991 (1984); Found. Phys. 14, 1095 (1984).
7). Nikolova, L. and Rizov, V.A., Dimensional reduction of invariant linear connections and tensor fields on multidimensional spacetime, submitted for publication to J. Math. Phys.
8). Trautman, A., Rep. Math. Phys. (Toruń) 1, 29 (1970).
9). Daniel, M. and Viallet, C.M., Rev. Mod. Phys. 52, 175 (1980).
10). Kobayashi, S. and Nomizu, K., Foundations of Differential Geometry, vol.I and vol.II, Interscience Publishers, New-York-London, 1963, 1969.

FIBER BUNDLES AND KALUZA-KLEIN THEORY

A. Jadczyk

Institute of Theoretical Physics, University of Wroclaw
ul.Cybulskiego 36, 50-205 Wrocław
POLAND

ABSTRACT

A new geometrical framework for dimensional reduction is presented. For a G/H Kaluza-Klein theory it predicts a reasonably big effective gauge group G_{eff} which (locally) is a product of N(H)|H and G|C, where N(H) is the normalizer of H and C is the center of G.

1. INTRODUCTION

The works of Kerner[1], Trautman[2], Cho[3], Cho and Freund[4], laid foundations for a geometrical understanding of a unified description of gravity and (non-Abelian) gauge theory. The idea developed in these papers was to generalize the 5-dimensional Kaluza-Klein theory, and to consider the principal bundle on which gauge fields live not as an auxiliary geometrical object but as a true arena where "real" physics takes place. Thus four-dimensional space-time was replaced by an "extended universe" which locally looked like a product M×G of space-time M and an internal space isomorphic to the group manifold of a compact Lie group G (however, non-compact groups may be also relevant). With an appropriate "Ansatz" or by imposing certain geometrical constraints one could show that gravity in 4+n dimensions (n=dimG) splits into gravity, Yang-Mills field and, possibly, some scalar fields in 4 dimensions. The effective gauge group of the resulting theory in four dimensions is again G (although Higgs mechanisms involving scalar fields could reduce it to a smaller one). That simple method of unification had to come to end with observation that there must be a reasonable compromise between the two opposing tendencies:n should be big enough to accomodate for at least SU(3)×SU(2)×U(1) gauge fields, but n should be small enough not to produce particles of spin higher than 2 in the effective particle spectrum of the dimensionally reduced theory. Witten[5] analysed the situation in supergravity and found that a viable compromise is possible by assuming that the internal space is a homogeneous space G/H rather than the group G itself[*]. From that time on quite a number of models of this type has been studied. What was lackig was a clear geometrical understanding of the "Ansatz", understanding comparable to that achieved in the principal bundle case. Percacci and Randjbar-Daemi[7] attempted to close this gap but their proposal did not seem natural. In [8] Coquereaux and the present Author proposed a simple and natural model of dimensional reduction of gravity. The machinery developed in that paper was later used[9,10] to improve the results of Forgacs and Manton[11] (see also [12,13]) on symmetric Yang-Mills fields. The only bad thing with [8] was that the predicted effective gauge group was not G as expected. The "coefficient of effectiveness" as measured by the ratio (dim.of the effective gauge group)/(dim.of the internal space) was in fact smaller than one. In this lecture a possible solution to this dilemma is proposed. I will suggest a new model, more general than that of ref.[8], which allows for the effective gauge group big enough to satisfy the needs: for a G/H internal space the effective gauge group is (locally) a product (N(H)|H)×(G|C).

2. AN ILLUSTRATIVE EXAMPLE: THE KLEIN BOTTLE AND THE TORUS

Consider the following two simple models of the extended universe: The torus and the Klein-bottle (see Fig.1). Both, the Klein-bottle and the torus, are S^1 fibrations over S^1. Internal spaces are circles and space-time (base of the fibration) is a circle in these simplest of the nontrivial examples. Both carry a flat Riemannian metric (ground-state metric) inherited from the piece of $\mathbb{R}^2$ from which they are glued (the gluings of the end circles in Fig.1. are supposed to respect the depic-

---

*) Already in 1959 Souriau[6] proposed to use S^2 as an internal space for SU(2) gauge field.

ted senses of rotation; that causes well know difficulties when attempted in $\mathbb{R}^3$). The local internal symmetry group ("internal" means it does not affect the base points) of the ground state metric is in both

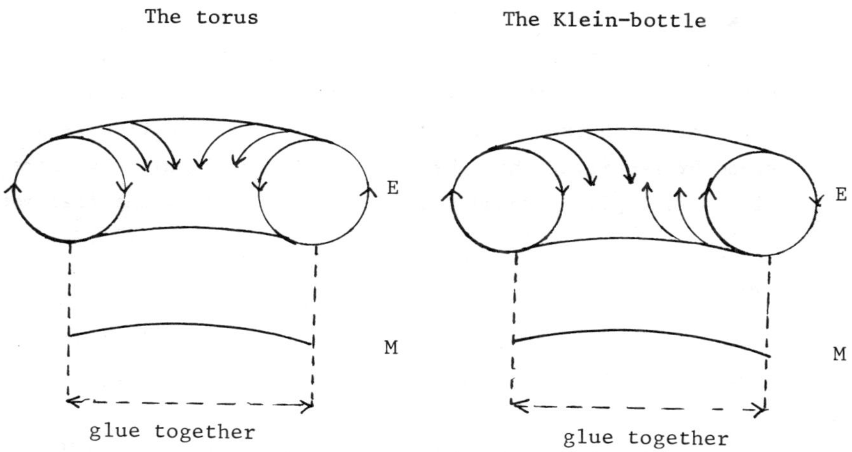

Fig.1.

Fig.1.

cases the same: it is U(1). However, and this is the difference which is crucial for us, *the U(1) acts globally on the torus but can not be made to act globally on the Klein-bottle* (the best proof of this fact is to try and see that it can not be done - see the arrows in the figure). Being locally indistinguishable from the torus, the Klein-bottle seems to be a legitimate candidate for a model of the extended universe. But it does not satisfy the most important assumption of ref.[8]: the requirement of a global group action.

Before we shall formulate assumptions of a scheme more general than that of ref.[8] let us first list some of the properties that both Klein-bottle and the torus have in common. Let E denote one or another and let M be the base circle; x and x' are points of M, and E_x denotes the fiber of E over x.

- E is a fibration over M

- each fiber E_x has a transitive group of isometries G_x
- the groups G_x and $G_{x'}$ acting on different fibers are isomorphic although a natural isomorphism need not to exist; in the Klein-bottle case there is no way to tell whether two rotations, one at x and another at x′ , have the same sense.
- it is however possible to set up a local convention and to identify the G_x-s with U(1) over an open neighborhood of every point of M; an example of such a local identification is provided by requiring that the U(1) should act locally by isometries of the given ground-state metric of E (e.g. the flat metric).

3. A NEW MODEL

Let us now make use of the lesson of the Klein-bottle case and let us try to formulate a general scheme on the basis of the properties listed above.

We assume E to be a fibration $\pi_E : E \to M$, and for each $x \in M$ let G_x be a (compact Lie) group acting transitively on E_x - the fiber of E over x. In application to Kaluza-Klein theory G_x will be the isometry group of a certain metric on E_x. All the groups G_x, $x \in M$, are assumed to be isomorphic to a standard one, say G, although no particular isomorphism is being distinguished at the beginning. In the Klein-bottle case, for example, there is a Z_2-ambiguity in identifying a rotation of a fibre with an element of U(1), and this ambiguity cannot be consistently removed by a smooth global convention. To describe the situation more precisely we introduce the *bundle of groups* $\mathbb{G} = \bigcup_x G_x$, and $\mathbb{G}$ is assumed to be locally trivial (see below). The actions of G_x on E_x, when x runs over M, give rise to a bundle map

$$E \times \mathbb{G} \to E$$

$$(y,a) \to ya, \qquad a \in E_x, \quad a \in G_x.$$

Locally E is assumed to look like $E \times (H \backslash G)$. The fact that locally we can describe the situation as acting with G on $H \backslash G$ is expressed by the following fundamental postulate:

Local Triviality: There is an open covering (U_α) of M and, for each α, there are

$$\phi_\alpha : \pi_E^{-1}(U_\alpha) \to U_\alpha \times (H\backslash G)$$

$$\psi_\alpha : \pi_{\mathbb{G}}^{-1}(U_\alpha) \to U_\alpha \times G$$

(ϕ_α and ψ_α are assumed to be diffeomorphisms and are called local trivializations of E and $\mathbb{G}$ respectively) such that ϕ_α restricts to group isomorphisms $G_x \to G$ of the fibers, and

$$\psi_\alpha(ya) = \psi_\alpha(y)\phi_\alpha(a) \qquad (3.1)$$

for all $y \in E_x$, $a \in G_x$, $x \in U_\alpha$. The product on the right hand side is understood as $(x,[a])(x,b)=(x,[a]b)$, where $[a]=Ha \in (H\backslash G)$ and $b \in G$.

On the intersection of two trivializing neighborhoods U_α and U_β we get transition maps $\psi_{\alpha\beta}$: $(H\backslash G) \to (H\backslash G)$ and $\phi_{\alpha\beta} : G \to G$. By the very definition $\phi_{\alpha\beta}$ is an automorphism of G and $\psi_{\alpha\beta}$ satisfies a condition analogous to (3.1). This motivates the following definition:

Definition A pair (ϕ,ψ) of maps $\phi : G \to G$ and $\psi : (H\backslash G) \to (H\backslash G)$ is called a *twisted automorphism* of $H\backslash G$ if

i) ϕ is an automorphism of G
ii) $\psi([a]b) = \psi([a])\phi(b)$.

The set of all twisted automorphisms of $(H\backslash G)$ is a group under composition of maps. This group will be denoted $\mathrm{Taut}(H\backslash G)$.

In the following we shall always assume that $H\backslash G$ is an *effective homogeneous space* i.e. that the action of G on $H\backslash G$ is effective. If this is the case then the map ϕ is completely determined by ψ and the property ii). According to the definition of a "coordinate bundle" by Steenrod[14), $\mathbb{G}$ and E are such bundles with structure groups AutG and

Taut(H\G) respect. In a canonical way one can construct then the associated principal bundles (see ref.[14,§8.1]) denoted (P,M,Aut) and (Q,M,Taut), so that $\mathbb{G}$ and E can be also considered as bundles associated to P and Q respectively: $\mathbb{G} = P\times_{AutG}G$, $E = Q\times_{Taut(H\backslash G)}(H\backslash G)$.

These are the rudiments of the new model. In the next Section we shall elaborate a little bit on some of its properties. But first let us see how does the rigid model of ref.[8] is a particular case of the one presented above. Assume P is a *trivial* principal bundle. It means it has a global cross-section $\sigma : M \to P$. This allows us to define the global (right) action R_σ of G on E by

$$R_\sigma(a)y \doteq y(\sigma(x)\cdot a), \qquad x\in\pi_E(y),$$

where $\sigma(x)\cdot a \in \mathbb{G} \cong P\times_{AutG}G$. (Of course with a change of σ the associated action of G on E will change accordingly).

4. THE EFFECTIVE GAUGE GROUP

We have seen how, starting with natural assumptions concerning the structure of E, a principal bundle Q with structure group Taut(H\G) can be constructed. It is this kind of analysis that was lacking in ref.[7] where structure group G was guessed*). The group Taut(H\G) introduced in Sec.3 is the biggest possible effective gauge group allowed by geometry. In the following we will denote it also as G_{eff}. Let us list the most important properties of the group G_{eff}; these are

1° N(H)|H - the effective gauge group of ref.[8] - is an invariant subgroup of G_{eff}. In fact we have the exact sequence

$$1 \to Aut(H\backslash G) \to Taut(H\backslash G) \to Aut_{(H)}G \to 1$$

where Aut(H\G)=N(H)|H (see ref.[8]), and $Aut_{(H)}G$ consists of those automorphisms ϕ of G for which $\phi(H)$ is conjugated to H (in particular

*) For another guess see e.g. ref.[15], where the Authors suggest the group G×H ! See however the end remark of this Section.

$Aut_{(H)}G$ contains all inner automorphisms). In fact the bundle P of Sec. 3 has structure group $Aut_{(H)}G$, and can be also constructed as the quotient of Q by $N(H)|H$.

2° Let $\bar{G}$ be the semidirect product of AutG and G

$$\bar{G} \doteq \mathrm{Aut}\,G \circledS G$$

and let $\bar{H} \doteq I \times H \subset \bar{G}$. Then we have the following natural isomorphisms

i) $\mathrm{Taut}(H\backslash G) \cong N(\bar{H})|\bar{H}$

ii) $\mathrm{Aut}\ G \times (H\backslash G) \cong \bar{H}\backslash\bar{G}$

iii) $H\backslash G \cong H\backslash\bar{G}/\mathrm{AutG}$ - a double coset

3° Locally we have

$$G_{eff} \cong \mathrm{Taut}(H\backslash G) \overset{loc}{=} (N(H)|H) \times (G|C)$$

where C is the center of G

The following comments are relevant:

Remark 1 There is a simple method which allows us to obtain the bundle E as discussed above with the help of constructions known already from ref.[8]. Nemely, having $\bar{G}$, $\bar{H}$, and Q which(by i) is a principal bundle with structure group $N(\bar{H})|\bar{H}$), we can construct, with the methods of ref.[8], the associated bundle $\bar{E}$ with typical fiber $\bar{H}\backslash\bar{G}$. The group $\bar{G}$ acts globally on $\bar{E}$ (from the right). This is the situation known from ref.[8]. Now, AutG is a subgroup of $\bar{G}$, and we can take the quotient bundle $\bar{E}/\bar{G}$. Then, by iii), this quotient bundle is naturally isomorphic to E:

$$E \cong \bar{E}/\mathrm{Aut}\,G\ .$$

This suggests us the class of interesting metrics on E: those which are images of $\bar{G}$-invariant metrics on $\bar{E}$. It is clear then from the results of ref.[8] that this class of metrics give rise to gauge fields with gauge group $N(\bar{H})|\bar{H} \equiv G_{eff}$ (recall that E is a bundle associated to Q)*). Thus G_{eff} is not only geometrically allowed, but also kinematically available.

Remark 2 If G is semisimple then G is (locally) contained in G_{eff} (see 3°). But even if G has central U(1) factors, if they are not in H (which is anyway excluded by the assumption of effectiveness of $H\backslash G$) then they contribute to G_{eff} via $N(H)|H$. Below are three simple examples of G, H and G_{eff} (all equalities are local)

a) $G = U(1)$, H trivial, the case of 5-dimensional Kaluza-Klein theory. Then Aut(G) is trivial($=Z_2$), $N(H)|H = U(1)$. Thus $G_{eff} = U(1)$

b) $G = SO(8)$, $H = SO(7)$, $H\backslash G = S^7$. $Aut(G) = SO(8)$, $N(H)|H$ is trivial ($=Z_2$). Thus $G_{eff} = SO(8)$ - the isometry group of the round seven-sphere

c) $G = U(2;H)$, $H = U(1;H)$, $H\backslash G = S^7$, $AutG = U(2;H) = SO(5)$, $N(H)|H =$ $= U(1;H) = SU(2)$. Thus $G_{eff} = SO(5)\times SU(2)$ - the isometry group of the squashed seven-sphere (see ref.[17-18]).

These few examples show us that the formula for G_{eff} given by 2° i) and 3° predicts a reasonable effective gauge group even without any particular dynamical model. As we have already remarked, for concrete models the Higgs mechanism can break a part of G_{eff}.

Remark 3 The bundle P will be, in general, non-trivial. Nevertheless one can expect that the ground-state metric of E will give rise to a *flat* principal connection in P. In that case although a *global* cross-section σ of P does not exist, there are *distinguished local* cross-sections: the *horizontal* sections. Transition functions between these sections are then *constant*. Thus different actions of G on E as defined by

---

*) In the particular case of trivial H, E becomes a "weak principal bundle", and the induceed G_{eff}-connection becomes a "weak principal connection", according to the terminology of ref.[16].

these cross-sections (see end of Sec.3) are related by *constant* group automorphisms. This enables one to apply the equivariant techniques of ref.[19] also to the present, more general, case.

Remark 4 Castellani, Romans and Warner [20] analysed Killing vectors on coset spaces and indicated N(H)|H×G (with a correction for common U(1)-factors) as a maximal gauge group for a given realization of a coset. This statement essentially agrees with G_{eff} as derived above by a careful analysis of fiber bundle structure of the Kaluza-Klein theory. To see the relation between the two groups observe that every pair (n,r) with $n \in N(H)|H$ and $r \in G$ determines an element $(\phi,\psi) \in Taut(H\backslash G)$ given by $\phi(a)=rar^{-1}$, $\psi([a])=[nar^{-1}]$, and in the connected neighborhood of the identity every element of Taut(H\G) is of this form.

ACKNOWLEDGMENTS

I am very pleased to thank L.Castellani, R.Coquereaux, M.Dubois-Violette and K.Pilch for discussions and useful comments at the early stage of development of the presented scheme.

5. REFERENCES

[1] Kerner, R. Ann.Inst.H.Poincaré 9,143(1968)

[2] Trautman, A. Rep.Math.Phys.1,29(1970)

[3] Cho, Y.M. J.Math.Phys.16,2029(1975)

[4] Cho, Y.M.,Freund, P.G.O. Phys.Rev.D12,1711(1975)

[5] Witten, E. Nucl.Phys.B186,412(1981)

[6] Souriau, J.M. in: Les Theories relativistes de la gravitation. Colloque de Royaumont 1959, Paris: CNRS 1962.

[7] Percacci, R., Randjbar-Daemi, S. J.Math.Phys.24,807(1983)

[8] Coquereaux, R., Jadczyk, A. Commun.Math.Phys.90,79(1983)

[9] Jadczyk, A., Pilch, K. Lett.Math.Phys.8,97(1984)

[10] Coquereaux, R., Jadczyk, A. Commun.Math.Phys.98,79(1985)

[11] Forgacs, P., Manton, N.S. Commun.Math.Phys.72,15(1980)

[12] Harnad, J., Schnider, S., Vinet, L. J.Math.Phys.21,2719(1980)

[13] Harnad, J., Schnider, S., Tafel, J. Lett.Math.Phys.4,107(1980)

[14] Steenrod, N. "The topology of fibre bundles", Princeton University Press, Princeton, New Jersey, 1951

[15] Volkov, D.V., Sorokin, D.P., Tkach, V.I. Theor.and Math.Phys.56, 171(1983), in Russian.

[16] Kolář, I. in: Global Differential Geometry and Global Analysis, Colloquium Berlin 1979, Lecture Notes in Mathematics, 838, Springer-Verlag, Berlin-Heidelberg-New York 1981

[17] Duff, M.J., Nilsson, B.E.W., Pope, C.N., Phys.Rev.Lett.50,2043 (1983), see also Preprint ICTP /82-83/29

[18] Bais, F.A., Nicolai, H., van Nieuwenhuizen, P.,CERN preprint TH 3577/83

[19] Coquereaux, R., Jadczyk, A. Preprint CERN TH 4023/84

[20] Castellani, L., Romans, L.J., Warner, N.P. Nucl.Phys.B241,429 (1984), see also Caltech preprint CALT-68-1057(1983)

RELATION BETWEEN GROUP CONTRACTION AND NON-LINEAR REALIZATIONS

E. Celeghini[+], M. Tarlini[°] and G. Vitiello[*]

+) Dipartimento di Fisica dell'Università, 50125 Firenze, Italia
I.N.F.N. - Sezione di Firenze

°) I.N.F.N. - Sezione di Firenze, 50125 Firenze, Italia

*) Dipartimento di Fisica dell'Università, 84100 Salerno, Italia
I.N.F.N. - Sezione di Napoli

ABSTRACT

Non linear realizations of the invariance group are recovered by means of contraction of group representations in spontaneously broken symmetry theories. A detailed analysis of Goldstone model, SO(n) vector model and SU(2) doublet model is given. Extension to gauge theories is finally presented.

1. INTRODUCTION

Aim of this paper is to report about some recent work [1] relating group theoretical aspects of spontaneous symmetry breakdown (s.s.b.) with effective Lagrangian theories and, more specifically, with non linear realizations [2-4] of invariance group in quantum field theory. From a general point of view , one reason of interest in spontaneously broken symmetry theories is that the dynamics is supplemented since the begin with informations about the ground state structure of the theory, which, in turn, provide relations among masses, coupling constants and other physically relevant parameters. In the study of spontaneous breakdown of symmetry one is thus naturally led to investigate the relations between the dynamical level (Heisenberg field equations) and the asymptotic fields representation of the theory (LSZ field equations). The informations provided by the spontaneous symmetry breakdown condition are indeed informations about the vacuum state for the asymptotic fields. A question to ask is therefore how the invariance of the dynamical Heisenberg field equations manifests itself at the level of asymptotic field equations. It is clear that knowledge of symmetry properties of asymptotic field equations is most relevant to phenomenology since observables are expressed in term of asymptotic fields. A central object of study is then the dynamical mapping [5-6] among Heisenberg fields, say $\psi(x)$ and asymptotic LSZ fields, say $\phi^{o}(x)$ (here ϕ^{o} means ϕ^{in} or ϕ^{out}):

$$\langle a|\psi(x)|b\rangle = \langle a|F(x;\phi^{o}(x))|b\rangle \tag{1.1}$$

where /a> and /b> are vectors of the Fock space for physical states, and F denotes a functional of normal ordered products of ϕ^{o} fields. Knowledge of the map (1.1) (which is called also the Haag expansion) is all what one needs to solve the dynamics since it provides all the matrix elements of the theory. When Heisenberg fields operators and their relevant functionals, e.g. the Hamiltonian, the generators of some internal symmetry group, the currents, etc., are expressed in terms of ϕ° fields, we say that the physical (or free) field picture is used (the quasi-particle picture in many body terminology).

Suppose a symmetry group G for the Heisenberg field equations and let G° be the symmetry group for the asymptotic field equations, then eq.(1.1) requires

$$\langle a | g\psi(x) | b \rangle = \langle a | F(x; g^{\circ}\phi^{\circ}(x)) | b \rangle \tag{1.2}$$

with $g \in G$ and $g^{\circ} \in G^{\circ}$. In the presence of s.s.b. G° is different from G. We say that a dynamical rearrangement of symmetry occurs and a general rule [7] is that G° is the Inönü-Wigner [8] group contraction of G. This rule has been found true under very general conditions and it is directly related to the content of the Goldstone theorem. A detailed analysis can be found in refs. 7,9-11. An interesting point is that the Abelian transformations belonging to G° give account of the condensation of the Goldstone bosons in the vacuum state. This condensation can be spacially homogeneous or else spacially non homogeneous. In this latter case, macroscopically behaving field configurations (soliton-like extended objects) can be obtained [12,13]. Moreover, low energy theorems are derived as infrared effects due to low momentum Goldstone bosons, the vacuum condensate structure thus being related to the low energy behaviour of the theory [7,9]. Main task of this paper is to show how the occurrence of group contraction naturally leads to non linear realizations of the invariance group G in effective Lagrangian theories. This will be done in Sec.3. Before, however, we present some relations obtained [12,13] in the study of Nambu-

-type model, which we shortly comment in order to clarify the above considerations about dynamical rearrangement of symmetry. This is the content of section 2.

2. HOMOGENEOUS AND NON HOMOGENEOUS CONDENSATION

Let $\phi(x)$ be a spin-$\frac{1}{2}$ field and $A_\mu(x)$ a chiral gauge field. Assume a Lagrangian invariant under chiral phase transformations

$$\begin{aligned} \phi(x) &\to \phi'(x) = e^{i\vartheta\gamma_5}\phi(x) \\ A_\mu(x) &\to A'_\mu(x) = A_\mu(x) \end{aligned} \tag{2.1}$$

as well as chiral gauge transformations

$$\begin{aligned} \phi(x) &\to \phi'(x) = e^{ie_0\lambda(x)\gamma_5}\phi(x) \\ A_\mu(x) &\to A'_\mu(x) = A_\mu(x) + \partial_\mu\lambda(x) \end{aligned} \tag{2.2}$$

where $\lambda(x)$ vanishes in the limit $|X_0| \to \infty$ and /or $|\bar{x}| \to \infty$.
We work in the Lorentz gauge $\partial_\mu A_\mu = 0$. The spontaneous breakdown condition is $(\bar{\phi} = \phi^\dagger\gamma_0)$:

$$\langle 0|\bar{\phi}(x)\phi(x)|0\rangle = \sqrt{2}\,\tilde{v} \neq 0 \tag{2.3}$$

with real $\tilde{V}$. By use of path-integral formalism we find [12] the following dynamical maps

$$\begin{aligned} \phi(x) = :\exp\left[i\frac{Z_\chi^{\frac{1}{2}}}{2\tilde{v}}\chi^0(x)\gamma_5\right] \times \\ \times\left[Z_\phi^{\frac{1}{2}}\phi^0(x) + F_\phi[\phi^0, U^0_\mu, \partial(\chi^0 - b^0)]\right]: \end{aligned} \tag{2.4}$$

$$A_\mu(x) = Z_3^{1/2} U^0_\mu(x) + \frac{Z_\chi^{1/2}}{2e_0\tilde{v}} \partial_\mu b^0(x) + + :F_\mu[\phi^0, U^0_\mu, \partial(\chi^0 - b^0)]: \tag{2.5}$$

where χ^0 is the Goldstone field which is found to be a bound state of fermion-antifermion, ϕ^0 is the fermion asymptotic field, U^0_μ in the massive asymptotic gauge field (a Proca field), b^0 is a massless ghost field which appears in the covariant Lorentz gauge. Z_χ, Z_ϕ and Z_3 are wave function renormalization constants. The asymptotic field equations are

$$(\not\partial + m)\phi^0(x) = 0 \tag{2.6}$$

$$(\Box - M^2) U^0_\mu(x) = 0 \quad , \quad \partial_\mu U^0_\mu = 0 \tag{2.7}$$

$$\Box \chi^0(x) = 0 \tag{2.8}$$

$$\Box b^0(x) = 0 \tag{2.9}$$

with

$$B^0(x) = \chi^0(x) - b^0(x) \tag{2.10}$$

and

$$B^{0(-)}(x)|phys\rangle = 0 \tag{2.11}$$

$B^{0(-)}(x)$ denotes the annihilation operators. Eq. (2.11) is the condition to be satisfied by the physical states of the theory and is similar to the Gupta-Bleuler condition in covariant QED. Eq. (2.11) expresses the fact that the Goldstone particle χ^0 is unobservable in physical states. The chiral transformations (2.1) and (2.2) of the Heisenberg fields $\phi(x)$ and $A_\mu(x)$ are induced by the following transformations of the asymptotic fields:

$$\begin{aligned} \chi^0(x) &\to \chi^0(x) + \frac{\tilde{v}}{Z_\chi^{1/2}}\, \delta f(x) \\ b^0(x) &\to b^0(x) \\ \phi^0(x) \to \phi^0(x)\,, &\quad U_\mu^0(x) \to U_\mu^0(x) \end{aligned} \tag{2.12}$$

which induce the global transformation (2.1), and

$$\begin{aligned} \chi^0(x) &\to \chi^0(x) + \frac{e_0 \tilde{v}}{Z_\chi^{1/2}}\, \lambda(x) \\ b^0(x) &\to b^0(x) + \frac{e_0 \tilde{v}}{Z_\chi^{1/2}}\, \lambda(x) \\ \phi^0(x) \to \phi^0(x)\,, &\quad U_\mu^0(x) \to U_\mu^0(x) \end{aligned} \tag{2.13}$$

which induce the local transformations (2.2). In eqs. (2.12) and (2.13) $f(x)$ and $\lambda(x)$ must be solutions of the χ^0-equation: $\Box f(x)=0$, $\Box\lambda(x)=0$; moreover eq. (2.12) induce the transformations (2.1) in the limit $f(x)\to 1$: $f(x)$ is needed in order to avoid infrared divergencies in the generator of χ^0 transformation (2.12). When (2.12) is implemented and $f(x)\to 1$ the Goldstone particle χ^0 is condensed in the ground state homogeneously. Local condensation is obtained by means of the boson transformation

$$\chi^0(x) \to \chi^0(x) + \frac{\tilde{v}}{Z_\chi^{1/2}}\, f(x) \tag{2.14}$$

$$\Box f(x) = 0\,. \tag{2.15}$$

To preserve the condition (2.11) and the field equation for $A_\mu(x)$ under (2.14), $U^\circ_\mu(x)$ must transform as

$$U^\circ_\mu(x) \rightarrow U^\circ_\mu(x) + u_\mu(x) \tag{2.16}$$

$$(\Box - M^2)\, u_\mu(x) = -\frac{M}{2e_0 Z_3^{1/2}} \partial_\mu f(x)$$

$$\partial_\mu u_\mu(x) = 0 \,. \tag{2.17}$$

The transformations (2.14) and (2.16) induce the vector potential $A^f_\mu(x)$ and the axial current $j^f_{\mu 5}(x)$. When contributions from F (cf.eq. (2.5)) are ignored we have[13].

$$(\Box - M^2)\, a_\mu(x) = -\frac{M^2}{2e_0} \partial_\mu f(x) \tag{2.18}$$

$$\langle 0 | A^f_\mu(x) | 0 \rangle = a_\mu(x) \tag{2.19}$$

$$\langle 0 | j^f_{\mu 5}(x) | 0 \rangle = j_\mu(x) =$$
$$= M^2 \left[a_\mu(x) - \frac{1}{2e_0} \partial_\mu f(x) \right] \tag{2.20}$$

Eqs. (2.18) - (2.20) are the relations for c-number ground state vector potential $a_\mu(x)$ and ground state current $j_\mu(x)$ induced by the local condensation (2.14). When $f(x)$ is a regular function eqs. (2.15) and (2.18) show that

$$a_\mu(x) = \frac{1}{2e_0} \partial_\mu f(x) \tag{2.21}$$

i.e. $j_\mu(x)=0$ and $F_{\mu\nu} = \partial_\mu a_\nu - \partial_\nu a_\mu = 0$, which means that no observable potential and current develope due to the boson condensation. It can be shown [13] indeed that (2.14) is equivalent to a gauge transformation of the kind (2.2) when f(x) is regular and $\frac{1}{2e_0} f(x) = \lambda(x)$. When, on the contrary, f(x) is a singular function $F_{\mu\nu}$ and j_μ do not vanish since eq.(2.21) does not hold. In eq. (2.20) the $a_\mu(x)$ term denotes then the Meissner current and the $\partial_\mu f(x)$ term the boson current. We thus conclude that classical gauge field $a_\mu(x)$ and vacuum current $j_\mu(x)$ exist only for Goldstone particle local condensation regulated by singular f(x). It can be shown that many extended objects, as vortices, dislocations, point defects, wall domains can be generated when f(x) carries topological singularities. Furthermore, those extended object with topological singularities can be created only by condensation of gapless modes; this explains why they exist only in a background which exhibits an ordered pattern, namely where gapless modes exist as Goldstone modes. To see this [6], let f(x) carry a topological singularity, i.e. let f(x) be path dependent such that $[\partial_\mu, \partial_\nu] f(x) \neq 0$, but also $[\partial_\mu, \partial_\nu] \partial_\rho f(x) = 0$, since $\partial_\mu f(x)$ is path independent being related to observable quantities (cf. eq. (2.20)). Assume that f(x) satisfies the equation

$$(\Box - m^2) f(x) = 0 \tag{2.22}$$

then

$$\partial_\nu f(x) = -\frac{1}{\Box - m^2}\partial_\mu[\partial_\mu,\partial_\nu]f(x) \tag{2.23}$$

Since $\partial^2[\partial_\mu,\partial_\nu]f(x) = 0$, (1.24) gives $\partial^2 f(x)=o$ and therefore from (2.22) we find $m=o$.

A detailed analisis of extended objects with topological singularities can be found in ref. 6,12, 13.

Let us finally observe that eq. (2.12) and (2.13) belong to the group of transformations for the asymptotic fields which is the contraction of the chiral transformation group for the Heisenberg fields $\phi(x)$ and $A_\mu(x)$ (see eqs. (2.1) and (2.2)). Examples of rearrangements leading to the contraction of SU(2), SU(3), chiral SU(2)xSU(2), SU(n)xSU(n), as well as general theorems for SO(n) and SU(n) have been studied in several models of physical interest by means of path integral formalism [7,9], projective geometry [4] and group theoretical methods [7,11]. In the following section we use the formalism of the contraction of group representations [15] to investigate the relation between group contraction and non linear realizations.

3. NON LINEAR REALIZATIONS

It may be useful to sketch first of all what is a group contraction and what are its very general properties [16]. We define the contraction (Inönü-Wigner contraction) on the Lie algebra g of the group G. Let X_i a set of basis vectors of the algebra and let Y_i a new set related to the X_i by

$$Y_j = U(\varepsilon)_j{}^i X_i \tag{3.1}$$

where the matrix $U(\varepsilon)$ is singular only for $\varepsilon = 0$.

Then the commutation relations with respect to the new basis are given by

$$[Y_i, Y_j] = C_{ij}{}^{k}(\varepsilon)\, Y_k \tag{3.2}$$

where

$$C_{ij}{}^{k}(\varepsilon) = U(\varepsilon)_i{}^{r}\, U(\varepsilon)_j{}^{s}\, C_{rs}{}^{t}\, U^{-1}(\varepsilon)_t{}^{k}$$

If the limit of $C_{ij}{}^{k}(\varepsilon)$ exists it characterizes a contracted Lie algebra. Of course the number of generators and the dimension of the group remain, in the contraction process, unchanged. In every contraction a subalgebra of g preserves its structure constants, moreover it is always present an abelian invariant subalgebra. The contracted algebra is then necessarily nonsemisimple. Coming back to the general connection between contraction and s.s.b. the stable subalgebra generates the subgroup H of the unbroken symmetry and the abelian part is spanned by the contraction of the broken generators; they plays the role of translations operators for the Goldstone bosons of the theory. The important point we want to stress is that it is possible to extend this limit process to the fields: the contraction of fields in the representation space of G gives the representation space of the asymptotic symmetry G^{in} (contraction of G), i.e. the phenomenological in - (out) - fields symmetry group. This connection between Heisenberg fields by way of the contraction limit is the first order of the standard non linear realization of s.s.b. theories [11]. Following our point of view the non linear realization is a group expansion [16]; this means to express the initial fields and the generators of their linear transformation group G in terms of the contracted fields (in - out - fields) and the contracted generators (algebra of G^{in}).

We will treat in detail two simple examples: the Goldstone type model with the generalization to SO(n) and the complex doublet of SU(2).

We start with the simple symmetry U(1) and with a complex field ϕ defined by $\phi = \frac{1}{\sqrt{2}}(\psi + i\chi)$.

The action of the generator J of U(1) on ϕ is

$$[J, \begin{pmatrix}\psi \\ \chi\end{pmatrix}] = \begin{pmatrix} & i \\ -i & \end{pmatrix}\begin{pmatrix}\psi \\ \chi\end{pmatrix} \tag{3.3a}$$

$$[\psi, \chi] = 0,\ [J, \psi] = i\chi,\ [J, \chi] = -i\psi \tag{3.3b}$$

We can see (3.3b) as the commutation relations of an **E** (2) algebra. The orbits of J in the representation space given by ($\begin{smallmatrix}\psi \\ \chi\end{smallmatrix}$) are labeled by the Casimir of such E (2):

$$\mathbf{C} = \psi^2 + \chi^2 \tag{3.4}$$

The potential of the theory will be a functional of C and if the minimum is characterized by $C \neq 0$ we have spontaneous symmetry breaking. We choose the direction of the breakdown given by ψ , thus we assume

$$\langle 0 | \psi | 0 \rangle = v \tag{3.5}$$

It is clear that the rotational invariance in the plane $\{\psi, \chi\}$ is broken. How the contraction of the E (2) can describe the same situation ? We will see that the symmetry in the contraction is broken automatically. If we want to draw the "limit" curve of a circle

(circle of minima in our case) we have to draw a tangent line, this means to choose the tangent point. Every choice is equivalent; coherently with (3.5) we choose $\left\{\phi = \frac{v}{\sqrt{2}}, 0\right\}$. The Inönü-Wigner contraction describing this choice is given by the following position:

$$\bar{J} = \varepsilon J \,, \quad \bar{\varphi} = \varphi \,, \quad \bar{\chi} = \varepsilon^{-1}\chi \tag{3.6}$$

The algebra for $\bar{J}$, $\bar{\varphi}$ and $\bar{\chi}$ is

$$[\bar{J},\bar{\varphi}] = \varepsilon^2 i\bar{\chi} \,, \quad [\bar{J},\bar{\chi}] = -i\bar{\varphi}, \quad [\bar{\chi},\bar{\varphi}] = 0 \tag{3.7}$$

The Casimir C is rewritten as

$$C = \bar{\varphi}^2 + \varepsilon^2\bar{\chi}^2 \tag{3.8}$$

It is clear from (3.8) that the rotational invariance is deformed by the ε presence.

The asymptotic behavior of this model is given in terms of the LSZ in - (out) Goldstone boson (whose existence is required by the breaking of the generator of the initial U(1) symmetry) and of the asymptotic limit of the field in the broken direction (with v.e.v. different to zero); e.g. see the superconductivity theory. We want to show how the contraction exactly gives the phenomenologic description of the model making explicit the low energy behaviour. In the aim of the contraction process we have to make the limit $\varepsilon \to 0$ keeping finite $\bar{J}$, $\bar{\varphi}, \bar{\chi}$; this means to obtain the following commutation relations

$$[J^{in}, \varphi^{in}] = 0 \,, \quad [J^{in}, \chi^{in}] = -i\varphi^{in}, \quad [\varphi^{in}, \chi^{in}] = 0 \tag{3.9a}$$

$$J^{in} = \lim_{\varepsilon\to 0} \bar{J}$$

$$\varphi^{in} = \lim_{\varepsilon\to 0} \bar{\varphi} \qquad (3.9b)$$

$$\chi^{in} = \lim_{\varepsilon\to 0} \bar{\chi}$$

The Casimir C results in the limit $\varepsilon \to 0$

$$C^{in} = (\varphi^{in})^2 \qquad (3.10)$$

At this point it is clear that the rotational symmetry is broken. The action of J^{in} does not transform φ^{in} (from (3.10) φ^{in} is now seen to be an invariant) and it translates χ^{in} by an ammount given by the invariant φ^{in}.

From (3.5)

$$\langle 0 | \varphi^{in} | 0 \rangle = v \qquad (3.11)$$

Thus taking the classical value of φ^{in} (φ^{in} is an invariant of the asymptotic symmetry, then is a constant on the representations) we have:

$$[J^{in}, \chi^{in}] = -iv \qquad (3.12)$$

This commutation relation suggests the interpretation of χ^{in} as the Goldstone boson of the theory, indeed (3.12) implies no mass terms for χ^{in}; to understand better this point we have to consider not only the abstract structure of the symmetry but also the representation space; i.e. to consider all the equalities between operators in a weak sense; i.e. as relations between matrix elements. The representation states of the initial $E(2)$ given by J, Ψ, χ are labeled by the Casimir eigenvalue and, for example, by the J eigenvalue. On this representation space we have

$$\begin{aligned}
\langle \nu h' | J | \nu h \rangle &= h\,\delta(h'-h) \\
\langle \nu h' | \Psi | \nu h \rangle &= \frac{\nu}{2}[\delta(h'-h-1)+\delta(h'-h+1)] \\
\langle \nu h' | \chi | \nu h \rangle &= i\frac{\nu}{2}[\delta(h'-h-1)-\delta(h'-h+1)] \\
\langle \nu h' | C | \nu h \rangle &= \nu^2\,\delta(h'-h)
\end{aligned} \tag{3.13}$$

where $\nu \in \mathbb{R}$ and h is integer or halfinteger. To recover the previous contraction we can define, this time on the representation parameters [15], a singular ε-dependent transformation given by

$$\begin{pmatrix} \nu \\ p \end{pmatrix} = M(\varepsilon)\begin{pmatrix} \nu \\ h \end{pmatrix} = \begin{pmatrix} 1 & \\ & \varepsilon \end{pmatrix}\begin{pmatrix} \nu \\ h \end{pmatrix} \tag{3.14}$$

This is equivalent to define a succession of representations

$$|\nu p\rangle_\varepsilon = \sum \langle \nu h | \nu p \rangle_\varepsilon \, |\nu h\rangle \tag{3.15}$$

where

$$\langle \nu h | \nu p \rangle_\varepsilon = [\det M(\varepsilon)]^{-1/2}\,\delta(h-\varepsilon^{-1}p) \tag{3.16}$$

On these states $|\nu p\rangle_\varepsilon$ the matrix elements of the operators $\bar{J}$, $\bar{\Psi}$, $\bar{\chi}$ and the Casimir given in (3.6) and (3.8) are:

$$
\begin{aligned}
{}_{\varepsilon}\langle \nu p'|\bar{\varphi}|\nu p\rangle_{\varepsilon} &= \frac{\nu}{2}\left[\delta(p'-p-\varepsilon)+\delta(p'-p+\varepsilon)\right] \\
{}_{\varepsilon}\langle \nu p'|\bar{\chi}|\nu p\rangle_{\varepsilon} &= -\frac{i\nu}{2\varepsilon}\left[\delta(p'-p-\varepsilon)-\delta(p'-p+\varepsilon)\right] \\
{}_{\varepsilon}\langle \nu p'|\bar{J}|\nu p\rangle_{\varepsilon} &= p\,\delta(p'-p) \\
{}_{\varepsilon}\langle \nu p'|C|\nu p\rangle_{\varepsilon} &= \nu^2\delta(p'-p)
\end{aligned}
\tag{3.17}
$$

Making the explicit limit of (3.17) we recover the representation of an asymptotic symmetry:

$$
\begin{aligned}
{}^{in}\langle \nu p'|\varphi^{in}|\nu p\rangle^{in} &= \nu\,\delta(p'-p) \\
{}^{in}\langle \nu p'|\chi^{in}|\nu p\rangle^{in} &= i\nu\,\delta'(p'-p) \\
{}^{in}\langle \nu p'|J^{in}|\nu p\rangle^{in} &= p\,\delta(p'-p) \\
{}^{in}\langle \nu p'|C^{in}|\nu p\rangle^{in} &= \nu^2\delta(p'-p)
\end{aligned}
\tag{3.18}
$$

From (3.18) it appears clear what we said looking at the algebra. The main point studying the representations is that it is possible to find what is the functional expression of J, φ, χ in terms J^{in}, φ^{in}, χ^{in}. We will show that in this way we recover the Coleman-Wess-Zumino non linear realization [2,3] in which the Goldstone mode is just χ^{in}. To achieve this let us define the ε dependence in $\bar{\varphi}$, $\bar{\chi}$ and $\bar{J}$ by the definitions:

$$
\begin{aligned}
{}^{in}\langle \nu p' | \bar{\mathcal{P}}(\varepsilon) | \nu p \rangle^{in} &\equiv {}_{\varepsilon}\langle \nu p' | \bar{\mathcal{P}} | \nu p \rangle_{\varepsilon} \\
{}^{in}\langle \nu p' | \bar{\mathcal{X}}(\varepsilon) | \nu p \rangle^{in} &\equiv {}_{\varepsilon}\langle \nu p' | \bar{\mathcal{X}} | \nu p \rangle_{\varepsilon} \qquad (3.19) \\
{}^{in}\langle \nu p' | \bar{J}(\varepsilon) | \nu p \rangle^{in} &\equiv {}_{\varepsilon}\langle \nu p' | \bar{J} | \nu p \rangle_{\varepsilon}
\end{aligned}
$$

In the r.h.s. of (3.19) (see (3.17))we have terms like $\delta(p'-p\pm\varepsilon)$; using a Taylor expansion we get

$$
\delta(p'-p\pm\varepsilon) = \sum_{n=0}^{\infty} \frac{(\pm\varepsilon)^n}{n!}\, \delta^n(p'-p) \qquad (3.20)
$$

where $\delta^n(p'-p)$ is the nth derivative of the $\delta(p'-p)$ distribution . Using (3.18) we can write:

$$
\delta(p'-p\pm\varepsilon) = \sum_{n=0}^{\infty} \frac{(\pm\varepsilon)^n}{n!}\, {}^{in}\langle \nu p' | \left[-i \frac{\mathcal{X}^{in}}{V} \right]^n | \nu p \rangle^{in} =
$$

$$
= \langle \nu p' | \exp\left(\mp i\varepsilon \frac{\mathcal{X}^{in}}{V} \right) | \nu p \rangle^{in} \qquad (3.21)
$$

Substituting (3.21) in (3.19) we get:

$$
\begin{aligned}
{}^{in}\langle vp'|\,\overline{\psi}(\varepsilon)\,|vp\rangle^{in} &= {}^{in}\langle vp'|\, v\cos\frac{\varepsilon\chi^{in}}{v}\,|vp\rangle^{in} \\
{}^{in}\langle vp'|\,\overline{\chi}(\varepsilon)\,|vp\rangle^{in} &= {}^{in}\langle vp'|\,\frac{v}{\varepsilon}\sin\frac{\varepsilon\chi^{in}}{v}\,|vp\rangle^{in} \\
{}^{in}\langle vp'|\,\overline{J}(\varepsilon)\,|vp\rangle^{in} &= {}^{in}\langle vp'|\, J^{in}\,|vp\rangle^{in}
\end{aligned}
\tag{3.22}
$$

This gives the non linear realization of the U(1) broken symmetry as weak relations. The inhomogeneous action of J^{in} on χ^{in} produces a linear rotation of χ and ψ under J. Eqs. (3.22) clarify the role Goldstone boson χ^{in}. v is the v.e.v. of the ψ field.

The example just treated has a natural generalization, the case of a SO(n) symmetry, in the vectorial representation, broken with respect to SO(n-1) subgroup. The presence now of an unbroken part of the invariance group leads to preserve in the contraction limit the generators of the unbroken SO(n-1). The analogy with the model studied above gives:

$$
\begin{aligned}
&\overline{\phi}_i = \varepsilon^{-1}\phi_i\ ; \qquad \overline{\phi}_0 = \phi_0\ ; \\
&\overline{J}_{0i} = \varepsilon J_{0i}\ ; \qquad \overline{J}_{ij} = J_{ij}\ ;
\end{aligned}
\tag{3.23}
$$

where J_{ij} are the generators of SO(n-1), J_{oi} are the remaining generators of SO(n) and $\{\phi_i, \phi_0\}$ is a vectorial field representation. The commutation relations of the operators defined in

(3.23) are the standard ones, except for

$$[\bar{J}_{0i}, \bar{J}_{0j}] = -i\varepsilon^2 \bar{J}_{ij} \tag{3.24}$$

$$[\bar{J}_{0i}, \phi_0] = -i\varepsilon^2 \bar{\phi}_i \tag{3.25}$$

that have an ε -dependence.

Let us note that eqs. (3.23) choosing J_{ij} as the generators for the unbroken SO(n-1) give the broken direction to ϕ_0 ; this is equivalent to assume $\langle 0|\phi_0|0\rangle = v \neq 0$. Indeed if we assume spontaneous symmetry breaking

$$\langle 0|C^2|0\rangle = \langle 0|\phi_0^2 + \sum_i \phi_i^2|0\rangle = v^2 \neq 0$$

in the limit

$$\langle 0|\lim_{\varepsilon\to 0} \phi_0^2 + \lim_{\varepsilon\to 0} \sum_i \varepsilon^2 \bar{\phi}_i^2|0\rangle = \tag{3.26}$$
$$= \langle 0|\lim_{\varepsilon\to 0} \phi_0^2|0\rangle = v^2$$

In terms of the asymptotic limits

$$\lim_{\varepsilon\to 0} \bar{J}_{0i} = T_i^{in} ; \quad \lim_{\varepsilon\to 0} \bar{J}_{ij} = J_{ij}^{in} ;$$
$$\lim_{\varepsilon\to 0} \bar{\phi}_0 = \phi_0^{in} ; \quad \lim_{\varepsilon\to 0} \bar{\phi}_i = \phi_i^{in} ; \tag{3.27}$$

the asymptotic algebra is the contraction $IO(n-1)$ of $SO(n)$ (see the limit of (3.24)). The commutators for the asymptotic fields are:

$$[J_{ij}^{in}, \phi_k^{in}] = i(\delta_{jk}\phi_i^{in} - \delta_{ik}\phi_j^{in});$$
$$[J_{ij}^{in}, \phi_0^{in}] = 0\ ;\ [T_i^{in}, \phi_j^{in}] = i\delta_{ij}\phi_0^{in} \tag{3.28}$$
$$[T_i^{in}, \phi_0^{in}] = 0$$

We note that ϕ_0^{in} is a group invariant, it has v.e.v. different from zero (see (3.26)); this proves the inhomogeneous action of T_i^{in} on ϕ_j^{in}. The unbroken $SO(n-1)$ generated by J_{ij}^{in}, as expected, acts linearly. Thus we found the following corrispondence between initial and asymptotic fields:

$$\begin{pmatrix} \bar{\phi}_i \\ \bar{\phi}_0 \end{pmatrix} \approx \begin{pmatrix} 0 \\ \phi_0^{in} \end{pmatrix} + \begin{pmatrix} \phi_i^{in} \\ 0 \end{pmatrix} \tag{3.29}$$

In terms of non linear realization, using (3.23) we get:

$$\begin{pmatrix} \bar{\phi}_i \\ \bar{\phi}_0 \end{pmatrix} = \exp\left(\begin{array}{c|c} & \xi_i \\ \hline -\varepsilon^2\xi_i & \end{array}\right)\varphi_0 \tag{3.30}$$

Comparing the limit $\varepsilon \to 0$ of (3.30) with (3.29) and replacing ϕ_0^{in} by its v.e.v. we have:

$$\begin{pmatrix} \bar{\phi}_i \\ \bar{\phi}_o \end{pmatrix} = \exp\left[\frac{1}{v}\left(\begin{array}{c|c} & \phi_i^{in} \\ \hline -\varepsilon^2 \phi_i^{in} & \end{array}\right)\right]\begin{pmatrix} 0 \\ v \end{pmatrix} \tag{3.31}$$

Using (3.23) it can be rewritten as:

$$\begin{pmatrix} \phi_i \\ \phi_o \end{pmatrix} = \left(\begin{array}{c|c} \varepsilon & \\ \hline & 1 \end{array}\right) \exp\left[\frac{1}{v}\left(\begin{array}{c|c} & \phi_i^{in} \\ \hline -\varepsilon^2 \phi_i^{in} & \end{array}\right)\right]\left(\begin{array}{c|c} \varepsilon^{-1} & \\ \hline & 1 \end{array}\right) \times$$

$$\times \begin{pmatrix} 0 \\ v \end{pmatrix} = \tag{3.32}$$

$$= \exp\left[\frac{\varepsilon}{v}\left(\begin{array}{c|c} & \phi_i^{in} \\ \hline -\phi_i^{in} & \end{array}\right)\right]\begin{pmatrix} 0 \\ v \end{pmatrix}$$

We thus have

$$\phi_i = v \frac{\phi_i^{in}}{|\phi^{in}|} \sin \varepsilon \frac{|\phi^{in}|}{v}$$

$$\phi_o = v \cos \varepsilon \frac{|\phi^{in}|}{v} \tag{3.33}$$

The form of the generators J_{ij} and J_{oi} on ϕ_i^{in} is given, using their standard linear form and (3.33), by:

$$J_{ij} = i\left(\phi_i^{in}\frac{\delta}{\delta\phi_j^{in}} - \phi_j^{in}\frac{\delta}{\delta\phi_i^{in}}\right)$$

$$J_{oi} = \frac{i}{2}|\phi^{in}|\,\mathrm{cotg}\,\frac{\varepsilon|\phi^{in}|}{v}\left[\delta_{il} + \left(\frac{v}{\varepsilon|\phi^{in}|}\times\right.\right. \tag{3.34}$$

$$\left.\left.\times\,\mathrm{tg}\,\frac{\varepsilon|\phi^{in}|}{v} - 1\right)\frac{\phi_i^{in}\phi_l^{in}}{|\phi^{in}|^2}\right]\frac{\delta}{\delta\phi_l^{in}} + h.c.$$

The action of J_{ij} is clearly linear, whereas the action of J_{oi} is not. As expected in the limit

$$T_i^{in} = \lim_{\varepsilon\to 0} \bar{J}_{oi} = iv\frac{\delta}{\delta\phi_i^{in}} \tag{3.35}$$

The non linear dynamical maps (3.33) between Heisenberg fields and asymptotic fields have to be understood in a weak sense; i.e. between matrix elements in the Fock space for the asymptotic fields. In the spirit of phenomenological Lagrangians it is easy to give an interpretation to the parameter ε ; let us consider a Lagrangian invariant under global SO(n) internal symmetry:

$$\mathcal{L}(\phi_i,\phi_o) = \frac{1}{2}\partial_\mu\phi\,\partial^\mu\phi - \frac{1}{2}m^2\phi^2 - \frac{g}{4}\phi^4 \tag{3.36}$$

where $\phi = \{\phi_i, \phi_0\}$ and $m^2 < o$ to allow s.s.b. Using (3.33) the effective Lagrangian is defined:

$$\mathcal{L}_\varepsilon(\xi^{in}, v) \equiv \varepsilon^{-2}\mathcal{L}\left[\frac{v\xi_i^{in}}{|\xi^{in}|}\sin\frac{\varepsilon|\xi^{in}|}{v}, v\cos\frac{\varepsilon|\xi^{in}|}{v}\right] \tag{3.37}$$

Note that for SO(4) and $\langle o/\phi_0/o\rangle = v \equiv F_\pi$ we reproduce, from (3.36) and (3.37), the chiral pion phenomenological lagrangian:

$$\mathcal{L} = \frac{1}{2}\partial_\mu\pi_i\partial^\mu\pi_i -$$

$$-\frac{1}{2}\left(1 - \frac{F_\pi^2}{\varepsilon^2|\pi|}\sin\frac{\varepsilon|\pi|}{F_\pi}\right)\left(\partial_\mu\pi_i\partial^\mu\pi_i - \partial_\mu|\pi|\partial^\mu|\pi|\right) \tag{3.38}$$

where π_i are the pion fields replacing ξ^{in} and F_π is the pion decay constant. From (3.37) we see that the ε-power in the expansion of $\mathcal{L}_\varepsilon$ is related with the number of loops L and external lines E by:

$$P = E + 2L - 2$$

L = 0 gives the tree approximation; the tree diagrams are thus characterized by P = E-2. If N_i is the number of lines departing from the ith vertex, the power of ε in that vertex is N_i-2; we find then no vertex in the $\varepsilon \to 0$ limit: this means the LSZ asymptotic limit of the physical fields as observed previously. So we see that the parameter which, by way of the contraction, gauges the strength of the s.s.b. is the same ε-parameter of the expansion of the phenomenological Lagrangian [2,3]. The last example we want to treat is peculiar in the sense that we find asymptotic fields as bounded states.

The model is the doublet complex field under a SU(2) global internal invariance:

$$\psi = \begin{pmatrix} \phi \\ \chi \end{pmatrix} \quad ; \quad \psi^{+} = (\phi^{*} \;\; \chi^{*})$$

$$[J_+, J_-] = 2J_3 \; ; \; [J_3, J_\pm] = \pm J_\pm \tag{3.39}$$

$$[J_\pm, \psi] = -\frac{\sigma_\pm}{2}\psi \; ; \; [J_3, \psi] = -\frac{\sigma_3}{2}\psi$$

where ψ is the doublet, $J_\pm$, J_3 are the generators of SU(2) and $\sigma_\pm = \sigma_1 \pm i\sigma_2$, σ_3 the Pauli matrices. To have s.b.s. of SU(2) respect to U(1) generated by J_3 we need to assume that the bilinear $\psi^{+}\frac{\sigma_3}{2}\psi$ has v.e.v. different to zero:

$$\langle 0|\psi^{+}\frac{\sigma_3}{2}\psi|0\rangle = \langle 0|\phi^{*}\phi - \chi^{*}\chi|0\rangle = M \neq 0 \tag{3.40}$$

The non invariance of the vacuum under $J_\pm$ leads to the existence of two Goldstone bosons. The assumption of the unbroken generator J_3 gives the following contraction to describe the s.b.s.:

$$\bar{J}_\pm = \varepsilon J_\pm \; ; \qquad \bar{J}_3 = J_3 \tag{3.41}$$

From (3.30) the only possibility is to define

$$\bar{\psi} = \begin{pmatrix} \bar{\phi} \\ \bar{\chi} \end{pmatrix} = \begin{pmatrix} 1 & \\ & \varepsilon^{-1} \end{pmatrix} \begin{pmatrix} \phi \\ \chi \end{pmatrix} \tag{3.42}$$

Other choices give in the limit trivial commutation relations. The eqs. (3.39) are now rewritten as:

$$\begin{aligned}
&[\bar{J}_+, \bar{J}_-] = 2\varepsilon^2 \bar{J}_3 \; ; \quad [\bar{J}_3, \bar{J}_\pm] = \pm \bar{J}_\pm \; ; \\
&[\bar{J}_+, \bar{\psi}] = -\varepsilon^2 \frac{\sigma_+}{2} \bar{\psi} \; ; \\
&[\bar{J}_-, \bar{\psi}] = -\frac{\sigma_-}{2} \bar{\psi} \; ; \\
&[J_3, \bar{\psi}] = -\frac{\sigma_3}{2} \bar{\psi} \; ;
\end{aligned} \tag{3.43}$$

The limit of (3.43) is trivial; we get that the asymptotic symmetry is E (2) and:

$$\begin{aligned}
&[J_+^{in}, \lim_{\varepsilon \to 0} \bar{\psi}] = 0 \\
&[J_-^{in}, \lim_{\varepsilon \to 0} \bar{\psi}] = -\frac{\sigma_-}{2} \lim_{\varepsilon \to 0} \bar{\psi}
\end{aligned} \tag{3.44a}$$

$$[J_3, \lim_{\varepsilon\to 0} \overline{\Psi}] = -\frac{\sigma_3}{2} \lim_{\varepsilon\to 0} \overline{\Psi} \qquad (3.44b)$$

The limit of (3.40) is

$$\lim_{\varepsilon\to 0} \langle 0| \frac{1}{2}(\overline{\phi}^* \overline{\phi} - \varepsilon^2 \overline{\chi}^* \overline{\chi})|0\rangle =$$

$$= \langle 0| \frac{1}{2} \phi^{*in} \phi^{in} |0\rangle = M \qquad (3.45)$$

where $\lim_{\varepsilon\to 0} \overline{\phi} = \phi^{in}$.

The bilinear $\phi^{*in}\phi^{in}$ is invariant under the asymptotic symmetry as we can check from (3.44), indeed the limit of the SU(2) invariant is:

$$\lim_{\varepsilon\to 0} \overline{\Psi}^{+}\overline{\Psi} = \lim_{\varepsilon\to 0} (\overline{\phi}^* \overline{\phi} + \varepsilon^2 \overline{\chi}^* \overline{\chi}) = \phi^{*in}\phi^{in}. \qquad (3.46)$$

The two Goldstone bosons are the asymptotic fields we have to find, they are identified by the contraction of the elements of the adjoint representation corresponding to the broken generators:

$$\lim_{\varepsilon\to 0} \overline{\Psi}^{+} \frac{\sigma_+}{2} \overline{\Psi} = \lim_{\varepsilon\to 0} \frac{1}{2} \overline{\phi}^* \overline{\chi} = i\,\alpha^{in}$$

$$\lim_{\varepsilon\to 0} \overline{\Psi}^{+} \frac{\sigma_-}{2} \Psi = \lim_{\varepsilon\to 0} \frac{1}{2} \overline{\chi}^* \overline{\phi} = -i\,\alpha^{*in} \qquad (3.47)$$

The transformations of α^{in} and α^{*in} are:

$$[J_-^{in}, \alpha^{in}] = iM \; ; \qquad [J_+^{in}, \alpha^{*in}] = iM \; ;$$

$$[J_+^{in}, \alpha^{in}] = 0 \; ; \qquad [J_-^{in}, \alpha^{*in}] = 0 \; ; \tag{3.48}$$

$$[J_3, \alpha^{in}] = 0 \; ; \qquad [J_3, \alpha^{*in}] = 0 \; ;$$

where we substituted the invariant $\frac{1}{2}\phi^{*in}\phi^{in}$ with its v.e.v. M. The relations (3.48) say that α^{in} and α^{*in} translate under J_-^{in} and J_+^{in}, respectively and are objects with "spin" one; they are the Goldstone modes we were looking for. Thus the asymptotic limit of $\bar{\chi}$ is:

$$\lim_{\varepsilon \to 0} \bar{\chi} = \frac{i}{M} \alpha^{in} \phi^{in} . \tag{3.49}$$

The relation between Heisenberg fields and asymptotic fields is then:

$$\lim_{\varepsilon \to 0} \begin{pmatrix} \bar{\phi} \\ \bar{\chi} \end{pmatrix} = \begin{pmatrix} \phi^{in} \\ 0 \end{pmatrix} + \frac{i}{M} \begin{pmatrix} 0 \\ \alpha^{in}\phi^{in} \end{pmatrix} \tag{3.50}$$

In terms of (3.43) we have the non linear realization

$$\bar{\Psi} = \exp\left[i \xi \frac{\sigma_-}{2} + i \xi^* \varepsilon^2 \frac{\sigma_+}{2} \right] \Psi_0 \tag{3.51}$$

We find in the limit the corrispondence:

$$\Psi_0 = \begin{pmatrix} \phi^{in} \\ 0 \end{pmatrix} \; ; \quad \xi = \frac{\alpha^{in}}{M} .$$

Then from (3.51) and (3.42)

$$\phi = \cos\frac{\varepsilon}{M}\sqrt{\alpha^{*in}\alpha^{in}}\ \phi^{in}$$

$$\chi = i\,\frac{\sin(\varepsilon/M)\sqrt{\alpha^{*in}\alpha^{in}}}{(\varepsilon/M)\sqrt{\alpha^{*in}\alpha^{in}}}\ \varepsilon\,\frac{\alpha^{in}}{M}\,\phi^{in} \tag{3.52}$$

The eqs. (3.52) can be rewritten as:

$$\psi = \exp\left[i\frac{\varepsilon}{M}\left(\alpha^{in}\frac{\sigma_-}{2} + \alpha^{*in}\frac{\sigma_+}{2}\right)\right]\begin{pmatrix}\phi^{in}\\ 0\end{pmatrix} \tag{3.53}$$

The effect of the presence of the contraction parameter affects only the Goldstone bosons and not the field ϕ^{in}. The parameter ε is the right expansion constant if we want to follow the rearrangement of symmetry and the fields limit from their dynamical level to the LSZ asymptotic structure.

Finally, we have to introduce the covariant derivative. We follow the notations of ref.3 and write

$$D_\mu \xi = p_\mu$$

$$D_\mu \psi = \partial_\mu \psi + V_\mu \cdot T\psi \tag{3.54}$$

Here ξ denote the Goldstone fields, ψ the other fields of the theory, T_i the representation of the generators V_i of the stability group H; p_μ and V_μ are defined by

$$e^{-i\varepsilon\xi A}\,\partial_\mu\, e^{i\varepsilon\xi A} = i\,p_{\mu i}A + i\,V_\mu\cdot V \tag{3.55}$$

where A_i are the generators of G :H. From eqs. (3.54) and (3.55) we see that $p_\mu = O(\varepsilon)$ and $V_\mu = O(\varepsilon^2)$ which shows the dependence of the covariant derivative on ε .

Extension to the case of a coordinate dependent group of transformation is achieved when gauge fields are introduced. Let us denote them as $\rho_{\mu i}$ and $a_{\mu \ell}$, which are associated to V_i and A_ℓ generators respectively. Their transformation properties are:

$$i\rho'_\mu \cdot V + i\varepsilon a'_\mu \cdot A = \\ = g\left(i\rho_\mu \cdot V + i\varepsilon a_\mu \cdot A\right)g^{-1} - f^{-1}(\partial_\mu g)g^{-1} . \tag{3.56}$$

The new covariant derivatives are now defined by

$$e^{-i\varepsilon\xi\cdot A}\left[\partial_\mu + if\left(\rho_\mu \cdot V + \varepsilon a_\mu \cdot A\right)\right]e^{i\varepsilon\xi\cdot A} = \\ = iV_\mu \cdot V + ip_\mu \cdot A \tag{3.57}$$

with

$$V_\mu = \rho_\mu + O(\varepsilon^2)$$

$$p_\mu = \partial_\mu \varepsilon\xi + f\varepsilon a_\mu + iD^{(h)}(\rho_\mu \cdot V)\varepsilon\xi + O(\varepsilon^3).$$

An explicit example of dynamical maps in the case of gauge field theory model is the one presented in sec. 2 for the case of Nambu-type model (cf. eqs. (2.4) and (2.5)).

In conclusion the authors would like to thank Professor J. Mozrzymas, Professor L. Michel and the Organizing Committee of the XXI Winter School of Theoretical Physics, Karpacz 1985, for the warm hospitality.

REFERENCES

1. Celeghini, E., Magnollay, P., Tarlini, M. and Vitiello, G., Non linear realizations and contraction of group representations, preprint 1985.

2. Coleman, S., Wess, J. and Zumino, B., Phys. Rev. 177, 2239 (1969).

3. Callan jr., C.G., Coleman, S., Wess, J. and Zumino, B., Phys. Rev. 177, 2247 (1969).

4. Salam, A. and Strathdee, J., Phys. Rev. 184, 1750 (1969).

5. Umezawa, H., in "Renormalization and Invariance in QFT", Ed. Caianiello, E.R., Plenum, N.Y. p.275 (1973).

6. Umezawa, H., Matsumoto, H. and Tachiki, M., Thermofield Dynamics and Condensed Matter, North-Holland (1982).

7. De Concini, C. and Vitiello, G., Nucl. Phys. 116B, 141 (1976).

8. Inönü,E. and Wigner, E.P., Proc. Nat. Sci. U.S. 39, 510 (1953).

9. Shah, M.N., Umezawa, H. and Vitiello, G., Phys. Rev. 10B, 4724 (1974).

10. Hongoh, M., Matsumoto, H. and Umezawa, H., Progr. Theor. Phys. 65, 315 (1981).

11. Celeghini, E., Tarlini, M. and Vitiello, G., Nuovo Cimento A84, 19 (1984).

12. Matsumoto, H., Papastamatiou, N.J., Umezawa, H. and Vitiello, G., Nucl. Phys. 97B, 61 (1975).

13. Matsumoto, H., Papastamation, N.J. and Umezawa, H., Nucl. Phys. 97B, 90 (1975).

Ezawa, Z.F. and Tze, H.C., Nucl.Phys., B96, 264 (1975).

14. De Concini, C. and Vitiello, G., Phys. Lett. 70B, 355 (1977).

15. Celeghini, E. and Tarlini, M., Nuovo Cimento 61B, 265 (1982); 65B, 172 (1982); 68B, 133 (1982).

16. Gilmore, R., Lie groups, Lie algebras and some of their applications, J. Wiley, New York (1974).

14. De [illegible] and C. de Villedi[illegible], [illegible] Phys. Lett. [illegible] (19[illegible]).

15. Celegnini, E. and [illegible], Nuovo Cimento [illegible] (1940);
658, [illegible] (1942) [illegible]

16. [illegible] "The [illegible] and [illegible] —
cations", [illegible] Wiley, New York (19[illegible]).

GEOMETRY OF SPONTANEOUS SYMMETRY BREAKING

Henri Ruegg
Département de Physique Théorique-Université de Genève
24, boulevard d'Yvoy, CH-1211 Genève 4
SWITZERLAND

ABSTRACT

We discuss the geometry of orbit space of a compact Lie group in terms of symmetry types and an invariant integrity basis. This is used to study minimas of Higgs potentials and their maximal or non maximal little groups.

1. INTRODUCTION

Spontaneous symmetry breaking plays an important role in different fields of physics : phase transitions, dynamical systems, elementary particles. We shall be interested in the latter. In this case, the broken symmetry is a gauge group based on a compact Lie group G, which acts on a representation space R^n (if the representation is complex, one adds the conjugate representation). The relevant mathematical object is the so-called Higgs potential $V(\phi)$, when $\phi \in R^n$. This is a G-invariant function of ϕ of degree 4, bounded from below and with its lowest value not at the origin. The symmetry is broken in the direction of the minimum $\overset{\circ}{\phi}$ of V, and the remaining symmetry is the little group (isotropy group) $H \subset G$, such that $H\overset{\circ}{\phi} = \overset{\circ}{\phi}$.

In general, the minimization is tedious. It is therefore important to use as much general information as possible. We have now a good understanding of the stationarity condition $\partial V = 0$ thanks to the pioneering work of Michel and Radicati [1][2]. This was followed by Abud and Sartori [3] who studied the relation between the tree isotropy subgroups and the stratification of orbit space. This space is spanned by the gradients of G invariant functions of ϕ.

Concerning the second derivative, Michel [2] made the following conjecture : for a representation which is irreducible on the real, the little group of the minimum is maximal. This means : if $V(\phi^\circ)$ is minimum, $H\overset{\circ}{\phi} = \overset{\circ}{\phi}$, then there exist no little group K ($K\overset{1}{\phi} = \overset{1}{\phi}$, $\overset{1}{\phi}\in R^n$) such that $H \subset K \subset G$.

It has been shown that the conjecture is fulfilled for the adjoint representation of SU(n) [4][5] and other examples, such as 2. rank tensors [6]. However there exist now counterexamples : the representations 75 of SU(5) [7] and 27 of SU(3) [8].

It therefore seems interesting to study the conditions of validity of the Michel conjecture. This has physical applications for model builders of spontaneously broken grand (or super) unified theories, especially since a non maximal little group is more "economic" [9]. We shall also see that discrete subgroups, which are usually neglected by model builders, play an important role. It may also be relevant for the "family" problem [8]. On the mathematical side, it is worthwhile to get a deeper understanding of the second derivative.

In this lecture we shall emphasize the geometrical approach of Abud and Sartori [3] which was extensively used in the papers of Abud et al. [7]. Next we give a short proof that the Michel's conjecture is satisfied by the adjoint representation of SU(n). We then show a simple example of a potential of degree 6 where the little group of the minimum is not maximal. Finally, we discuss in detail the application of the geometric method to the representation 75 of SU(5).

2. GEOMETRY OF THE ORBIT SPACE [1][3]

Let G be a compact connected Lie group acting linearly on a representation space R^n ($\phi\in R^n$). The set of all points $g\phi$, $g\in G$, is called the orbit Ω_ϕ of ϕ. If G_ϕ is the little group of ϕ, all points on Ω_ϕ have little groups conjugated to G_ϕ : one says that they have the same symmetry type $[\Omega]$. To characterize symmetry types it is useful to introduce the retraction of ϕ' on Ω, $\phi'\notin\Omega$: it is the point $\rho_\Omega(\phi')\in\Omega$ for which the distance of ϕ' to Ω is minimum (if ρ_Ω is unique, otherwise the retraction is not defined). Because the distance is an invariant, $\rho_\Omega(g\phi') = g\rho_\Omega(\phi')$. Hence, if $h\in G_\phi$, then $h\rho_\Omega(\phi') = \rho_\Omega(h\phi') = \rho_\Omega(\phi')$ and so $h\in G_{\rho_\Omega}$, and therefore

$$G_{\rho_\Omega} \supseteq G_{\phi'}, \qquad \text{or} \quad [\Omega] \supseteq [\Omega'] \tag{2.1}$$

if Ω' is in the domain of $\rho_\Omega(\cdot)$. The hierarchy introduced by (2.1) in symmetry types can be visualized by the following example : let Ω be a point ϕ, Ω' the circle around ϕ, and $\phi'\in\Omega'$. Then $\rho_\Omega(\phi')$ is unique, but not $\rho_{\Omega'}(\phi)$, and hence $[\Omega] \supset [\Omega']$.

The stratum Σ_ϕ containing ϕ is the set of all points with same symmetry $[\Omega_\phi]$, that is the little groups are conjugated to G_ϕ. For example $\lambda\phi$ is not on the orbit Ω_ϕ, but has same G_ϕ. One should notice that two points whose little groups are isomorphic are on the same stratum iff the isomorphism is a conjugation. The orbit space is the space R^n/G where every orbit is collapsed into a point. This is done by a canonical projection π, which maps the stratum Σ on $\hat{\Sigma}$.

Since the symmetry type is a property of a stratum, the hierarchy introduced in (2.1) applies to strata. We now show that this is related to the dimension of $\hat{\Sigma}$. For this we need the tangent plane. If $\breve{G}_\phi$ is the Lie algebra of G_ϕ, and T_ϕ the tangent plane to Ω_ϕ at ϕ, then $T_\phi = \{t\phi, t\in\breve{G}-\breve{G}_\phi\}$. Hence $\dim \Omega_\phi = \dim G - \dim G_\phi$. Call N_ϕ the normal complement of T_ϕ in R^n, N_ϕ^0 the subspace of G_ϕ invariant vectors orthogonal to T_ϕ, N_ϕ^1 the rest, i.e. $R^n = T_\phi \oplus N_\phi^0 \oplus N_\phi^1$. Then one can prove the following theorem :

$$T_\phi(\Sigma_\phi) = T_\phi(\Omega_\phi) \oplus N_\phi^{(0)} . \qquad (2.2)$$

From this follows :

$$\dim \hat{\Sigma}_\phi = \dim N_\phi^0 . \qquad (2.3)$$

This shows that the dimension of $\hat{\Sigma}_\phi$ is given by the number of G_ϕ singlets not in T_ϕ.

We give a sketch of the proof [3]. Let $C_\phi(\omega)$ be a curve in Σ_ϕ through ϕ, $\phi'(\omega)\in C_\phi$, $\rho_\Omega(\phi')$ the retraction of ϕ' on Ω_ϕ. Then from (2.1) $G_{\rho_\Omega} \supseteq G_{\phi'}$. Since ϕ' is on the same stratum as ρ_Ω, it follows that $G_{\rho_\Omega} = G_{\phi'}$, for ϕ' near enough to ϕ. The tangent vector τ_ϕ to C_ϕ at ϕ is given by :

$$\tau_\phi = \lim_{\omega\to 0} \frac{\phi'(\omega)-\phi}{\omega} = \lim_{\omega\to 0} \frac{\phi'(\omega)-\rho_\Omega(\phi'(\omega))}{\omega} + \lim_{\omega\to 0} \frac{\rho_\Omega(\phi')-\phi}{\omega} . \qquad (2.4)$$

In the limit $\omega \to 0$, $G_{\phi'}(\omega) = G_{\rho_\Omega}(\omega) = G_\phi$, and hence the first vector in (2.4) is in N_ϕ^0, the second in T_ϕ.

One can now classify strata by their dimension. On one end one has the generic stratum Σ_p. One says that Ω is generic iff $[\Omega'] = [\Omega]$ for all Ω' in the domain of ρ_Ω, that is near enough to Ω. One can show that Ω is generic iff N_ϕ^1 = zero for all $\phi\in\Omega$. All generic orbits lie in a unique stratum Σ_p. Hence, $\dim \Sigma_\phi = n$. This means that almost all points of R^n have the same symmetry type. Lower dimensional strata are called singular and are at the border of the generic stratum. More generally, each stratum has at its border a union of strata of lower dimension. Remarkably, the lower the dimension of a stratum, the larger is its little group. Hence one arrives [1][2] at a stratification of R^n by the type of symmetry of its points.

According to an important result due to Bierstone [10] there is a one-to-one correspondence between the strata of R^n classified according to the symmetry type and the algebraic varieties of R^n/G, the latter being given in terms of G-invariant functions $I(\phi) = I(g\phi)$, all $g \in G$, all $\phi \in R^n$. We use the theorem of Hilbert, that the set $P^G_{(R^n)}$ of all real polynomials in $\phi \in R^n$, invariant under the linear action of G, admits a finite integrity basis (IB), such that $P(\phi) = \hat{P}'(\theta_1(\phi),\ldots,\theta_q(\phi))$. The IB separates the orbits, that is if Ω and Ω' are distinct, there exists at least one θ_i which takes different values on Ω and Ω'. One now defines an orbit map $\theta : R^n \to S \subset R^q$, $\phi \to \theta(\phi) = (\theta_1(\phi),\ldots,\theta_q(\phi))$. One can introduce a stratification of R^q and show [11][10] that these strata are related in a one-to-one way to the strata $\hat{\Sigma}$ previously introduced. In fact the dimension of $\hat{\Sigma}$ is given by the number of algebraically independent θ_i or by the number of linearly independent grad θ_i (these latter are obviously in N_ϕ^0). For a one-dimensional stratum, all gradients are parallel, say to ∂Q, where $Q = ||\phi||^2$. Then for every C^∞ invariant function I one can write $\partial I = (\partial \hat{I}/\partial Q_i)\partial\theta_i$ and on the sphere Q = const. one has $\partial I = 0$. This is the theorem of Michel and Radicati [1].

In conclusion, we can use the following strategy. Given a compact connected Lie group and a linear representation D in R^n, we first construct all little groups of G and for each of them we count the number of singlets of the representation D when restricted to the subgroup. This gives a first stratification of R^n according to symmetry type. We then construct an integrity basis $(\theta_1,\ldots,\theta_q)$ and the subspace $S \in R^q$ corresponding to the range of θ_i. Points of S with smallest little group correspond to the generic stratum. Adding algebraic relations on θ_i, that is linear relations on grad θ_i, we get strata on the periphery with larger symmetries, ending with dimension one and a maximal little group.

3. PROOF OF MICHEL'S CONJECTURE FOR THE ADJOINT REPRESENTATION OF SU(n) [4][12]

Let ϕ be a $n \times n$ hermitean, traceless matrix. Under $g \in SU(n)$, ϕ transforms as $\phi' = g\phi g^\dagger$, $g^\dagger = g^{-1}$, det g = 1. ϕ can be diagonalized, its eigenvalues being a_i $(i = 1,\ldots,n)$, with $\sum_{i=1}^{n} a_i = 0$. Consider the little group of a particular ϕ. It is maximal if only two eigenvalues a_i are different, non maximal otherwise. The most general Higgs potential can be written as

$$V = -\frac{\mu^2}{2}\mathrm{tr}\phi^2 + \frac{a}{4}(\mathrm{tr}\phi^2)^2 + \frac{b}{2}\mathrm{tr}\phi^4 + d\,\mathrm{tr}\phi^3 . \qquad (3.1)$$

A necessary condition for V to have a minimum is to be minimal with respect to four arbitrarily chosen values a_i, the $n - 4$ others being fixed. Then one can consider

$$F = -\frac{\mu^2}{2}\varphi + \frac{a}{4}\varphi^2 + \frac{b}{2}\sum_{i=1}^{4} a_i^4 + d\sum_{i=1}^{4} a_i^3 , \qquad (3.2)$$

$$\sum_{i=1}^{4} a_i^2 = \varphi \; ; \quad \sum_{i=1}^{4} a_i = -\sum_{i=5}^{n} a_i = \sigma . \qquad (3.3)$$

A necessary condition for a minimum is

$$\frac{\partial F}{\partial a_i} = 0 \quad ; \quad \frac{\partial^2 F}{\partial a_i \partial a_j} \geq 0 \qquad i,j = 1,\ldots,4 . \qquad (3.4)$$

The first equation (3.4) is cubic, so there are at most three different eigenvalues. (3.3) and (3.4) can be solved explicitly and one gets four types of solutions :

$$\binom{1}{2} \quad a_1 = a_2 = a_3 = \frac{1}{4}(\sigma \pm 2\sqrt{3}x); \; a_4 = \frac{1}{4}(\sigma \mp 6\sqrt{3}x) \qquad \binom{3.5}{3.6}$$

$$(3) \quad a_1 = a_3 = \frac{1}{4}(\sigma + 6x); \; a_2 = a_4 = \frac{1}{4}(\sigma - 6x) \qquad (3.7)$$

$$(4) \quad a_1 = a_2 = \frac{1}{4}\sigma + \frac{3}{2}y; \; \binom{a_3}{a_4} = \frac{1}{4}\sigma - \frac{3}{2}y \pm 3\left(\frac{x^2-y^2}{2}\right)^{1/2} \qquad (3.8)$$

$$x = \frac{(4\varphi - \sigma^2)^{1/2}}{6} \qquad y = \frac{d}{b} + \frac{1}{2}\sigma .$$

Solution (3.8) corresponds to a non maximal little group. It exists only if $x^2 > y^2$.

We can now apply an idea of Erick Weinberg [12] using Morse theory [2] : equation (3.3) defines a 2-sphere. Then Morse theory implies that the number of minimas + number of maximas − number of saddle points = 2. Define $b_\alpha = 1$ if solution α is a minimum or a maximum (α numbers the solutions (3.5) to (3.8)). Define $b_\alpha = -1$ if solution α is a saddle point. Define $b_4 = 0$ if solution 4 does not exist. We have to take into account the permutations of the a_i , namely four for $\alpha = 1,2$, six for $\alpha = 3$ and 12 for $\alpha = 4$.

Then Morse theory tells us that

$$4b_1 + 4b_2 + 6b_3 + 12b_4 = 2 . \qquad (3.9)$$

There are only two solutions :

$$b_1 = b_2 = 1; \; b_3 = -1; \; b_4 = 0 ; \qquad (3.10)$$

$$b_1 = b_2 = b_3 = 1; \; b_4 = -1 . \qquad (3.11)$$

There is no solution with $b_4 = +1$. Thus solution (4), whose little group is non maximal, corresponds to a saddle point.

In this proof, it is crucial that the degree of the Higgs potential is not larger than four. Otherwise, there exist an infinity of counter examples. For instance, consider SU(6) and the invariant potential

$$V = \mathrm{tr}[\phi - D(a_1 a_1 0000)]^2 [\phi - D(00 a_2 a_2 00)]^2 [\phi - D(0000 a_3 a_3)]^2$$

where D is a diagonal 6×6 matrix. Then V has an absolute minimum at

$$\overset{\circ}{\phi} = D(a_1 a_1 a_2 a_2 a_3 a_3) \ , \quad \sum_{i=1}^{3} a_i = 0 \ .$$

The little group is obviously not maximal.

4. NON MAXIMAL LITTLE ALGEBRAS AND NON MAXIMAL LITTLE GROUP [7]

We shall now treat in detail the representation 75 of SU(5), which has some virtues for model builders [9]. The group acts on the tensor ϕ^{cd}_{ab}

$$\phi^{cd}_{ab} = -\phi^{dc}_{ab} = -\phi^{cd}_{ba} \quad ; \quad \phi^{ad}_{ab} = 0 \ . \tag{4.1}$$

There are four algebraically independent invariants of degree not larger than four :

$$Q = \phi^{ab}_{cd}\phi^{cd}_{ab} \quad ; \quad C = \phi^{ab}_{ef}\phi^{cd}_{ab}\phi^{ef}_{cd}$$
$$K_2 = \phi^{ab}_{gh}\phi^{cd}_{ab}\phi^{ef}_{cd}\phi^{gh}_{ef} \quad ; \quad K_3 = \phi^{ab}_{cg}\phi^{cd}_{ab}\phi^{ef}_{dh}\phi^{gh}_{ef} \ . \tag{4.2}$$

The most general Higgs potential is

$$V = -\mu^2 Q + c\mathbb{C} + \lambda_1 Q^2 + \lambda_2 K_2 + \lambda_3 K_3 \ . \tag{4.3}$$

A sufficient condition for V to be positive when $Q \to \infty$ is $\lambda_1 > \sum_{i=2,3} |\lambda_i|$.

We first construct the non trivial part of the tree of subalgebras of su(5). Isomorphic subalgebras are distinguished by the decomposition of the fundamental representation 5. The generators of u(1) algebras (denoted x_i) are obtained by the decomposition of the adjoint representation 24 (see figure).

We then count the number of singlets in the decomposition of the representation 75 with respect to the subalgebras. In this way we find four <u>maximal little algebras</u> with one singlet. They are

Figure : Tree of subalgebras

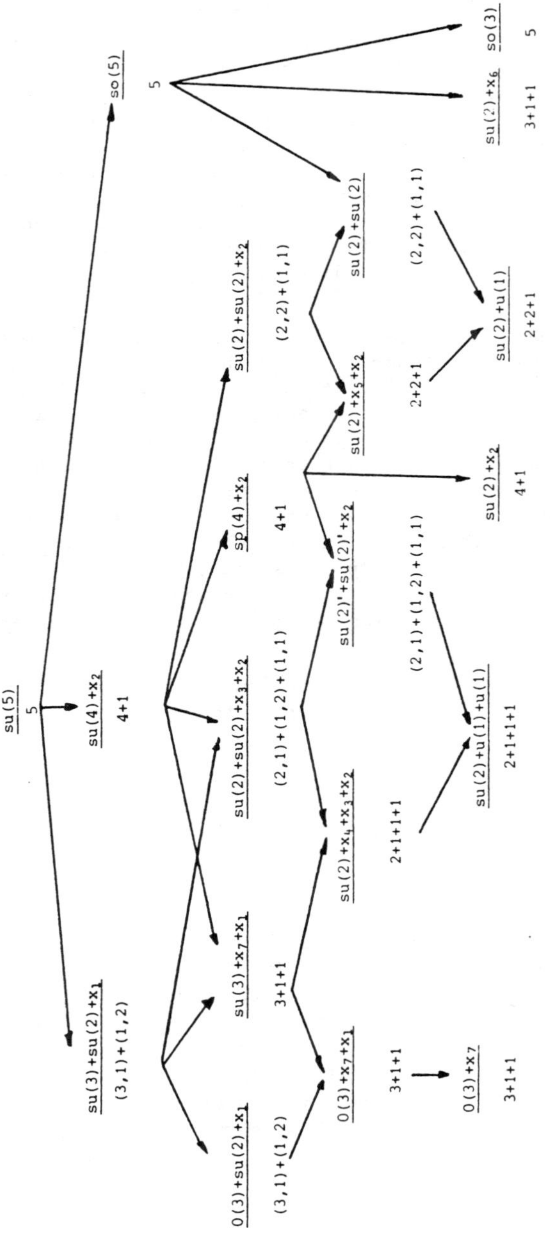

$g_1 = su(3) + su(2) + x_1$; $g_2 = sp(4) + x_2$; $g_3 = su(2) + su(2) + x_2$; $g_4 = so(3)$. After that, we shall consider the non maximal little algebras $g_5 = su(2) + su(2) + x_3 + x_2$ with two singlets and $g_6 = su(2) + x_5 + x_2$ with three singlets.

To study the structure of the strata, we now look for relations between the invariants. The integrity basis of R^{75} has at least $75 - 24 = 51$ elements. We take into account only the four invariants (4.2). This, however, will already yield useful information. For the four maximal subalgebras with one singlet we find indeed that all the gradients of the invariants are parallel, as it should be for one-dimensional strata. There is then one equation $\partial V = 0$ for an extremum, which is easily found. We find numerically that in each case there is an open set of parameters of V for which the second derivative is positive (or zero), corresponding to a minimum.

We next consider the non maximal subalgebra $g_5 = su(2) + su(2) + x_3 + x_2$ with two singlets. The invariants can now be written in terms of two real parameters a and b as follows : $Q = 24(2a^2 + b^2)$, $C = 96a(a^2 + b^2)$, $K_2 = 96(6a^4 + 8a^2b^2 + b^4)$, $K_3 = 1/6(Q^2 + 2K_2)$. There are two independent gradients. Taking generic points in this two-dimensional stratum, we find no minimum. We now look for relations between gradients at the border of the stratum. We find again the previous case g_1 for $a^2 = b^2$, but a new case when $b = 0$. The corresponding singlet is invariant under g_5 and an additional discrete group $Z_2 \in SU(5)$. This is an example when the little algebra is non maximal, but the little group is maximal, because Z_2 doesn't belong to the maximal little groups previously found. Interestingly, we find a set of parameters for which the potential V has a minimum.

So far, everything agrees with Michel's conjecture. But we find a counter-example in the three-dimensional stratum with little algebra $g_6 = su(2) + x_5 + x_2$. We parametrize the invariants with three parameters, and find :

$$Q = 24(\alpha^2 + \beta^2 + \gamma^2), \quad C = 48(-\alpha^2\beta - \alpha^2\gamma - 3\beta\gamma^2 + \gamma^3),$$

$$K_2 = 96(\alpha^4 + 2\alpha^2\beta^2 + 2\alpha^2\gamma^2 + 4\alpha^2\beta\gamma + \beta^4 + 6\beta^2\gamma^2 - 4\beta\gamma^3 + 3\gamma^4),$$

$$K_3 = 16(8\alpha^4 + 16\alpha^2\beta^2 + 16\alpha^2\gamma^2 + 8\alpha^2\beta\gamma + 9\beta^4 + 18\beta^2\gamma^2 + 9\gamma^4) .$$

Looking for relations between gradients, we recover previous cases and a new one : $\alpha^2 = 21\gamma^2$, $\beta = -4\gamma$. In this case, there are two non parallel gradients, ∂Q and ∂K_2 , ∂C being parallel to ∂Q. So we have a line in a three-dimensional stratum with two independent invariants. On this line, $\partial V = 0$ gives rise to two conditions and we find that V has a minimum. The little algebra g_6 is obviously not maximal. But do we get additional discrete groups on the line ? To answer this question, we consider the mass spectrum (eigenvalues of the second derivative of V). It has the correct subgroup structure, with the right number $24 - 5 = 19$ zero eigenvalues (Goldstone Bosons). The important point is that there is no degeneracy beyond the one required by Schur lemma as would be the case if there were further discrete symmetry. We therefore think that the little group of the minimum on the line is non maximal.

5. CONCLUSION

We think that the geometric method outlined in Section 2 is very useful when analyzing complicated subgroup structures. With its help we have found minima with non maximal little algebras and non maximal little groups.

To understand better the conditions of validity of Michel's conjecture and therefore the structure of the second derivative, it would be worthwhile to study the common features of the counter-examples [7][8].

REFERENCES

[1] Michel, L. and Radicati, L., Ann. of Phys. 66, 758 (1971).

[2] Michel, L., in : Regards sur la physique contemporaine, p. 157, CNRS, Paris (1980); CERN-TH 2716 (1979) and references therein.

[3] Abud, M. and Sartori, G., Phys. Lett. 104B, 147 (1981) and Ann. of Phys. 150, 307 (1983).

[4] Ruegg, H., Phys. Rev. D22, 2040 (1980).

[5] Murphy, T. and O'Raifeartaigh, L., Nucl. Phys. B229, 509 (1983).

[6] Li, L.-F., Phys. Rev. D9, 1723 (1974).

[7] Abud, M., Anastaze, G., Eckert, P. and Ruegg, H., Phys. Lett. 142B, 371 (1984) and Ann. of Phys., to appear (1985).

[8] Burzlaff, J., Murphy, T. and O.Raifeartaigh, L., Dublin Preprint 1985.

[9] Georgi, H., Phys. Lett. 108B, 283 (1982).

[10] Bierstone, E., Topology 14, 254 (1975) and "The Structure of Orbit Spaces and the Singularities of Equivalent Mappings", Inst. de Matemática Pura e Aplicada Publications, Rio de Janeiro (1980).

[11] Schwarz, G.W., Topology 14, 63 (1975).

[12] Weinberg, E., private communication.

CAN THE PRINCIPLE OF MAXIMUM SPEED IMPLY LORENTZ INVARIANCE?

H.-J. Borchers

Institut für Theoretische Physik
der Universität Göttingen
Bunsenstr. 9
D-3400 Göttingen
FEDERAL REPUBLIC OF GERMANY

1. INTRODUCTION

Looking at the development of physics one could get the impression, especially from the particle physics, that group theory plays a central role. Even more, all important theories in particle physics which came into existence during the last few decades seem to be at first glance nothing else then group theory with a little bit extra.

If this is the truth, and no doubt there is something to it, then one should wonder about the relation between group theory and nature. During this lecture I will try to look at these relations and it will turn out that this connection becomes more and more vague the closer one is trying to look at it.

In the next section we will discuss some of the so-called symmetries and we will realize that there exist no symmetry in nature except those which are build in into the concept of physics. If there are no symmetries in physics then one should ask why nature is pretending that there are symmetry groups present. What I am suggesting is that one should understand the reason for the pretention just mentioned.

In the third section I will have a closer look at the homogeneous part of the Lorentz group. The discuss in section 2 will show us that in quantum field theory, except for academic models, the homogeneous part of the Lorentz group will not be a symmetry or at least it will be a spontaneously broken symmetry. However, we will show that the principle of maximal velocity c has drastic consequences. One knows from classical theory that the invariance-group of this concept is the group of conformal transformations[1]. Without assuming that Lorentz transformations are a symmetry we will show that the spectrum of the translations is located on a Lorentz invariant set in momentum space. This shows that the experimentally well established relation $p^2 = m^2$ for isolated particles follows from Einsteins principle of maximal velocity alone without using any invariance under Lorentz transformation. (For the mass relation see e.g.[2]).

2. DISCUSSION OF SOME SYMMETRY GROUPS

2.1 Space-time translations

In a physical theory one has to deal with two sets of objects:

the observables 0,

the states S

and a mapping of $S \times 0 \to R$ called measurement. The two sets are such that each one is separating the other.

A symmetry is then a pair of mappings

$$\alpha: 0 \to 0$$

$$\beta: S \to S$$

such that for $A \in 0$ and $\omega \in S$ one has

$$(\beta\omega, \alpha A) = (\omega, A)$$

where (ω, A) denotes the image of the pair $\{\omega, A\}$ under the map called measurement.

In physical theories considered so far 0 and S are both embedded in linear structures $L(0)$ and $L(S)$ and α and β are such that they can be extended to linear mappings $\tilde{\alpha}$ and $\tilde{\beta}$ on the linear spaces $L(0)$ and $L(S)$. In this situation $L(0)$ and $L(S)$ is a pair in duality and $\tilde{\alpha}$ and $\tilde{\beta}$ are the transposed mappings of each other.

Natural sciences and especially physics is based on the fact that experiments can be repeated at other places and other times. Since a repetition of an experiment means that one must prepare the state and that one has to do the measurement. If it was a good physical experiment then also the outcome is the same. But this implies that one has a symmetry in the above sense. This, however, makes some assumption namely that one has an underlying space and that this space is homogeneous. But if we deal with a theory where this assumption is made then the space time translations must be a symmetry of the theory.

If one drops the assumption of homogeneity of the space as it is done in general relativity then the laws of physics become dependent on space and time. In this case there is no longer a translational symmetry and even more, no experiment becomes verifiable at least in the naive sense. However I do not want to discuss general relativity. I rather prefer to stay in realm of theories with a pre-given space-time.

The space time translations form an active symmetry in the sense that we have to rebuild the experiment. For the sake of simpler discussion assume one needs two pieces, one the machinery which prepares the state, and the other one the observable which analyses the state. If we translate the two parts, then one can not be sure that the two parts have been shifted exactly the same amount. Therefore one must have

$$(\omega,A)-(\beta_x\omega,\beta_{x'}A) = o(|x-x'|)$$

in order that the translations become an observable symmetry. This is of course true for every active symmetry and we obtain.

> An active symmetry can be observed as a symmetry if the symmetry group acts continuously.

There are not many active symmetries. Besides the translations there are also the relations which are a symmetry because the space is assumed to be isotropic.

Recall again that in the definition of the symmetry appears the shift of the observable and the shift of the state. In the presence of an infinite amount of matter one usually is interested in the local deviations from the background which is assumed to be in a kind of equilibrium. Shifting the local excitation is not the same as shifting also the background and hence it does not necessarily define a symmetry in the usual sense.

2.2 Lorentz boosts

The Lorentz-group has been known before Lorentz as the invariance group of the wave equation. Woldemar Voigt[3)] has used it for instance as a tool for computing the Doppler effect of the sound. But nowadays it is known as invariance-group of special relativity, which is a classical theory. Quantum mechanics of finite number of degrees of freedom is in contrast to this a theory which can not be transformed into a Lorentz invariant theory. The incorporation of special relativity into quantum mechanics is only possible in field theory, this means in a theory of infinitely many degrees of freedom, at least in some models. However, when it comes to more realistic models then again there are problems. It has been observed by Fröhlich, Marchio and Strocchi[4)] that in quantum electrodynamics the Lorentz symmetry must be broken in every charged sector as a consequence of Gauss law.

If we accept this result then it means that one can have Lorentz invariance at most in the vacuum sectors. If one wants to apply a Lorentz transformation to a charged sector then one gets to an inequivalent representation for the observables. This puts us into an unpleasant situation because we are no longer allowed to argue with help of Lorentz invariance. Even if we start with a theory which is formally (on the algebraic level) invariant under such transformations we are not able to derive from this e.g. the mass relation $p^2 = m^2$ for the electron.

Every physicist is, with great success, pretending that Lorentz transformations are symmetries also for charged particles. But this discussion shows that there is need for some explanation that one can do what one does. It is not enough just to remark that the two things do not contradict each other.

2.3 Internal "symmetries"

Much more mysterious to me becomes the situation if one wants to discuss internal symmetries of particle physics. For discussing the situation we stay with the simplest case, the isospin. Here the different components of a multiplet carry different charges. Which means nothing else that isospin is a broken symmetry. The usual philosophy is that one can describe physics by a Lagrangean theory and that the Lagrange function is the sum of different terms which describe different interactions, and which have different coupling constants. If now one lets the coupling constant in front of the breaking term tend to zero then one obtains a theory which is invariant.

I personally have strong objection against this procedure. It is generally accepted in physics that the set of observables shall separate the states. If one stays with this principle than it is impossible to distinguish a proton from a neutron if the electromagnetic interaction is switched off. This means that it is not only allowed to switch off the electromagnetic coupling constant but that one has to deform the symmetry group SU(2) at the same time in such a way that it disappears together with the electromagnetic interaction.

The disappearance of the group means that it becomes abelian and consequently the parameters of the Lie algebra of the symmetry group must be functions of the interaction which breaks it and they should go to zero together with the breaking.

Some people do not like the idea of internal symmetries being there only because they are broken. They usually argue that the scattering of two nucleons should exhibit the singlet and the triplet state. But this only can be observed when the "proton" and the "neutron" appear in different ratios. But since no interaction distiguishes the two nucleons we can not expect that they can be prepared in two different states. If this is the case then also the singlet and the triplet state will appear as a fixed superposition and consequently as one state and not as four different states.

If it is so that the internal symmetries can only be observed because they are broken then one should conclude that they are only present because they are broken. Such an attitude seems to be describable in Lagrangean field theory, but, if one is interested in the general background then one obtains problems. Whatever approach to field theory you like you are faced with the problem that you use symmetries for computations and you do not understand why you are allowed to do so.

I can not answer this problem of internal symmetries in particle physics. But for the problem of the Lorentz boosts there are some results which give at least the answer to some of the questions raised before. These results will be discussed in the next section.

3. THE PRINCIPLE OF MAXIMAL VELOCITY AND SOME OF ITS CONSEQUENCES

The frame for our discussion will be the theory of local observables. Doing experiments then they are done in a laboratory with some instruments which are activated for some time. This means for every real physical instrument we can associate a region in space-time. An abstraction of this situation is that one associates to every observables A a domain D(A) in which A can be measured. Then one puts

(A) In every bounded open region 0 in Minkowski space we denote by

$A(0)$ = algebra generated by the set of observables which can be measured in 0

(B) By this definition we must have $0_1 \subset 0_2$ implies $A(0_1) \subset A(0_2)$.

Furtheron we will assume that $A(0)$ are C^*-algebras and moreover

A denotes the C^*-inductive limit of $\bigcup_0 A(0)$.

(C) Since we want to have a theory on Minkowski-space we know from the discussion in the beginning of section 2 that the translations in space and time must be a symmetry of our theory this means for every $x \in R^d$ exists anautomorphism α_x of A such that $\alpha_x A(0) = A(0+x)$.

Since the representations π of the algebra A are considered as special situations one has to require that for physically reasonable representations the expressions $\pi(\alpha_x(A))$ shall be contiuous on R^d.

What we call reasonable depends on the situation one wants to describe. For example if π shall describe a thermodynamic system then we expect the continuity in the time alone. If we deal with particle physics then we require the continuity for space and time translations.

Definition: A representation π on the Hilbertspace H is called a particle representation if

(i) On H exists a continuous unitary representation $U(x)$ of the translation group R^d.

(ii) $U(x)$ implements the automorphism α_x this means

$$U(x)\pi(A)U^*(x) = \pi(\alpha_x(A))$$

(iii) The spectrum of the representation of $U(x)$ is contained in the forward light cone $\overline{V^+}$. This means there exists a spectral measure $E(\Delta)$ with

$$U(x) = \int_{\overline{V^+}} e^{i(p,x)} dE(p)$$

Since one believes that all physics is contained in the laws of particle physics one has to require that the observable algebra denoted by $\{A(0), A, R^d, \alpha\}$ permits a faithful particle representation.

In our set of axioms we have not required that the group of space-time translations U(x) does belong to the von Neumann algebra generated by $\pi(A)$. In the case of temperature states it is known that the time translations U(t) do not belong to this algebra. Therefore it is unclear in this situation in what sense the energy is an observable. In the case of particle physics the situation is much more favourable.

(D) The principle of maximal speed means that there is no particle or no information which can travel with a speed larger then the speed of light. This implies if we have two observables A and B and if their associated domains D(A) and D(B) are space-like separated then the two measurements can not influence each other. In quantum mechanics we know that two observables commute if the measurement of the one does not influence the measurement of the other and vice versa. But this implies:

If 0_1 and 0_2 are spacelike separated, and if $A \in A(0_1)$ and $B \in A(0_2)$ then $[A,B] = 0$.

(E) Finally the last condition deals with the characterization of those representations, which shall be associated to particle physics.

In the description of particles we want to guarantee that they can obtain a velocity at most that of light. This means the four momenta of the particle can take values only in the forward light cone. Since the forward light cone is an additive set we want that the common spectrum of energy-momentum operator takes values only in the forward light cone V^+.

There is however a problem, namely energy and momentum are not observables in our algebra of observables A, but they are so-called global observables and therefore they must be associated to the representations π of the algebra A. As usual in quantum theory the energy and momentum operators will be identified with the generators of the time and space translations. This leeds to the following

I. Theorem[5)]: Let $\{A,R^d,\alpha\}$ be a C*-dynamical system and let $V\subset R^d$ be a convex closed proper cone with interior points and, with dual cone V'. Assume (π,H) is a representation of A on H such that:

(i) On H exist a continuous unitary representation $U(x)$ of the group R^d.

(ii) $U(x)$ implements the automorphism α_x this means:

$$\pi(\alpha_x(A)) = U(x)\pi(A)U^*(x).$$

(iii) The spectrum of $U(x)$ is contained in V'.

Then there exist $U'(x)$ in $\pi(A)''$ with

(i) $U'(x)$ is a continuous unitary representation of the group R^d and $U'(x)\in\pi(A)''$.

(ii) $U'(x)$ implements α_x this means

$$U'(x)\pi(A)U'^*(x) = \pi(\alpha_x(A)).$$

(iii) Spectrum of $U'(x)$ is contained in V'.

This theorem tells us that in particle physics one can consider the energy and the momentum as limits of local observables.

But, there is still one problem namely the above conditions do not fix the spectrum uniquely, since the multiplication of $U(x)$ by a phase-factor implies a shift of the origin in momentum space. If we do not fix $U(x)$ then one can not speak about energy and momentum as absolute quantity, and in particular we are not allowed to compare the energies of two different particle representations. As an example in such situation we cannot speak about the mass of charged particle. But this is a situation which can not be excepted.

The result of Theorem I is a purely kinematical result and it holds for any convex cone V'. On the other hand our arguments about energy and momentum are taken from the real world this means that we cannot expect a unambiguous definition of energy and momentum without any imput of dynamics. Indeed it turns out that the principle of maximal velocity in configuration space, this means the locality condition is sufficient for the definition of energy and momentum. The result is the following.

II. Theorem[6)]: Let $\{A(0),A,R^d,\alpha\}$ be a theory of observables and let $\{\pi,H,U^1(x)\}$ be a particle representation that is $U^1(x)$ implements the translation, it belongs to $\pi(A)''$ and its spectrum is contained in the forward light cone V^+. Then there exists a unique minimal representation of the translation group $U(x)$ with:

(i) $U(x)$ is continuous,

(ii) $U(x)$ implements α_x,

(iii) Spectrum $U(x)$ is contained in V^+.

(iv) $U(x)$ is minimal in the following sense: Let $U(x) = e^{i(x,P)}$, and let $U^1(x)$ fulfill (i) - (iii), and $U^1(x) = e^{i(x,Q)}$, then $(t,P) \leq (t,Q)$, for every $t \in V^+$.

(v) $U(x)$ is unique and belongs to $\pi(A)''$.

The generators of the unique representation can now be indentified with energy and momentum of the theory and it can be given an absolute meaning. If we do so then we are in accordance with the usual field theory. If $\{\pi,H,U(x),\Omega\}$ is a vacuum representation and if we fix as usual $U(x)$ by the requirement that $U(x)\Omega = \Omega$ then $U(x)$ coincides with the above unique minimal representation. If the algebra of observables can be imbedded in a field algebra $\{F(0),F,R^d,\alpha\}$ and if $\{\pi,H,U(x),\Omega\}$ is a vacuum representation of F, then the restriction of π to the algebra of observables need no longer be irreducible. Then $\{\pi(A),H,U(x)\}$ is no longer a vacuum representation, but this $U(x)$ coincides with the unique minimal representation defined in Theorem II.

For the results we have obtained so far we have used the principle of maximal velocity but we have not assumed that Lorentz transformations are a symmetry of the theory. Therefore one might think that any subset of the forward light cone V^+ can appear as the support of the spectrum of the translations. But it turns out that this is not the case. By a miraculous interplay of the spectrum condition and the locality condition only a limited number of sets can appear as supports of the spectrum of the translations. The result is the following

III. Theorem[7]: Let $\{A(0),A,R^d,\alpha\}$ be a theory of local observables and let $\{\pi,H,U(x)\}$ be a particle representation of A with $U(x)$ the unique minimal representation of the translation described in Theorem II. Then spectrum $U(x)$ is a set which is invariant under Lorentz-transformations.

4. CONCLUSIONS

It is my personal belief that group theory is an invention of mankind but not of the gods. The assumption that there are continuous symmetries in nature seems not to withstand a close inspection. But nevertheless most approximating theories are full of symmetry groups. Therefore group theory is powerful theory for obtaining results.

This indicates that one is in a very paradoxical situation. If one does not accept that nature is leading us by the nose then one has to find reasons why nature is mimicking symmetry groups.

In special relativity the principle of maximal speed has the conformal group as its invariance group. But from this it is not allowed to conclude that it is also the symmetry group of every theory obeying the principle of maximal velocity. Nevertheless we have seen that every quantum mechanical particle theory has energy-momentum spectrum which is located on a Lorentz invariant subset of the forward light cone. From this one sees if the spectrum contains an isolated manifold than it is the manifold $p^2 = m^2$. Up to now we do not know all consequences of the interplay of the locality condition in configuration space and the spectrum condition in momentum space. But from what we known we must conclude that it would be very hard or almost impossible to device experiment showing the breaking of the Lorentz-symmetry.

The experience with the Lorentz group makes me hope that also for other symmetry groups exist simple principles which do not imply these symmetries, but which have consequences imitating the existence of a symmetry group. If one finds such principles one can hope that they have a much wider range of applications. The principle of maximal speed is also valid in general relativity and I am convinced that in future theory which combines quantum mechanics with general relativity the consequences of that principle are as drastic as in the flat case.

Having methods at hand which permit to evaluate the consequences of such principles then one might obtain results which are not consequences of a group symmetry. As an example in the theory which has been discussed before in details one obtaines:

<u>IV. Theorem[8)]</u>: Let $\{A(0),A,R^d,\alpha\}$ be a theory of local observables and assume $\{\pi,H,U(x)\}$ is a particle representation of A where U(a) is the unique minimal representation. Assume the spectrum of U(x) starts at m_0 and assume there is a gap in the spectrum which is between m_1 and m_2. Then one has

$$m_2 \leq 3m_0.$$

This tells us if the spectrum starts at m_0 then above $3m_0$ there are no gaps in the spectrum. It is assumed that π is not the vacuum-representation because for the vacuumsector we know[9)] that the continuum starts at least at $2m_0$. The physical interpretation of the above result is the following. Usually one is believing that a particle representation of the observable algebra is associated to some charge quantum number or some other global quantum number q. So since the spectrum starts at m_0 in our theory should exist particles with mass m_0 and charge q. Moreover one believes that to every particle should exist an antiparticle. If this is true then one has neutral particle--antiparticle pairs with masses $\geq 2m_0$. Since we can add to the particle with mass m_0 and charge q a neutral pair we see that in the same "charge sector" we have states with mass $\geq 3m_0$.

For the proof of the above theorem one does not have to leave the representation space H. This means charges and antiparticles do not appear in the proof. It indicates that the principle of maximal velocity implies the existence of antiparticles. R. Jost's proof of the T.C.P. theorem makes heavilyuse of the Lorentz invariance. But it might turn out that one day the T.C.P. theorem can be proved without the use of Lorentz invariance.

5. REFERENCES

1) Zeeman, E.C., "Causality Implies the Lorentz Group", J. Math. Phys. 5, 490-493 (1964).
Borchers, H.J. and Hegerfeldt, G.C., "The Structure of Space-Time Transformations", Commun. Math. Phys. 28, 259-266 (1972).
Borchers, H.J. and Hegerfeldt, G.C., "Über ein Problem der Relativitätstheorie: Wann sind Punktabbildungen des R^n linear?", Nachrichten der Göttinger Akademie der Wissenschaften Jahrgang 1972, Nr. 10, 205-229.

2) Faragó, P.S. and Jánossy, L., "Review of the Experimental Evidence for the Law of Variation of the Electron Mass with Velocity", Nuovo Cimento 5, 1411-1436 (1957).

3) Voigt, W., "Doppler's Princip", Göttinger Nachrichten (1887).

4) Fröhlich, J., Marchio, G. and Strocchi, F., "Infrared Problem and Spontaneous Breaking of the Lorentz Group in QED", Phys. Lett. 89B, 61 (1979).

5) Borchers, H.J., "Energy and Momentum as Observables in Quantum Field Theory", Commun. Math. Phys. 2, 49-54 (1966).
Borchers, H.J., "Translation Group and Spectrum Condition", Commun. Math. Phys. 96, 1-13 (1984).

6) Borchers, H.J. and Buchholz, D., "The Energy-Momentum Spectrum in Local Field Theories with Broken Lorentz-Symmetry", Preprint (1984).

7) See ref[6] and
Borchers, H.J., "Locality and Covariance of the Spectrum", Preprint (1984).

8) Borchers, H.J., "Additivity Properties of the Spectrum of the Translations", Preprint (1985).

9) Wightman, A.S., Trieste Lectures 1962, Theorem 12.

DYNAMICAL TREATMENT OF CONSTRAINTS IN GAUGE THEORIES: INTEGRATION OVER ALL POTENTIALS

Iwo Białynicki-Birula
Institute for Theoretical Physics, Polish Academy of Sciences
Lotników 32/46, 02-668 Warsaw, Poland

ABSTRACT

Two inequivalent approaches to the description of dynamical systems with constraints are known. In the first approach, the system is not allowed to leave the surface of constraints and the equations of constraints are used to reduce the number of the degrees of freedom. In the second approach, the constraints are enforced dynamically, by introducing appropriate forces, which prevent the system from departing too far from the surface of constraints. In these lectures, I explore the consequences of using the analog of the second approach in quantum theory of gauge fields. I shall argue that a natural way to dynamically enforce the constraints is to integrate over all potentials in the Feynman path integrals. The generating functionals obtained in this manner are explicitely gauge covariant, because no gauge-fixing terms are ever introduced. Two main disadvantages of this new approach are: the necessity to keep nonvanishing background sources and an apparent lack of renormalizability when the standard criteria are applied.

1. INTRODUCTION

Gauge theories play such an important role in modern theoretical physics that every effort leading to a better understanding of their properties is justified, even if this understanding does not yield any tangible results. The need for a better understanding of gauge theories is made more acute by two facts. i) We are still lacking a fully satisfactory formulation of gauge theories, especially of non-Abelian gauge theories, that would combine gauge and relativistic invariance with all the essential features of the quantum theory. ii) Gauge theories exhibit new, unique features not found in non-gauge theories; color confinement, monopoles and asymptotic freedom are three examples of such features, which require new methods and new formulations.

The best proof that a fully satisfactory formulation of gauge theories at the quantum level has not yet been found is a continuing proliferation of competing formulations. These formulations differ in many ways, offering various advantages, but none of them exhibits manifest gauge invariance. What is even more disturbing is the fact that a transition from one gauge to another often drastically changes the theory, for example, leading from a renormalizable to a nonrenormalizable theory. In my opinion, this state of affairs is highly unsatisfactory since the mathematical elegance and the physical beauty of gauge invariance and gauge transformations are the strongest arguments in favor of gauge theories.

In my lectures, I would like to describe a novel approach to the problem of the quantization of gauge field theories that puts the principle of manifest gauge invariance in the forefront and fully conforms to this principle. The possibility of a novel approach is rooted in an ambiguity of the quantization procedure for the theories with constraints.

Every field theory with gauge invariance describes a constrained system. The dynamics of constrained systems, even at the classical level, exhibits certain ambiguities. At the quantum level of description, these ambiguities are amplified and certainly deserve a thorough study.

In classical and quantum mechanics and in statistical mechanics the problems encountered in the description of constrained systems have been noticed and discussed in the literature. For a lucid review and a list of references, I refer the readers to a recent article by N.G. van Kampen and J.J. Lodder[1]. The results of these studies may be summarized as follows. There exist two approaches to the description of constrained systems, depending on whether one treats the constraints kinematically or dynamically. In the first approach, the space of allowed states is reduced by the equations of constraints. In the second approach, the motion of the system is influenced by an additional potential in the form of a potential gully, infinitely narrow in the limit, which is forcing the system to move along the surface of constraints. It was surprising to learn that these two approaches are not equivalent.

In the standard approach to gauge field theories, the constraints are treated kinematically. While at the classical level the dynamical treatment of constraints would be rather artificial, at the quantum level two distinct approaches corresponding to the kinematical and to the dynamical treatment are possible. The differences between these two approaches are best described in terms of Feynman path integrals. In the kinematical treatment of constraints, the path integral is extended only over those trajectories which obey the equations of constraints. In the dynamical treatment, the integration is to be extended over <u>all trajectories</u>, including those that do not satisfy the equations of constraints. Only in the simplest case, namely, for quadratic Lagrangians, these two approaches are equivalent, but even then they conceptually differ. It seems to me that for interacting fields, when the two approaches are inequivalent, the second method should also be tried and all its consequences should be explored. An attempt in this direction has been made in our earlier publication[2] and in the Ph.D. thesis of Jerzy Przeszowski[3]. In these lectures I shall outline the results of our investigations.

2. QUADRATIC LAGRANGIANS

As I have mentioned in the Introduction, quadratic Lagrangians have certain exceptional properties, when it comes to the discussion of constraints. I shall explain these properties using the simplest case of a quantum-mechanical particle moving in one dimension. As has been already shown in 1951 by Feynman[4], the path-integral representation of the propagator can be written in the Hamiltonian form

$$G(q,t;q_0,t_0) = \int Dq\, Dp \exp\{ i\int dt\, (p\dot{q} - H(p,q))\} \quad , \tag{1}$$

where the integration is extended over all trajectories in the phase space that originate at q_0 and end at q. The symbols Dp and Dq include appropriate normalization factors. For the Hamiltonians of the form

$$H(p,q) = (p + A(q))^2/2m + V(q) \quad , \tag{2}$$

the integration over p is Gaussian and can be easily performed. After this integration, we obtain tha Lagrangian form of the path integral,

$$G(q,t;q_0,t_0) = \int Dq \exp\{ i\int dt\, L(q,\dot{q})\} \quad , \tag{3}$$

where

$$L(q,\dot{q}) = m\,\dot{q}^2/2 - \dot{q}\,A(q) - V(q) \quad . \tag{4}$$

In contrast to the situation in classical mechanics, the equivalence of the Hamiltonian and the Lagrangian forms in quantum mechanics is not universal. This equivalence holds <u>only</u> when the integration over the momenta in (1) is tantamount to the elimination of the variable p from the expression $p\dot{q} - H$ with the help of the inverse Legendre transformation, i.e., by solving the equation

$$\dot{q} = \partial H/\partial p \tag{5}$$

with respect to p. For non-quadratic Hamiltonians the integration over p and the classical elimination of p through the formula (5) do not give the same results. Only for the Hamiltonians of the form (2), or equivalently for the Lagrangians of the form (4), these two procedures are equivalent.

The two procedures described above are the direct counterparts of the two treatments of the constrained systems, which were discussed in the Introduction. The kinematical approach, characterized by the use of the classical equations of constraints to reduce the number of the degrees of freedom, corresponds to the elimination of the momenta with the help of the classical Legendre transformation. In this approach the momenta and the velocities obey the classical relations (5). The dynamical approach, in which the trajectories that wander off the surface of constraints are allowed, corresponds to the integration over all trajectories in the phase space. For most such trajectories, the classical relations (5) do not hold.

This discussion applies, mutatis mutandis, to quantum field theory. The observed equivalence of the Hamiltonian and the Lagrangian form finds its analog in the often used equivalence of the first-order formalism and the second-order formalism. For example, in quantum electrodynamics one may either use the action W_1,

$$W_1 = \int d^4x \left(\tfrac{1}{4} f_{\mu\nu} f^{\mu\nu} - \tfrac{1}{2} f^{\mu\nu} (\partial_\mu A_\nu - \partial_\nu A_\mu) \right) \quad , \tag{6}$$

and integrate over all f's and all A's, or use the action W_2,

$$W_2 = \int d^4x \left(-\tfrac{1}{4} f_{\mu\nu} f^{\mu\nu} \right) \quad , \tag{7}$$

and integrate over all A's, treating the f's as being defined in terms of the potentials by the standard relation

$$f_{\mu\nu} = \partial_\mu A_\nu - \partial_\nu A_\mu \quad . \tag{8}$$

This relation is now the counterpart of the equations of constraints or the equation (5). For non-quadratic Lagrangians the equivalence of the first-order formulation and the second-order formulation does not hold in quantum theory. For example, in nonlinear electrodynamics of the Born-Infeld type[5)] the two forms of the action that correspond to W_1 and W_2 lead to equivalent classical theories, but give two different results when inserted into the path integrals.

3. GAUGE FIELDS

The ambiguities encountered in the quantization by path integrals are a mere reflection of the well known fact that most classical theories do not have unique quantum counterparts. The problem of finding a quantization procedure , which would unambiguously produce for every classical theory its quantum counterpart, does not seem to be very important. Most classical theories do not deserve their quantum versions. Who would need the quantum description of a gearbox or an avalanche? The problem of finding the correct method of quantization is important only for those few classical theories that are believed to be of fundamental significance. Gauge theories certainly fall into this cathegory.

Classical theory of gauge fields is understood now quite well, even in the non-Abelian case, with all the intricacies of its fiber bundle structure, topological invariants, monopole solutions, etc. Quantum theory, however, due to the enormous complexity is still in its infancy. Being unable to make precise predictions, we do not even know whether the canonical quantization procedure, borrowed essentially from nonrelativistic quantum mechanics, is working. In the Abelian case in quantum electrodynamics or in the non-Abelian case when the gauge symmetry is spontaneously broken, as in the Glashow-Salam-Weinberg theory of electroweak interactions, we are getting right answers, but even then all known formulations of the quantum theory lack elegance - a clear indication that something has gone wrong. Since there is little doubt that the classical version of QED or the GSW theory are the right starting point, the fault presumably lies in the quantum formulation. It is true that for ordinary relativistic field theories without the gauge group the canonical quantization prescription gives (almost) a unique result. Gauge theories are , however, quite different. Gauge invariance does not find a natural expression within the canonical framework, since the gauge parameters are not affected by the action of the physical generators, such as the energy, momentum and angular momentum[6]. This results in the loss of explicit gauge invariance and relativistic invariance. Path integrals are being now commonly used to bypass the complications of the canonical formulation in gauge theories, but it is always assumed that at the end we must reproduce the results of the canonical scheme. However, once we acknowledge the inherent ambiguity of the quantization procedure, we should start a search for new methods better suited for gauge theories, paying more attention to the internal consistency and elegance than to its compatibility with the canonical quantization. Integration over all potentials in Feynman path integrals should, in my opinion, be explored as one plausible candidate for such a new method. The integration over all potentials has been practiced in the past, but it was then always combined with an alteration of the Lagrangian resulting in the destruction of the gauge invariance. As a matter of fact, this method preceded the integration over the potentials restricted by the gauge condition. The restriction, in the form of a δ-function was introduced in QED in my paper[7] and later extended to non-Abelian gauge theories by Faddeev and Popov[8]. Such a restriction is indeed necessary for integrands that are fully invariant under gauge transformations, since the integration over the group parameters leads then to an infinite factor.

In the presence of external sources the integrand is not invariant under the gauge transformations, because the external sources, in general, distinguish certain directions in the gauge space. As a result, for nonvanishing external sources we may obtain a nonsingular result, provided the external sources completely fix the gauge. This interesting possibility has been overlooked by Sommerfield[9], who was the first to consider the integration over all potentials for gauge invariant Lagrangians. The simplest case of the free electromagnetic field, considered by Sommerfield, is pathological in this context, because the external current J^{μ} does not fix the gauge. The generating functional in this theory has a δ-function factor and, therefore, can not be properly normalized. It is sufficient, however, to couple the electromagnetic field to charged fields and to introduce nonvanishing sources of these fields in order to obtain a regular generating functional, without the δ-function singularity. In what follows, I will show how this program is working in the Abelian case of scalar electrodynamics and in the non-Abelian case of the Yang-Mills field coupled to the scalar field with the SU(2) gauge group.

4. SCALAR QUANTUM ELECTRODYNAMICS

The simplest case that allows for an implementation of the program outlined above is the theory of the charged scalar field coupled to the electromagnetic field - scalar electrodynamics. In this case the gauge invariant action functional has the form

$$S[A,\phi,\phi^*] = \int d^4x \,\{-\tfrac{1}{4} f_{\mu\nu}f^{\mu\nu} + (D_\mu\phi)^* D^\mu\phi - m^2\phi^*\phi\} \quad . \tag{9}$$

Following the procedure introduced in Ref.2, I shall expand the functional (9) around the field ϕ_0 generated by the constant source η_0,

$$m^2\phi_0 = \eta_0 \quad , \tag{10}$$

The quadratic part of the expanded action is

$$S^{(2)} = \int d^4x \,\{\tfrac{1}{2} A_\mu(g^{\mu\nu}\Box - \partial^\mu\partial^\nu)A_\nu - \tilde{\phi}^*(\Box + m^2)\tilde{\phi} + e^2\phi_0^*\phi_0 A_\mu A^\mu$$

$$+ ie\,\phi_0^*\partial^\mu A_\mu\tilde{\phi} - ie\,\tilde{\phi}^*\partial^\mu A_\mu\phi_0\} \quad , \tag{11}$$

where

$$\tilde{\phi} = \phi - \phi_0 \quad . \tag{12}$$

This quadratic form is not diagonal, but it can be diagonalized by simply shifting $\tilde{\phi}$ by the following A-dependent term

$$\tilde{\phi}(x) \to \tilde{\phi}(x) + ie\int d^4y\, \Delta_F(x-y)\, \partial^\mu A_\mu(y)\phi_0 \quad , \tag{13}$$

where Δ_F denotes the Feynman propagator of the scalar field. After the shift, the quadratic part of the action becomes

$$S^{(2)} = \int d^4x\,\{\tfrac{1}{2} A_\mu g^{\mu\nu}(\Box + \mu^2)A_\nu - \tilde{\phi}^*(\Box + m^2)\tilde{\phi}\}$$

$$+ \tfrac{1}{2}\int d^4x \int d^4y\, \partial^\mu A_\nu(x)\{\delta(x-y) + \mu^2\Delta_F(x-y)\}\partial^\nu A_\nu(y) \quad , \tag{14}$$

where $\mu^2 = 2e^2|\phi_0|^2$ plays the role of the mass term for the vector field, just as in the Higgs model with the spontaneously broken symmetry. However, this time the symmetry is broken by the external source η_0. This induced symmetry breaking has an interesting feature: the gauge parameters (the phase of the ϕ field in this case) adjust themselves automatically to the gauge chosen for the external source. As a result, the generating functional becomes gauge covariant, adjusting itself to the changes of the gauge for external sources. In Section 6 I shall derive explicit formulas showing the transformation properties of the generating functional under gauge transformations.

The propagators of the scalar field $\Delta^{(s)}$ and of the vector field $\Delta^{(v)}_{\mu\nu}$, that follow from the expression (14), have the following form in momentum space

$$\Delta^{(s)}(k) = (m^2 - k^2)^{-1} \quad , \tag{15}$$

$$\Delta^{(v)}_{\mu\nu}(k) = (\mu^2 - k^2)^{-1}(-g_{\mu\nu} + k_\mu k_\nu/\mu^2) + k_\mu k_\nu/\mu^2 m^2 \, . \tag{16}$$

The mass μ of the quanta of the A field has the same origin as the mass of the plasmons: the electromagnetic field is screened by the charges of the sources η and η^* of the scalar field.

Having determined the propagators of both fields, we may develop a perturbation theory expanding the trilinear and the quadrilinear terms in the exponent into the power series. The only remaining problem is that of renormalizability. The form of the propagator of the vector field indicates lack of renormalizability, at least in the conventional sense, since the propagator grows quadratically with k. On the other hand, it is hard to believe that a theory which is known to be renormalizable, in the gauge covariant form would loose this property. This problem certainly deserves a careful study, but I shall not undertake it in these lectures.

5. YANG-MILLS FIELD COUPLED TO A SCALAR FIELD

I shall restrict myself to the simplest case of the non-Abelian gauge group - the SU(2) group. The scalar field will be taken in the fundamental representation. In this theory also, as in scalar electrodynamics, every nonvanishing source of the scalar field completely fixes the gauge. This induced breakdown of the gauge symmetry again leads to the generation of the mass of vector particles. In order to see this, we should expand the action fuctional around the classical solution generated by a constant source term. All the calculations are practically the same as in the previous case and I shall not repeat them here. The propagator of the gauge field turns out to have the same form (16), except for the Kronecker symbol δ_{ab} with respect to the internal symmetry indices. Thus, all the quanta of the gauge field acquire the mass μ, with $\mu^2 = 2g^2 |\phi_0|^2$. This is achieved because the symmetry group is completely broken. Should there be a subgroup of unbroken symmetry left, the quanta of the vector field corresponding to this subgroup would remain massless. For example, if the scalar field is taken in the adjoint representation of SU(2) then a constant source of this field does not completely break the symmetry. There remains a U(1) symmetry subgroup (the little group that leaves the source field invariant) and one out of the three quanta of the gauge field does not acquire the mass.

The general discussion of this problem can be patterned after an analogous discussion in the theory of spontaneously broken symmetries, which has been described in detail by Louis Michel and other lecturers at this School.

In view of the necessity to keep in my approach the external sources different from zero at all times, there arrises the problem of the normalization of the generating functional. In the standard approach one normalizes this functional to 1 at the point where all the sources vanish. Such a simple normalization is ruled out because as we have seen this turns out to be a singular point. Instead, I shall normalize Z to 1 at the point where the source of the scalar field takes on a prescribed constant value. I expect that one may impose also here the natural normalization prescription at the zero values, but after the Legendre transformation from the source variables (J,η,η^*) to the mean field variables $(\langle A\rangle,\langle\phi\rangle,\langle\phi^*\rangle)$. The reason why it is possible that the effective action $\Gamma[\langle A\rangle,\langle\phi\rangle,\langle\phi^*\rangle]$ is finite at the zero point, where the generating functional is singular, is the singularity of the Legendre transformation at this point. The vector-field propagator exists and the potentials can be uniquely determined in terms of the external sources only when the sources of the scalar field do not vanish.

6. GAUGE COVARIANCE OF THE GENERATING FUNCTIONAL

The generating functional defined by the path integral over all potentials has a very simple dependence on the gauge parameters of the external sources. Under an arbitrary gauge transformation of the sources it changes only by a phase factor, very much like the Schrödinger wave function. I shall determine this phase factor in the more complicated, but at the same time more interesting case of the non-Abelian gauge theory. The Abelian case of scalar electrodynamics will be mentioned at the end.

Let us compare the values of the generating functional for two sets of the external sources (J,η,η^*) and $('J,'\eta,'\eta^*)$ that differ by a gauge transformation, i.e.,

$$'J^\mu_a(x) = T_{ab}(x)\, J^\mu_b(x) \quad , \tag{17a}$$

$$'\eta_\alpha(x) = S_{\alpha\beta}(x)\, \eta_\beta(x) \quad , \tag{17b}$$

where T_{ab} and $S_{\alpha\beta}$ represent the same element of the SU(2) group, but in two different representations: in the adjoint representation for the vector field and in the fundamental representation for the scalar field. In the path integral representation of the generating functional we may change the integration variables A and ϕ, transforming them according to the same gauge transformation (17). Since the volume element $dA\, d\phi\, d\phi^*$ and the action functional are gauge invariant, we get at the end the following relation

$$Z[J,\eta,\eta^*] = N \int dA\, d\phi\, d\phi^*\, e^{i(S[A,\phi,\phi^*] + J\cdot A + \eta^\dagger\cdot\phi + \phi^\dagger\cdot\eta)}$$

$$= e^{(-i\Omega/g)}\, Z['J,'\eta,'\eta^*] \quad , \tag{18}$$

where

$$\Omega = \tfrac{1}{2} \int d^4x\, J^\mu_a(x)\, \varepsilon^{abc}\, (\tilde{T}(x)\partial_\mu T(x))^{bc} \quad , \tag{19}$$

ε^{abc} denotes the Levi-Civita symbol and $\tilde{T}$ is the transpose of the T matrix. The phase factor arises from the inhomogeneous term in the transformation formula for the potentials,

$$A^a_\mu \to T^{ab}\, A^b_\mu + (2g)^{-1}\, \varepsilon^{abc}\, (\tilde{T}\partial_\mu T)^{bc} \quad . \tag{20}$$

The transformation formula (18) for the generating functional is the space-time analog of the formula obtained by Goldstone and Jackiw[10) for the wave functional.

The formulas (18) and (19) determine the dependence of the generating functional on the gauge parameters of the sources, since we may view the primed variables as being fixed in some particular gauge; for example, in the gauge in which the source of the scalar field has only one real component. Differentiating Z with respect to the gauge parameters (a rather tedious calculation that was explicitely performed in Ref.3) we may establish the validity of the following three equations

$$[\partial_\mu J^\mu_a + ig\, \varepsilon^{abc}\, J^\mu_b \delta/\delta J^\mu_c + ig(\eta_\beta \tau^a_{\alpha\beta}\delta/\delta\eta_\alpha + \eta^*_\alpha \tau^a_{\alpha\beta}\delta/\delta\eta^*_\beta)]Z[J,\eta,\eta^*] = 0 \tag{21}$$

These are the generating equations for the Ward-Fradkin-Takahashi identities in their purest form. These identities may also be obtained directly from the path integral (18) by performing an infinitesimal gauge transformation of the integration variables.

For scalar electrodynamics, the analog of Eqs.(18)-(20) are much simpler. The phase is given by

$$\Omega = \int d^4x\, J^\mu(x)\partial_\mu \alpha(x) \tag{22}$$

and the generating equation for the WFT identities is

$$[\partial_\mu J^\mu(x) - e(\eta(x)\delta/\delta\eta(x) - \eta^*(x)\delta/\delta\eta^*(x))]\; Z[J,\eta,\eta^*] = 0 \quad . \tag{23}$$

In the formula (22) α denotes the phase of the source , $\eta = |\eta| e^{-i\alpha}$.

Eqs.(21) and (23) are functional analogs of the current conservation and their validity in an unadulterated form is a clear indication that my formulation is gauge invariant. The significance of Eqs. (21) and (23) may be further elucidated by exploring the analogy between the generating functional and the wave functional, which has been mentioned above. This analogy is based on the following set of the substitution rules:

$$J^{\mu} \rightarrow -E^{i}\ , \quad -i\delta/\delta J^{\mu} \rightarrow A^{i}\ , \quad \eta \rightarrow \phi\ , \quad -i\delta/\delta\eta \rightarrow \pi^{\dagger},\quad \Psi \rightarrow Z\ , \qquad (24)$$

where E, A, ϕ and π are the canonically conjugate pairs of field operators and Ψ is the state vector represented in the Schrödinger representation by a functional of the field variables. Applying these rules to Eqs.(21) and (23), we obtain the Gauss law in scalar electrodynamics and in the Yang-Mills theory, imposed as a subsidiary condition on the state vector. Since any ad hoc modification of the Gauss law will play havoc with the fundamental properties of the gauge theory, we may expect that the modifications of Eqs.(21) and (23) are equally dangerous.

6. CONCLUSIONS

The formulation of quantum gauge theories presented in my lectures is certainly not complete. Its explicit gauge covariance is its greatest asset, but the apparent lack of renormalizability is its serious drawback. What remains to be done is to test the real value of this formulation by calculating the effective action beyond the tree approximation. If the radiative corrections can be made finite by some extension of the renormalization procedure, then we may have the first fully relativistic and gauge covariant formulation of gauge theories. At this point, I can only speculate on possible modifications of the renormalization procedure that could lead to the removal of the infinities. For example, it may become necessary to relate the physical source fields to the bare sources in a more complicated way than in the standard approach. There is no a priori reason, why the physical source fields should differ from the bare sources merely by multiplicative factors. More complicated (in general nonlinear) relations could provide for enough flexibility to conceal all infinities in these relations.

REFERENCES

1. van Kampen, N.G. and Lodder, J.J., Amer. J. Phys. 52, 419 (1984).
2. Bialynicki-Birula, I. and Przeszowski, J., Breaking of gauge symmetry by external sources and integration over all potentials, to appear in *Quantum Field Theory and Statistical Physics*.
3. Przeszowski, J., Ph.D. thesis (unpublished).
4. Feynman, R. P. Phys. Rev. 84, 108 (1951).
5. Born, M. and Infeld, L., Proc. Roy. Soc. A144, 425 (1934).
6. Bialynicki-Birula, I. and Bialynicka-Birula, Z. *Quantum Electrodynamics*, Pergamon, Oxford 1975, p.122.
7. Bialynicki-Birula, I., J. Math. Phys. 3, 1094 (1962).
8. Faddeev, L. D. and Popov, V. N., Phys. Lett. B 25, 30 (1967).
9. Sommerfield, C. M., Physica A 96, 309 (1979).
10. Goldstone, J. and Jackiw, R., Phys. Lett. B 74, 81 (1978).

HARMONIC SUPERSPACE: A NEW APPROACH TO EXTENDED SUPERSYMMETRY

E.Ivanov
Laboratory of Theoretical Physics,
Joint Institute for Nuclear Research

1. INTRODUCTION

The subject of these lecture is the new superfield (SF) approach to extended SUSY which is working out now at Dubna by Galperin, Ogievetsky, Kalitzin, Sokatchev and myself. It is based on the novel concept of harmonic superspace (SS) and seems to be capable enough to solve, on uniform geometric grounds, the longstanding problem of constructing uncounstrained off-shell SF formulations of all the theories with extended SUSY. To date, we have succeeded to describe along this line the N=2 matter, super Yang-Mills (SYM) and supergravity (SG) theories[1] as well as the N=3 SYM[2]. For the first two cases, a complete quantization procedure in harmonic SS has been recently developed[3].

Before starting, let me recall why are so important the unconstrained SF formulations. First, these ensure a deep insight into the extraordinary geometric structure of the SUSY theories. Second, and it is a more practical reason, such formulations drastically simplify the analysis of the divergency cancellations. In particular, the newly discovered remarkable property of certain models with extended SUSY, the property of ultraviolet finiteness (see e.g.[4,5b]) is expected to become manifest after passing to unconstrained SFs (this is just what comes about in the harmonic SS approach).

2. THE FUNDAMENTALS OF HARMONIC SUPERSPACE

2.1. Preliminaries. Any SS (like the ordinary Minkowski space), from the group-theoretical point of view is a coset space of the relevant supergroup over its certain subgroup. The problem of setting up a manifestly covariant SF description of a given SUSY theory is actually reduced to searching for the SS where the basic entities of the theory live as unconstrained SFs with a clear geometric meaning and where its fundamental gauge group gets the most simple and natural presentation. In the N=1 case, there are only two possibilities, viz. the real and chiral SSs :

$$\mathbb{R}^{4|4} = \frac{\{L_{\mu\nu}, P_m, Q_\alpha, \bar{Q}_{\dot{\alpha}}\}}{\{L_{mn}\}} = \{x^m, \theta^\alpha, \bar{\theta}^{\dot{\alpha}}\} \quad (2.1)$$

$$\mathbb{C}^{4|2} = \frac{\{L_{mn}, P_n, Q_\alpha, \bar{Q}_{\dot{\alpha}}\}}{\{L_{mn}, \bar{Q}_{\dot{\alpha}}\}} = \{x_L^m, \theta_L^\alpha\} \quad (2.2)$$

Both of them are utilized in the N=1 SUSY theories: the SF prepotentials of the N=1 SYM and SG are unconstrained functions over $\mathbb{R}^{4|4}$ while the matter SFs and the parameters of the SYM and SG gauge groups live in $\mathbb{C}^{4|2}$.

For N$\geq$2, the list of possible SS's is much richer due to the presence, in the algebra of the corresponding flat SUSY, of new purely internal generators associated with groups of automorphisms of spinor charges (besides, the central charges can appear). The straightforward generalizations of (2.1) and (2.2) are

$$\mathbb{R}^{4|4N} = \frac{\{L_{mn}, P_m, Q_\alpha^i, \bar{Q}_{\dot{\alpha}j}\}}{\{L_{mn}\}} = \{x^m, \theta_i^\alpha, \bar{\theta}^{\dot{\alpha}j}\} \quad (2.3)$$

$$\mathbb{C}^{4|2N} = \frac{\{L_{mn}, P_m, Q_\alpha^i, \bar{Q}_{\dot{\alpha}j}\}}{\{L_{mn}, \bar{Q}_{\dot{\alpha}j}\}} = \{x_L^m, \theta_{Li}^\alpha\} \quad (2.4)$$

which correspond to placing all the automorphism group generators into the stability subgroup. The number of θ -monomials increases as 2^{4N} or 2^{2N} with growing N, so the SFs defined on these SSs are highly reducible with respect to the superspin. To kill the superfluous superspins, one is forced to impose constraints or to allow for complicated gauge invariances of the nongeometric nature[5]. In view of this, it seems difficult to achieve satisfactory geometric unconstrained SF formulations of the $N \geq 2$ SUSY theories staying within $\mathbb{R}^{4|4N}$ or $\mathbb{C}^{4|2N}$ (with a finite number of SFs). Moreover, in a number of cases (e.g. for the N=2 Fayet-Sohnius hypermultiplet[6], for the $N \geq 3$ SYM) the no-go theorems exist[7,8] which state that it is not possible at all.

Fortunately, there is a way out, and it consists in going over to SSs of a more general type, to harmonic SSs. One has to include into play the automorphism group SU(N) of N-extended SUSY. Instead of considering, as in (2.3),

$$\frac{\text{Super - Poincaré}}{\text{Lorentz}} = \frac{\text{Super - Poincaré} \ltimes \text{SU(N)}}{\text{Lorentz} \times \text{SU(N)}}$$

one may replace SU(N) in the denominator by its certain subgroup H. This procedure leads to a <u>harmonic superspace</u>

$$Hr^{4+n|4N} = \frac{\text{Super-Poincaré} \ltimes \text{SU(N)}}{\text{Lorentz} \times H}, \quad H \in \text{SU(N)}, \quad n = \dim\frac{\text{SU(N)}}{H} \qquad (2.5)$$

with the 4+n - dimensional even part

$$M^4 \times \frac{\text{SU(N)}}{H} \qquad (2.6)$$

It is remarkable that such an extension of the even subspace allows us to single out in (2.5) analytic superspaces with a smaller Grassman dimensionality. The usual constraints become now the Grassman analyticity conditions[9] which are identically satisfied if the corresponding SF lives in the analytic subspace. The standard no-go theorems no longer work as these SFs contain infinite number of gauge and (or) auxiliary degrees of freedom. The harmonic SSs adequate to the N=2 and N=3 cases contain the following even subspaces[1,2]:

$$M^4 \times \frac{SU(2)}{SU(1)} \qquad \text{(6 dimensions) for } N=2$$

$$M^4 \times \frac{SU(3)}{U(1)\times U(1)} \qquad \text{(10 dimensions) for } N=3$$

where M^4 is ordinary Minkowski space. Further in this Section we concentrate on the case N=2.

2.2. Harmonic calculus on S^2. The internal sphere $S^2 \sim \frac{SU(2)}{U(1)}$ plays a crucial role in our construction. So, I explain first how to handle it.

We use a parametrization-independent description of S^2. The basic quantities are "zweibeins" $u_i^{\pm}$ which satisfy the condition

$$u^{+i}u_i^- = 1 \qquad (u^{\pm i} = \varepsilon^{ik}u_k^{\pm}) \tag{2.7}$$

and so form an SU(2)-matrix

$$u = \begin{pmatrix} u_1^+ & u_1^- \\ u_2^+ & u_2^- \end{pmatrix} \in SU(2), \quad u\bar{u} = I, \quad \det u = 1 \tag{2.8}$$

When SU(2) acts on u from the left, $u_i^{\pm}$ transform as an SU(2)-doublet with respect to index i. However, $u_i^{\pm}$ contain three independent parameters, and functions of $u_i^{\pm}$ are, in general, harmonic series on $S^3 \sim SU(2)$:

$$f(u) = \sum_{\substack{n=0 \\ m=0}}^{\infty} f^{(i_1 \dots i_n j_1 \dots j_m)} u^+_{(i_1} \dots u^+_{i_n} u^-_{j_1} \dots u^-_{j_m)} \tag{2.9}$$

where the parentheses mean symmetrization with the weight $1/(n+m)!$. The irreducible products of $u^{\pm}$ in the r.h.s. of (2.9) are nothing but the spherical harmonics, $u_i^{\pm}$ being the basic ones.

To pass to the coset SU(2)/U(1) with only two coordinates means to restrict properly the set of functions (2.9). To this end, let us introduce an auxiliary group U(1) with the right action on $u^{\pm}_i$, which are assumed to have the U(1)-charges $\pm$. This U(1) can, in fact, be extended to a whole right SU(2) with the generators:

$$D^{++}=u^{+i}\frac{\partial}{\partial u^{-i}}\,,\; D^{--}=u^{-i}\frac{\partial}{\partial u^{+i}}\,,\; D^{0}=[D^{++},D^{--}]=u^{+i}\frac{\partial}{\partial u^{+i}}-u^{-i}\frac{\partial}{\partial u^{-i}}; \tag{2.10}$$

$$D^{++}u^{+i}=0\,,\; D^{--}u^{+i}=u^{-i},\; D^{0}u^{\pm i}=\pm u^{\pm i} \tag{2.11}$$
$$D^{++}u^{-i}=u^{+i},\; D^{--}u^{-i}=0\,,$$

These generators commute with the initial SU(2) and, therefore, can be regarded as the covariant derivatives with respect to $u^{\pm}_i$. Being invariant under the action of SU(2) they can be used to reduce the set (2.9). Let us consider a subclass of (2.9) having a definite U(1) charge, i.e. being eigenfunctions of D^0 :

$$D^{0}F^{(q)}(u)=qF^{(q)}(u);\; U(1): F^{(q)\prime}(u')=e^{i\alpha q}F^{(q)}(u) \tag{2.12}$$

$$F^{(q)}(u)=\sum_{n=0}^{\infty}f^{(i_1\ldots i_{n+q}j_1\ldots j_n)}u^{+}_{(i_1}\ldots u^{+}_{i_{n+q}}u^{-}_{j_1}\ldots u^{-}_{j_n)} \tag{2.13}$$

The difference from the general expansion (2.9) is in the conservation of the U(1) charge throughout (2.13). One may easily see that the functions (2.13) effectively depend only on two real coordinates of S^2= SU(2)/U(1). Indeed, the SU(2)-matrix u (2.8) can always be divided into the product of the SU(2)/U(1) and U(1)-factors, and the latter can be absorbed into the overall U(1)-phase. Thus, to be left with the coset SU(2)/U(1), one should consider only those harmonic functions on SU(2) which are eigenfunctions of D^0 . Note that after imposing (2.12) we have two independent harmonic derivatives D^{++}, D^{--} , in accord with the dimensionality of S^2.

This approach simplifies considerably the integration rules on S^2. Instead of introducing an explicit parametrization of S^2 and then integrating over the manifold of parameters, we define the following (equivalent) integration rules

$$\int du \cdot 1 = 1$$
$$\int du\, (u^+)^{(m}(u^-)^{n)} = 0\,, \quad m+n>0 \qquad (2.14)$$

where

$$(u^+)^{(m}(u^-)^{n)} \equiv u^{+(i_1} \cdots u^{+i_m} u^{-j_1} \cdots u^{-j_n)} \qquad (2.15)$$

So, the integration on S^2 has become an algebraic operation (it simply singles out a u -independent part of integrand). An important rule due to the conservation of the U(1) charge on S^2 is that the integral of a charged quantity is always zero

$$\int du\, F^{(q)}(u) = 0\,, \quad q \neq 0 \qquad (2.16)$$

The rule for integration by parts is based on the property

$$\int du\, D^{++} F^{(-2)}(u) = \int du\, D^{--} F^{(+2)}(u) = 0$$

following from (2.13) and (2.14).

Useful technical tools are the harmonic δ-functions and the harmonic analogs of distributions of the type $\frac{1}{x^n}$. The defining relation of δ-functions is as follows[3]:

$$\int du_2\, \delta^{(q,-q)}(u_1,u_2) F^{(p)}(u_2) = F^{(q)}(u_1)\, \delta^{pq} \qquad (2.17)$$

whence

$$\delta^{(q,-q)}(u_1,u_2) = \sum_{n=0}^{\infty} (-1)^{n+q} \frac{(2n+q+1)!}{n!\,(n+q)!} (u_1^+)_{(n+q}(u_1^-)_{n)} (u_2^+)^{(n}(u_2^-)^{n+q)} \qquad (2.18)$$

Note that one needs to define separate δ -functions for each value of the U(1)-charge of the smearing function. Various properties of $\delta^{(q,-q)}(u_1,u_2)$ can be derived from eqs. (2.17), (2.18). In particular,

$$\delta^{(q,-q)}(u_1,u_2)=(u_1^+u_2^-)^q\,\delta^{(0,0)}(u_1,u_2) \tag{2.19}$$

where $\delta^{(0,0)}(u_1,u_2)$ is the ordinary δ-function of a special argument:

$$\delta^{(0,0)}(u_1,u_2)=2\,\delta\left[(u_1^+u_2^+)(u_1^-u_2^-)\right] \tag{2.20}$$

The second class of distributions which proves to be important in constructing harmonic Green functions[3] is given by the following harmonic expansions:

$$\frac{1}{(u_1^+u_2^+)^n}=\frac{1}{n!}\sum_{k=0}^{\infty}(-1)^{n+k}\,\frac{(2k+n+1)!}{k!\,(k+1)!}\,\frac{n}{n+k}\cdot (u_1^+)_{(k}(u_1^-)_{k+n)}(u_2^+)^{(k}(u_2^-)^{k+n)} \qquad (n\geq 0) \tag{2.21}$$

The notation $(u_1^+u_2^+)^{-n}$ is justified by the following properties of (2.21) (these all can be checked straightforwardly):

$$\begin{aligned}
(u_1^+u_2^+)^k\frac{1}{(u_1^+u_2^+)^n}&=\frac{1}{(u_1^+u_2^+)^{n-k}} \qquad k\geq 0\\
\frac{1}{(u_1^+u_2^+)^n}&=(-1)^n\frac{1}{(u_2^+u_1^+)^n}\\
D_1^{--}\frac{1}{(u_1^+u_2^+)^n}&=-n\,\frac{u_1^-u_2^+}{(u_1^+u_2^+)^{n-1}}
\end{aligned} \tag{2.22}$$

However, D^{++} has an unexpected action:

$$D_1^{++}\frac{1}{(u_1^+u_2^+)^n}=\frac{1}{(n-1)!}(D_1^{--})^{n-1}\delta^{(n,-n)}(u_1,u_2) \tag{2.23}$$

Of course, like the standard distributions, $(u_1^+u_2^+)^{-n}$ are ill defined at coinciding argements $u_1^+=u_2^+$ and in general, cannot be multiplied.

The last remark concerns the possibility to have real functions on S^2. The functions (2.13) cannot be real under the ordinary conjugation as $\overline{F^{(q)}}=\overline{F}^{(-q)}$. But it is possible to define another involution compatible with the sphere condition (2.7), namely a combination of the complex conjugation and the antipodal map of S^2:

$$(u^{\pm}_i)^* = u^{\pm i}\ ,\ (u)^{**} = -u \tag{2.24}$$

Now one can impose the reality condition

$$\left(F^{(q)}(u)\right)^* = F^{(q)}(u) \tag{2.25}$$

for even charges q. In particular, the analytic SFs (see sect. 2.3) can be real too.

2.3. <u>N=2 harmonic and analytic superspaces.</u> According to the general definition (2.5), the harmonic N=2 SS has the following coordinates

$$\mathbb{H}_2^{4+2|8} = \{x^m, \theta_{\alpha i}, \bar{\theta}^j_{\dot{\alpha}}, u^{\pm}\} \tag{2.26}$$

We will refer to the set (2.26) as to the <u>central basis</u> of $\mathbb{H}_2^{4+2|8}$. In this basis, SU(2)-covariance is explicit. The standard N=2 SFs defined in $\mathbb{R}^{4|8}$ are singled out by the evident condition

$$D^{++}\Phi\ (x,\theta,\bar{\theta},u) = 0 \Rightarrow \Phi = \Phi(x,\theta,\bar{\theta})$$

One may define in $\mathbb{H}_2^{4+2|8}$ also the other, <u>analytic basis</u>

$$\{x^m_A = x^m - 2i\theta^{(i}\sigma^m\bar{\theta}^{j)}u^+_i u^-_j,\ \theta^{\pm}_{\alpha} = u^{\pm}_i\theta^i_{\alpha},\ \bar{\theta}^{\pm}_{\dot{\alpha}} = u^{\pm}_i\bar{\theta}^i_{\dot{\alpha}},\ u^{\pm}_i\} \tag{2.27}$$

Now SU(2) is hidden, while the additional U(1) is explicit. The crucial point is that the set

$$\{x^m_A, \theta^+_{\alpha}, \bar{\theta}^+_{\dot{\alpha}}, u^{\pm}_i\} \equiv \{\zeta^M_A, u^{\pm}_i\} \tag{2.28}$$

is closed under the action of N=2 supersymmetry:

$$\begin{aligned} &\delta x^m_A = -2i\,(\varepsilon^i\sigma^m\bar{\theta}^+ + \theta^+\sigma^m\bar{\varepsilon}^i)u^-_i \\ &\delta\theta^+_{\alpha} = \varepsilon^i_{\alpha}u^+_i\ ,\ \delta\bar{\theta}^+_{\dot{\alpha}} = \bar{\varepsilon}^i_{\dot{\alpha}}u^+_i\ ,\ \delta u^{\pm}_i = 0 \end{aligned} \tag{2.29}$$

and thus forms a subspace $\mathbb{A}^{4+2|4}$ in $\mathbb{H}_2^{4+2|8}$. We call $\mathbb{A}^{4+2|4}$ the <u>analytic superspace</u> and the subclass of harmonic SFs restricted to $\mathbb{A}^{4+2|4}$ the <u>analytic</u>

superfields. These SFs are superextensions of the harmonic SU(2)/U(1)-functions (2.13), so they can carry the U(1)-charge q (along with the standard Lorentz indices). They satisfy the Grassman analyticity conditions having the most simple form in the basis (2.27)

$$D_\alpha^+ \phi^{(q)} = \frac{\partial}{\partial\theta^{-\alpha}} \phi^{(q)} = 0, \quad \bar{D}_{\dot\alpha}^+ \phi^{(q)} = -\frac{\partial}{\partial\bar\theta^{-\dot\alpha}} \phi^{(q)} = 0 \qquad (2.30)$$

It turns out that all the N=2 theories are formulated most naturally in terms of unconstrained superfunctions over $\mathbb{A}^{4+2|4}$.

How to handle the analytic SFs? Let us write down the θ-expansion of $\phi^{(q)}(\zeta_A, u^\pm)$:

$$\begin{aligned}\phi^{(q)}(\zeta_A, u) = {} & F^{(q)}(x_A, u) + \theta^{+\alpha}\psi_\alpha^{(q-1)}(x_A, u) + \\ & + \bar\theta^+_{\dot\alpha}\bar\varphi^{\dot\alpha\,(q-1)}(x_A, u) + \theta^+\theta^+ M^{(q-2)}(x_A, u) + \bar\theta^+\bar\theta^+ N^{(q-2)}(x_A, u) + \\ & + \theta^+\sigma^a\bar\theta^+ A_a^{(q-2)}(x_A, u) + \bar\theta^+\bar\theta^+\theta^{+\alpha}\xi_\alpha^{(q-3)}(x_A, u) + \\ & + \theta^+\theta^+\bar\theta^+_{\dot\alpha}\bar\chi^{\dot\alpha\,(q-3)}(x_A, u) + \theta^+\theta^+\bar\theta^+\bar\theta^+ D^{(q-4)}(x_A, u)\end{aligned} \qquad (2.31)$$

One immediately observes that the highest spin in $\Phi^{(q)}$ is carried out by the vector field and is one. So, $\phi^{(q)}$ for any q have zero superspin $Y = 0$ [x)]. On the other hand, each component in (2.31) is an SU(2)/U(1)-harmonic function of the type (2.13). Thus, $\phi^{(q)}$ contain infinite sets of ordinary fields with increasing isospins and, hence, infinite set of superisospins. The superisospin content of $\phi^{(q)}$ is specified by the formula[1]:

$$I = \left|\frac{q}{2} - 1\right| + n, \quad n = 0, 1, 2, \ldots \qquad (2.32)$$

Note that in the central basis $\phi^{(q)}$ is equivalent to an infinite tower of constrained ordinary N=2 SFs.

x) Extension to analytic SFs with Lorentz indices is straightforward .

We close this Section by several notes. First, D^{++} commutes with $D^+_\alpha, \bar{D}^+_{\dot\alpha}$ so $D^{++}\varphi^{(q)}$ is again an analytic SF. In the analytic basis D^{++} acquires a non-trivial x, θ - part. As applied to analytic SFs:

$$D_A^{++} = D^{++} - 2i\,\theta^+\sigma^m\bar\theta^+\partial_m^A \tag{2.33}$$

The second remark concerns the reality properties. The operation $\overset{*}{}$ admits a straigthforward extension to the supercase:

$$\overline{x_A^m}^* = x_A^m\,,\ \overline{(\theta^\pm_\alpha)}^* = \bar\theta^\pm_{\dot\alpha}\,,\ \overline{(\bar\theta^\pm_{\dot\alpha})}^* = -\theta^\pm_\alpha \tag{2.34}$$

The analytic SS (2.28) is closed under (2.24), (2.34) so we may define <u>real analytic SFs</u>

$$\overline{\varphi^{(q)}}^*(\zeta_A, u) = \varphi^{(q)}(\zeta_A, u)\,,\ q=2n \ \text{(cf.(2.25))} \tag{2.35}$$

This fact is crucial for the unconstrained SF description of N=2 SYM (Sect. 4).

Finally, we note that the standard notions of ordinary superspace (Berezin integral, Grassman δ-function, etc) have a simple extension to the present case[3]. An unusual feature is that the integration measure in $\mathbb{A}^{4+2|4}$ $d\zeta_A^{(-4)}du = d^4x_A d^2\theta^+ d^2\bar\theta^+ du$ has the U(1)-charge (-4) (since the Grassman integration is equivalent to differentiation). Therefore, the relevant δ-function has a nontrivial U(1)-assignment:

$$\int d\zeta_2^{(-4)}du_2\,\delta_A^{(q,4-q)}(\zeta_1,u_1|\zeta_2,u_2)F^{(p)}(\zeta_2,u_2) = \delta^{pq}F^{(q)}(\zeta_1,u_1)$$

$$\delta_A^{(q,4-q)}(\zeta_1,u_1|\zeta_2,u_2) = (D_2^+)^4\,\delta^{12}(z_1-z_2)\,\delta^{(q,-q)}(u_1,u_2) =$$

$$= (D_1^+)^4\,\delta^{12}(z_1-z_2)\,\delta^{(q-4,4-q)}(u_1,u_2) \tag{2.36}$$

Here $D^+_{\alpha(\dot\alpha)} = u^+_i D^i_{\alpha(\dot\alpha)}$, $(D^+)^4 \equiv \frac{1}{16}(D^{+\alpha}D^+_\alpha)(\bar D^+_{\dot\alpha}\bar D^{+\dot\alpha})$, $\delta^{12}(z_1-z_2)$ is the ordinary δ-function of $\mathbb{R}^{4|8}$ and $\delta^{(q,-q)}$

is defined by (2.18). The equivalence of two representations of $\delta_A^{(q,4-q)}$ can be easily checked using the basic properties of $\delta^{(q,-q)}(u_1,u_2)$.

Now we are ready to discuss concrete N=2 theories.

3. N=2 MATTER IN HARMONIC SUPERSPACE

3.1. Hypermultiplets. We begin with the Fayet-Sohnius hypermultiplet[6]. It has $Y=0$, $I=\frac{1}{2}$ and contains on-shell a scalar SU(2)-doublet $\varphi^i(x)$ and Dirac spinor $\psi_\alpha(x)$, $\bar{\rho}^{\dot\alpha}(x)$. A suitable analytic superfield is $q^+(\zeta_A, u)$ whose superisospin array starts just with $I=\frac{1}{2}$ (see eq. (2.32)). The free action and the equations of motion are

$$S_0^q = \int d\zeta_A^{(-4)} du\, \overset{*}{q}{}^+ D_A^{++} q^+ = \overset{*}{S}{}_0^q = \bar{S}_0^q \qquad (3.1)$$

$$D_A^{++} q^+ = 0 \qquad (3.2)$$

Using the explicit form of D_A^{++} (2.33) and expanding q^+ as in (2.31) one may easily see that the components with superisospins $> \frac{1}{2}$ do not propagate and thus are auxiliary, while for the physical fields the standard equations of motion appear. Thus we observe a principally new phenomenon consisting in the infiniteness of the number of auxiliary degrees of freedom. It has been proved recently that an off-shell formulation of q -hypermultiplet is possible if and only if the set of auxiliary fields is infinite[7].

One may easily promote (3.1) to the interacting theory. The requirement of SU(2)-invariance together with the conservation of U(1)-charge rather severely restrict a form of possible self-couplings:

$$S^q = S_0^q + S_{int}^q = \int d\zeta^{(-4)} du \left[\mathcal{L}^{(+4)}(\zeta,u) + \overset{*}{\mathcal{L}}{}^{(+4)}(\zeta,u)\right] \qquad (3.3)$$

$$\mathcal{L}^{(+4)} = \frac{1}{2}\overset{*}{q}{}^{+} D_A^{++} q^{+} + \lambda_1 (\overset{*}{q}{}^{+})^2 (q^{+})^2 + \lambda_2 \overset{*}{q}{}^{+} (q^{+})^3 + \lambda_3 (q^{+})^4 \quad (3.4)$$

The equation of motion is

$$D_A^{++} q^{+} = -4\lambda_1 \overset{*}{q}{}^{+} (q^{+})^2 - \lambda_2 (q^{+})^3 + 3\bar{\lambda}_2 q^{+} (\overset{*}{q}{}^{+})^2 + 4\bar{\lambda}_3 (\overset{*}{q}{}^{+})^3 \quad (3.5)$$

The auxiliary supermultiplets are expressed now in terms of the physical one to yield for the physical fields the Lagrangian of the σ-model type[1]. The bosonic part of the component action has the following generic form:

$$S_B = \int d^4x \{ g_{ik}(\varphi,\bar{\varphi}) \partial_m \varphi^i \partial_m \varphi^k + \bar{g}_{ik}(\varphi,\bar{\varphi}) \partial_m \bar{\varphi}^i \partial_m \bar{\varphi}^k + h_{ik}(\varphi,\bar{\varphi}) \partial_m \varphi^i \partial_m \bar{\varphi}^k \} \quad (3.6)$$

Based on the general theorem[10] one may conjecture that the metric in (3.6) is always hyperkählerian. To the present we explicitly know it only for the simplest interaction in (3.4) (with λ_1). As was expected, it is the hyperkählerian metric, namely, the Taub-NUT one[11]. It would be interesting to investigate the other cases too. Note that the much more general self-couplings become possible if one gives up the requirement of SU(2)-invariance[3]. The additional internal symmetries are easily introduced by attaching to q^+ appropriate group indices (q^+ may be assigned to complex representations without doubling the number of physical degrees of freedom).

Another simple representation of N=2 SUSY is the ω-hypermultiplet. It has $Y=0, I=1$ and reduces on-shell to the scalar singlet and isotriplet fields $\varphi(x)$, $\varphi^{(ij)}(x)$ and a doublet of Majorana spinors $\psi^{\alpha i}(x)$, $\bar{\psi}^{\dot{\alpha}}{}_i(x)$. Its first off-shell formulation, in terms of several constrained N=2 SFs, has been given by Howe, Stelle and Townsend[12]. The formulation in the harmonic SS seems to be more natural despite the infinite number of auxiliary fields. The corresponding SF is the scalar analytic SF $\omega(\zeta_A, u)$ with the following free action

$$S_0^{\omega} = \int dz^{(-4)} du \, \tfrac{1}{2} D_A^{++}\omega D_A^{++}\omega \;, \; \overset{*}{\omega} = \omega \tag{3.7}$$

whence

$$(D_A^{++})^2 \omega = 0 \tag{3.8}$$

Again, (3.8) kills in ω all superisospins except I=1. Nontrivial SU(2)-invariant interaction can be constructed only if the number of ω is two or more (otherwise it can be removed by an equivalence redefinition of ω):

$$S^{\omega} = \int dz^{(-4)} du \, g^{ik}(\omega) D_A^{++}\omega_i D_A^{++}\omega_k \,, \left(g^{ik} = \delta^{ik} + f^{ik}(\omega)\right) \tag{3.9}$$

In contrast to the q^+ hypermultiplet, putting ω in a complex representation of internal symmetry group would mean doubling the number of physical fields.

3.2. q-ω duality. In fact, q and ω hypermultiplets are not quite independent, being related to each other by a kind of duality transformation. To see this, let us come back to the free action (3.1). Besides the automorphism group SU(2), it is invariant under other $\widetilde{SU}(2)$ with respect to which q^+ and $\overset{*}{q}{}^+$ form a self-conjugated doublet:

$$q^{+i} = (q^+, \overset{*}{q}{}^+) \;, \; \overset{*}{q}{}^+_i = \varepsilon_{ik} q^{+k} \tag{3.10}$$

$$S_0^q = \int dz^{(-4)} du \left\{ \tfrac{1}{2} q^{+i} D_A^{++} q^+_i \right\} \tag{3.11}$$

Using the completeness property of u^{+i}, u^{-i} one may decompose q^{+i} as follows

$$q^{+i} = u^{+i}\omega + u^{-i} f^{++} \;, \; \overset{*}{\omega} = \omega, \; \overset{*}{f}{}^{++} = f^{++} \tag{3.12}$$

Insering (3.12) into (3.11) one obtains

$$S_0^q = \int dz^{(-4)} du \, \tfrac{1}{2} (\omega D_A^{++} f^{++} - f^{++} D_A^{++}\omega - f^{++} f^{++}) \tag{3.13}$$

The SF f^{++} is nonpropagating, so it can be eliminated by its equation of motion:

$$f^{++} = -D_A^{++}\omega \tag{3.14}$$

Substituting it back into (3.13) yields

$$S_0^q \Rightarrow \int dz^{(-4)} du \frac{1}{2} D_A^{++}\omega D_A^{++}\omega = S_0^\omega \tag{3.15}$$

In case of self-interacting q -hypermultiplet eq. (3.14) is modified by nonlinear terms but remains algebraic. Note that $\widetilde{SU}(2)$ is necessarily broken in interactions, however, this does not forbid the use of the decomposition (3.12). Thus, any self-interaction of q-hypermultiplets admits a dual form in terms of ω -ones.

3.3. <u>N=2 linear multiplet.</u> One more matter representation of N=2 SUSY is the N=2 linear multiplet[13]. Its physical field content is similar to that one of the ω -hypermultiplet. A crucial difference is, however, in that it contains a gauge antisymmetric field A_{mn} (notoph) instead of an isoscalar φ in ω . On-shell this difference disappears (A_{mn} has one physical degree of freedom on-shell[14]) but it results in different theories off-shell. In the harmonic SS language linear multiplet is described by a twofold charged analytic SF L^{++} with the constraint:

$$D^{++}L^{++} = 0 \tag{3.16}$$

and the free action

$$S_0^L = \int dz^{(-4)} du \, L^{++}L^{++} \tag{3.17}$$

It is easy to see that the condition (3.16) implies in particular, the standard notoph constraint for the vector field $A_\mu(x)$ (the first coefficient of the monomial $\theta^+\sigma^m\bar{\theta}^+$ in L^{++}):

$$\partial^m A_m = 0 \Rightarrow A_m = \varepsilon_{mnsp}\partial_s A_{np} \tag{3.18}$$

As is known, notoph is dual to the scalar field[14]. In the analytic SF language, this correspondence manifests itself as the duality between L^{++} and ω, the latter being introduced as the Lagrange multiplier for the constraint (3.16):

$$S_0^{L\prime} = \int dZ^{(-4)} du \left(L^{++} L^{++} + \omega D_A^{++} L^{++} \right) \tag{3.19}$$

Varying (3.19) with respect to ω we obtain just (3.16). On the other hand, varying with respect to L^{++} we have

$$L^{++} = -\frac{1}{2} D_A^{++} \omega \tag{3.20}$$

Putting it back into (3.19) produces the kinetic action for ω. This consideration immediately extends to the case of self-interacting L^{++}, demonstrating that any self-interaction of L^{++} is at the same time a self-interaction for ω. Note that the linear multiplet, as opposed to ω and q-hypermultiplets, is inconvenient to use to represent the N=2 matter because it admits no simple coupling with N=2 SYM (a covariantized version of the constraint (3.16) is unsolvable). At the same time, for ω and q-hypermultiplets such a coupling is constructed straightforwardly, by lengthening D_A^{++} in the actions (3.3), (3.9) (see Sect. 4.3).

3.4. <u>Quantization</u>. The quantization procedure in harmonic SS includes the construction of the SF Green functions and derivation of the relevant Feynman rules.

The Green functions for q and ω hypermultiplets are defined by the following equations:

$$D_1^{++} \langle \overset{*}{q}{}^{+}(1)\, q^{+}(2) \rangle = \delta_A^{(3,1)}(1|2) \tag{3.21}$$

$$D_1^{++} \langle \omega(1)\, \omega(2) \rangle = \delta_A^{(4,0)}(1|2) \tag{3.22}$$

where $\delta_A^{(q,4-q)}$ are analytic δ-functions defined in eqs. (2.36). It is not difficult to see that the following manifestly analytic expressions[3)]

$$\langle \overset{*}{q}{}^{+}(1)\, q^{+}(2)\rangle = -\frac{1}{\Box_1}(D_1^+)^4 (D_2^+)^4 \delta^{12}(z_1 - z_2)\frac{1}{(u_1^+ u_2^+)^3} \tag{3.23}$$

$$\langle \omega(1)\, \omega(2)\rangle = -\frac{1}{\Box_1}(D_1^+)^4 (D_2^+)^4 \delta^{12}(z_1 - z_2)\frac{u_1^- u_2^-}{(u_1^+ u_2^+)^3} \tag{3.24}$$

solve the equations (3.21), (3.22). In checking it, one may exploit a useful identity

$$\frac{1}{2}(D^+)^4 (D^{--})^2 \Phi(\zeta, u) = -\Box\, \Phi(\zeta, u) \tag{3.25}$$

for any analytic SF Φ .

The Green functions (3.23), (3.24) form a basis of the perturbation theory for N=2 quantum matter. As a simple example, let us consider the case of q hypermultiplet with the self-coupling $\sim (\overset{*}{q}{}^{+})^2 (q^+)^2$. By passing to the momentum representation one obtains the following Feynman rules

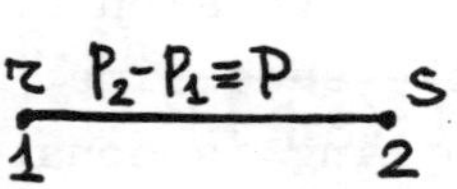

$$\frac{i}{p^2}\frac{(D_1^+)^4 (D_2^+)^4}{(u_1^+ u_2^+)^3}\delta^8(\theta_1 - \theta_2)\,\delta_s^r \tag{3.26}$$

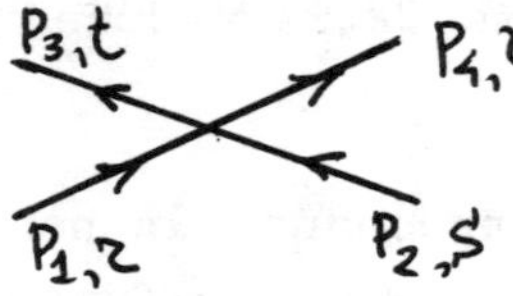

$$i\lambda\, \delta_t^{(r} \delta_v^{s)} (2\pi)^4 \delta(P_1 + P_2 - P_3 - P_4) \tag{3.27}$$

Fig. 1

where indices $r, s, \ldots$ refer to a representation of internal symmetry group to which q^+ may belong. At each vertex one integrates over all the internal momenta (with the measure $(2\pi)^{-4} d^4p$). Besides, an integration $\int d^4\theta^+ du$ (remaining from $\int d\zeta^{(-4)} du$ in the momentum representation) is also implied. One sees from (3.26) that at each analytic vertex there are factors $(D^+)^4$

which can always be used to restore the full Grassman measure $d^8\theta$ at the vertex. This important feature is common for the N=2 supergraph technique.

An illustrative example of N=2 supergraph calculations with the rules (3.26), (3.27) is the one-loop correction to the 4-point function

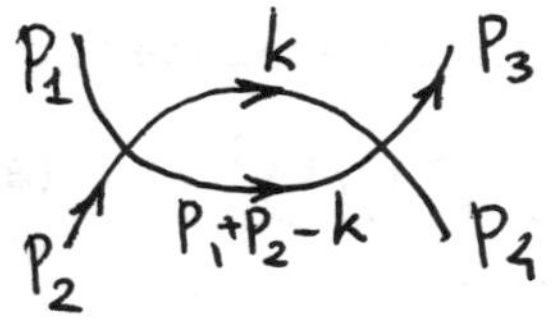

Fig. 2

The corresponding analytic expression is

$$\Gamma = \lambda^2 \int \frac{d^4p_1 \cdots d^4p_4\, d^4k}{(2\pi)^{16}} d^4\theta_1^+ d^4\theta_2^+ du_1 du_2\, \delta(p_1+p_2-p_3-p_4) \times$$

$$\times\, q^+(p_1,\theta_1,u_1)\, q^+(p_2,\theta_1,u_1)\, \overset{*}{q}{}^+(p_3,\theta_2,u_2)\, \overset{*}{q}{}^+(p_4,\theta_2,u_2) \times$$

$$\times\, \frac{(D_1^+)^4(D_2^+)^4}{(u_1^+u_2^+)^3}\, \delta^8(\theta_1-\theta_2)\, \frac{(D_1^+)^4(D_2^+)^4}{(u_1^+u_2^+)^3}\, \delta^8(\theta_1-\theta_2)\, \frac{1}{k^2(p_1+p_2-k)^2} \qquad (3.28)$$

By means of simple algebraic manipulations it is reduced to the form

$$\Gamma = \lambda^2 \int \frac{d^4p_1 \cdots d^4p_4\, d^4k}{(2\pi)^{16}} d^8\theta\, du_1 du_2\, \delta(p_1+p_2-p_3-p_4) \times$$

$$\times\, \frac{q^+(p_1,\theta,u_1)\, q^+(p_2,\theta,u_1)\, \overset{*}{q}{}^+(p_3,\theta,u_2)\, \overset{*}{q}{}^+(p_4,\theta,u_2)}{(u_1^+u_2^+)^2\; k^2\; (p_1+p_2-k)^2} \qquad (3.29)$$

Note that the harmonic distribution remaining in (3.29) does not lead to new divergences. Indeed, if the external lines are on-shell $(D^{++}q^+ \doteq 0)$, the u_2 integral in (3.29) can be computed using the identity

$$q^+(u_1)\, q^+(u_1) = \frac{1}{2} D_1^{++} D_1^{--} \left(q^+(u_1)\, q^+(u_1)\right)$$

and then integrating by parts with the help of the relation (2.23). After that the u-integral in (3.29) turns out to be

$$\frac{1}{2}\int du(\overset{*}{q}{}^+)^2 (D^{--})^2(q^+)^2$$

and does not contain any harmonic singularities. The momentum integral diverges logarithmically. Its divergent part is local in x-space (and thus in superspace):

$$\Gamma_\infty = -2\, c_\infty \lambda^2 \int d\zeta^{(-4)} du\, (\overset{*}{q}{}^+)^2 \Box (q^+)^2 \qquad (3.30)$$

Obviously Γ_∞ differs from the initial action (3.3) and the theory is nonrenormalizable (which is not surprising since $[\lambda] = M^{-2}$).

It is remarkable that in a three-dimensional space-time the graph in Fig. 2 is convergent. Moreover, in $d=2$ we may easily prove that the theory of the self-interacting q hypermultiplet is finite off-shell (the same applies to the ω-hypermultiplet as well as to more general couplings breaking SU(2)). In $d=2$ $[\lambda]=M^0$ and $[q^+(p,\theta,u)]=M^{-2}$. The n-particle contribution to the effective action has the generic form

$$\Gamma_n = \int d^8\theta\, du\, (d^2p)^{n-1} [q(p,\theta,u)]^n\, I(p)$$

The fact that the θ-integral has the full measure $d^8\theta$ follows from the Feynman rules, as explained above. We see that the momentum integral $I(p)$ has dimension M^{-2} and hence is convergent. As we have already mentioned, the self-interacting hypermultiplets reduce on-shell to the hyperkählerian σ-models[10)] (N=2 in d=4, N=4 in d=2). Thus, the manifestrly covariant supergraph technique provides a simple general proof of the finiteness of N=4, d=2 hyperkählerian σ-models. Note that the finiteness of some class of such models has been proved in[15)] at the component level by completely different methods.

4. N=2 GAUGE THEORY

4.1. λ-and τ-frames, analytic connection and all that. A natural starting point of geometric SF description of N=2 SYM is the familiar constraints in the ordinary N=2 SS[16]:

$$\{\mathcal{D}_\alpha^{(i}, \mathcal{D}_\beta^{j)}\} = \{\bar{\mathcal{D}}_{\dot\alpha}^{(i}, \bar{\mathcal{D}}_{\dot\beta}^{j)}\} = \{\mathcal{D}_\alpha^{(i}, \bar{\mathcal{D}}_{\dot\beta}^{j)}\} = 0 \tag{4.1}$$

$$\mathcal{D}^i_{\alpha(\dot\alpha)} = D^i_{\alpha(\dot\alpha)} + i A^i_{\alpha(\dot\alpha)}\ , \quad A^i_{\alpha(\dot\alpha)} \equiv A^{i\,b}_{\alpha(\dot\alpha)} T^b \tag{4.2}$$

where T^b are generators of gauge group. Harmonic N=2 SS comes into play in the process of solving these constraints. In terms of $\mathcal{D}^\pm_{\alpha(\dot\alpha)} = u^\pm_i \mathcal{D}^i_{\alpha(\dot\alpha)}$ the latter can be represented as the commutativity conditions[x)]:

$$\{\mathcal{D}^+_\alpha, \mathcal{D}^+_\beta\} = \{\bar{\mathcal{D}}^+_{\dot\alpha}, \bar{\mathcal{D}}^+_{\dot\beta}\} = \{\mathcal{D}^+_\alpha, \bar{\mathcal{D}}^+_{\dot\alpha}\} = 0 \tag{4.3}$$

which, as usual, imply for A^+_α, $\bar{A}^+_{\dot\alpha}$ a pure gauge form:

$$A^+_{\alpha(\dot\alpha)} \equiv A^i_{\alpha(\dot\alpha)} u^+_i = -i e^{-iv} D^+_{\alpha(\dot\alpha)} e^{iv} \tag{4.4}$$

$$\overset{*}{\overline{A^+_\alpha}} = -\bar{A}^+_{\dot\alpha} \quad \rightarrow \quad \overset{*}{\bar{v}} = v \tag{4.5}$$

The spinor connections $A^+_{\alpha(\dot\alpha)}$ by construction are linear in u^+_i that can also be expressed as the commutativity condition

$$[D^{++}, \mathcal{D}^+_{\alpha(\dot\alpha)}] = 0 \Rightarrow D^{++} A^+_{\alpha(\dot\alpha)} = 0 \Rightarrow A^+_{\alpha(\dot\alpha)} = A^i_{\alpha(\dot\alpha)} u^+_i \tag{4.6}$$

It follows from (4.6) that the harmonic prepotential $v(x, \theta, \bar\theta, u)$ is not arbitrary (in contrast to the N=1 case) but is subject to the constraints:

---

x) Such an interpretation of the N=2 SYM constraints has been suggested for the first time by Rosly[17].

$$\mathcal{D}^+_\alpha [e^{iv} D^{++} e^{-iv}] = \bar{\mathcal{D}}^+_{\dot\alpha} [e^{iv} D^{++} e^{-iv}] = 0 \tag{4.7}$$

These are easily recognized as the analyticity conditions for the SF object

$$V^{++} = -i e^{iv} D^{++} e^{-iv} \tag{4.8}$$

with no further constraints. Thus, we have solved (4.1) in terms of the pre-preportential V^{++} which is unconstrained SF over the analytic N=2 SS (2.28). To understand its geometric meaning, we note first that the solution (4.4) implies the existence of two different gauge groups in N=2 SYM, τ- and λ- ones:

$$e^{iv'} = e^{i\lambda} e^{iv} e^{-i\tau} \tag{4.9}$$

a) $D^{++}\tau = 0$ b) $D^+_{\alpha(\dot\alpha)}\lambda = 0 \; , \; \bar\tau = \tau \, , \; \overset{*}{\lambda} = \lambda$ (4.10)

The SF parameter τ does not depend on $u^\pm$ and is thus a standard real N=2 SF, whereas the parameter λ is a real analytic SF. In the representation (4.4) the τ - group is explicit: connections and matter fields transform with τ -parameters and due to eq. (4.10a) harmonic derivative D^{++} is covariant by itself. We call it the τ-representation or the τ -frame. On the other hand, there exists the other representation of the N=2 SYM, the λ -frame, where explicit is the λ-group. Passing to the λ -frame is accomplished by e^{iv} which thus plays the rôle of a "bridge" between two frames:

$$\mathcal{D}^+_{\alpha(\dot\alpha)\,(\lambda)} = e^{iv} \mathcal{D}^+_{\alpha(\dot\alpha)\,(\tau)} e^{-iv} = D^+_{\alpha(\dot\alpha)} \tag{4.11}$$

$$\mathcal{D}^{++}_{(\lambda)} = e^{iv} D^{++}_{(\tau)} e^{-iv} = D^{++} + iV^{++} \tag{4.12}$$

$$V^{++\prime} = e^{i\lambda}(V^{++} - iD^{++}) e^{-i\lambda} \tag{4.13}$$

Eqs. (4.11) show that the analyticity is manifest in the λ -frame, while the notion of u -independence becomes covariant (in the τ -frame the situation is opposite). The meaning of V^{++} is clear now: it is nothing but the gauge connection for the λ -group with respect to purely harmonic directions. To be convinced that its physical content corresponds exactly to N=2 gauge supermultiplet with $Y=0$, $I=0$ it is instructive to pass to the Wess-Zumino gauge where V^{++} is represented as

$$V^{++}_{W.Z.}(\zeta_A, u) = \theta^+\theta^+[M(x_A)+iN(x_A)]+\bar{\theta}^+\bar{\theta}^+[M(x_A)-iN(x_A)]+$$
$$+i\theta^+\sigma^a\bar{\theta}^+A_a(x_A)+\bar{\theta}^+\bar{\theta}^+\theta^{+\alpha}\psi^i_\alpha(x_A)u^-_i+$$
$$+\theta^+\theta^+\bar{\theta}^+_{\dot{\alpha}}\bar{\psi}^{\dot{\alpha}i}(x_A)u^-_i+\theta^+\theta^+\bar{\theta}^+\bar{\theta}^+D^{(ij)}(x_A)u^-_iu^-_j$$

Thus the infinite "tail" of superisospins beyond $I=0$ in V^{++} is purely gauge.

4.2. The analytic SF structure of N=2 SYM action. The action of N=2 SYM is known to be composed of the strength tensor $\bar{W}$ [16] introduced by:

$$\{\mathcal{D}^+_\alpha, \mathcal{D}^-_\beta\} = -2i\,\varepsilon_{\alpha\beta}\bar{W} \tag{4.14}$$

where, in the τ -frame

$$\mathcal{D}^-_\beta = D^-_\beta + iA^-_\beta\,,\quad A^-_\beta = (A^+_\beta)^* = -ie^{-i\bar{v}}D_\beta e^{i\bar{v}} \tag{4.15}$$

$$[D^{++}, \mathcal{D}^-_\beta] = \mathcal{D}^+_\beta \tag{4.16}$$

Using (4.4), (4.15) one gets

$$\bar{W} = -\frac{i}{4}e^{-i\bar{v}}\{D^-_\alpha[e^{i\bar{v}}e^{-iv}(D^{+\alpha}e^{iv}e^{-i\bar{v}})]\}e^{i\bar{v}} \tag{4.17}$$

or, after some algebraic manipulations

$$\bar{W} = -\frac{i}{4}e^{-iv}[(D^+)^2(e^{iv}D^{--}e^{-iv})]e^{iv} \tag{4.18}$$

The tensor $\bar{W}$ possesses a number of important properties. First, it directly follows from (4.14), (4.3), (4.6) and

(4.17) that $\overline{W}$ does not depend on $u^{\pm}$:

$$D^{++}\overline{W} = D^{--}\overline{W} = 0 \tag{4.19}$$

Second, $\overline{W}$ is covariantly chiral

$$\mathcal{D}^{+}_{\alpha}\overline{W} = \mathcal{D}^{-}_{\alpha}\overline{W} = 0 \tag{4.20}$$

It stems from eqs. (4.18), (4.6), (4.15). Finally, $\overline{W}$ obeys the irreducibility condition

$$\mathcal{D}^{+\alpha}\mathcal{D}^{-}_{\alpha}W = \bar{\mathcal{D}}^{+}_{\dot{\alpha}}\bar{\mathcal{D}}^{-\dot{\alpha}}\overline{W} = 0 \tag{4.21}$$

which is a direct consequence of Bianchi identities and can be checked to be compatible with (4.17), (4.18).

Due to (4.19), (4.20) the trace $tr(\overline{W}\overline{W})$ does not depend on $u^{\pm}$, is chiral and hence can be used to write the action[16] as an integral over chiral N=2 SS:

$$S_{SYM} = \frac{1}{2}\int dz_R \, tr(\overline{W}\overline{W}) \quad \left(z^M_R = (x^m_R = x^m - i\theta^k\sigma^m\bar{\theta}_k, \bar{\theta}^i_{\dot{\alpha}})\right) \tag{4.22}$$

For actual calculations one needs to know the expansion of S in terms of unconstrained analytic connection V^{++}. The variation of S can be found straightforwardly, using the properties (4.19)-(4.21) of $\overline{W}$. It turns out to be an integral over the analytic SS:

$$\delta S = -i\int dz^{(-4)} du \, tr\left[\delta V^{++}(D^{+})^4\left(e^{iv}D^{--}e^{-iv}\right)\right] \tag{4.23}$$

It then follows that the SF equations of motion are

$$(D^{+})^4\left(e^{iv}D^{--}e^{-iv}\right) = 0 \quad \text{or} \quad \bar{\mathcal{D}}^{+}\bar{\mathcal{D}}^{+}\overline{W} = 0 \tag{4.24}$$

To learn the V^{++}-expansion of S, one should find its next variations. It can be done with the help of the formula

$$e^{-iv}\delta e^{iv} = -i\int du_1 \frac{u^{+}u^{-}_1}{u^{+}u^{+}_1}(\delta V^{++})_{\tau}(z, u_1) \tag{4.25}$$

$$(\delta V^{++})_{\tau} \equiv e^{-iv}\delta V^{++}e^{iv}$$

the proof of which is based on the properties of harmonic distributions (2.19), (2.23). By eq. (4.25), all the higher-order variations of e^{iv} are expressed in terms of the first-order one. The second-order variation is, e.g.

$$\delta_2\delta_1 e^{iv} = e^{iv}\int du_1 du_2 \frac{u^+u_1^-}{u^+u_1^+}\Big\{\frac{u_1^+u_2^-}{u_1^+u_2^+}\big[(\delta_2 V^{++})_\tau,(\delta_1 V^{++})_\tau\big] - \frac{u^+u_2^-}{u^+u_2^+}(\delta_2 V^{++})_\tau(\delta_1 V^{++})_\tau\Big\} \quad (4.26)$$

Using these formulas it is a direct exercise to compute S to any order in V^{++}. Let us quote the kinetic part and the $(V)^3$ -interaction vertex[3)]:

$$S_{kin} = \frac{1}{2}\int dz^{(-4)} du_1 du_2 \, tr\Big[V^{++}(1)(D_1^+)^4 \frac{1}{(u_1^+u_2^+)^2} V^{++}(2)\Big] \quad (4.27)$$

$$S^3_{int} = \frac{g}{3!} f_{abc}\int dz^{(-4)} du_1 du_2 du_3 V_a^{++}(1)(D_1^+)^4 \frac{V_b^{++}(2)\, V_c^{++}(3)}{(u_1^+u_2^+)(u_1^+u_3^+)(u_2^+u_3^+)} \quad (4.28)$$

We see that the singularities of various factors of the denominator do not coincide (this property persists for the higher-order terms too). It is remarkable that there are no derivatives in the vertices obtained (unlike the case N=1). The harmonic nonlocalities prove to disappear in supergraph calculations.

4.3. Coupling to matter. Couplings of N=2 SYM with analytic N=2 matter are introduced on the pattern of N=0 case: by changing

$$D^{++} \to \mathcal{D}^{++} = D^{++} + iV^{++} \quad (4.29)$$

in the matter actions (3.3), (3.9). As an instructive example we consider the minimal gauge coupling of ω - hypermultiplet in the adjoint representation of gauge group

$$S^g_\omega = \int dz^{(-4)} du \, \frac{1}{2}\omega^a\big[(D^{++}+iV^{++})^2\big]^{ab}\omega^b$$
$$(V^{++})^{ab} = V^{++c}(T^c)^{ab}, \quad tr\, T^cT^m = \delta^{cm} \quad (4.30)$$

One may check that the sum $S = S_{SYM} + S^{g}_{\omega}$ possesses two additional supersymmetries

$$\delta V^{++} = \varepsilon^{i\alpha} u^{+}_{i} \theta^{+}_{\alpha} \omega + \overline{(\quad)}^{*}$$
$$\delta \omega = \frac{1}{2} (D^{+})^{4} (\varepsilon^{i\alpha} u^{-}_{i} \theta^{-}_{\alpha} e^{iv} D^{--} e^{-iv}) + \overline{(\quad)}^{*} \tag{4.31}$$

completing N=2 SUSY to N=4 SUSY. Thus this sum is identified as the N=4 SYM action written in terms of N=2 SFs.

4.4. Quantum N=2 SYM. Quantization of N=2 SYM in harmonic SS goes along the standard lines for gauge theories: one fixes the gauge, introduces the Faddeev-Popov ghosts etc. Due to nonlocality in harmonics this procedure is rather subtle in some points, but nevetherless it can be carried through completely. Let us omit details of computation and present the final answer for the kinetic part of the full N=2 SYM quantum action with the gauge fixing and ghost terms[3]:

$$\begin{aligned} S_{SYM} + S_{gf} + S_{FP} &= \frac{1}{2\alpha} \mathrm{tr} \int d\zeta^{(-4)} du\, V^{++} \Box V^{++} + \\ &+ \frac{1}{2}\left(1 + \frac{1}{\alpha}\right) \mathrm{tr} \int d\zeta^{(-4)} du_1 du_2\, V^{++}(1) (D^{+}_{1})^{4} \frac{1}{(u^{+}_{1} u^{+}_{2})^{2}} V^{++}(2) + \\ &+ \mathrm{tr} \int d\zeta^{(-4)} du\; i\, F\, D^{++} (D^{++} + i V^{++}) P + \ldots \end{aligned} \tag{4.32}$$

Here F^{a}, P^{a} are the ghost analytic SFs belonging to the adjoint representation of gauge group and α is a gauge-fixing parameter. Note that the kinetic operator in (4.32) is nondegenerate, thus avoiding the ghost-for-ghost problem[18].

After the gauge has been fixed, one can derive the Green function for the gauge superfield

$$G^{(2,2)}(1|2) = \langle V^{++}(1) V^{++}(2) \rangle$$

Its equation follows from (4.32). After some algebra it can be represented as

$$\int dz_2^{(-4)} du_2 \left[-\Box_1 \Pi^{(2,2)}(1|2) + \frac{1}{\alpha}\Box_1 \left(\delta_A^{(2,2)}(1|2) - \Pi^{(2,2)}(1|2) \right) \right] G^{(2,2)}(2|3) = \tag{4.33}$$
$$= \delta_A^{(2,2)}(1|3)$$

where

$$\Pi^{(2,2)}(1|2) = -\frac{(D_1^+)^4 (D_2^+)^4}{\Box_1} \frac{1}{(u_1^+ u_2^+)^2} \delta^{12}(z_1 - z_2)$$

is the projection operator for superisospin 0 for the analytic SFs. Using the projection property

$$\int dz_2^{(-4)} du_2 \Pi^{(2,2)}(1|2) \Pi^{(2,2)}(2|3) = \Pi^{(2,2)}(1|3)$$

we solve (4.33) as

$$G^{(2,2)}(1|2) = \frac{\alpha}{\Box_1} \delta_A^{(2,2)}(1|2) - \frac{1+\alpha}{\Box_1} \Pi^{(2,2)}(1|2) \tag{4.34}$$

The form of this propagator is reminiscent of that for the ordinary (N=0) gauge theories in the familiar α gauges. At $\alpha=0$ one obtains the Landau-Lorentz gauge, at $\alpha=-1$ - the Fermi-Feynman one which is preferred, being simplest and leading to a better infrared behaviour. The Green functions for the Faddeev-Popov ghosts are the same as for the ω - hypermultiplet (3.24).

The first examples of manifestly invariant N=2 supergraph calculations on the base of this quantization procedure are given in[3]. In particular, the one-loop pure Yang-Mills contribution to the V^{++} self-energy has been calculated and shown to cancel out the analogous contribution coming from the ω -hypermultiplet in the adjoint representation, in agreement with the finiteness property of N=4 SYM theory. The following step in developing the harmonic supergraph techniques is to work out a suitable variant of background field formalism which is known to be a powerful method of quantum calculations.

Closing this section we would like to emphasize that the main lesson we have drawn for the time being from the

N=2 supergraph quantum calculations is that these are not more difficult than in the case N=1. No new divergences associated with harmonic integrals appear. When the external lines are placed on-shell, all such integrals can be computed by simple algebraic manipulations, and dependence on harmonic coordinates is absent in the final answer. The crucial advantage of harmonic supergraph techniques is the preservation of manifest N=2 SUSY at each step of calculation.

5. OVER THE N=3 BARRIER

Harmonic superspace approach has been successfully applied to solve the problem of constructing an off-shell SF formulation of the N=3 SYM theory[2] (on-shell it is equivalent to N=4 SYM). The so-called "N=3 barrier", the "no-go" theorem about the absence of $N \geq 3$ SUSY gauge theories off-shell[8], has been circumvented due to the infinite number of auxiliary fields.

5.1. N=3 harmonic SS and its analytic subspace. Harmonic N=3 SS adequate to N=3 SYM turns out to have as an internal part the 6-dimensional coset[2]

$$SU(3)/U(1)\times U(1)$$

The N=3 harmonics can be gathered into an SU(3)-matrix

$$u = \begin{pmatrix} u_1^{(1,1)} & u_1^{(-1,1)} & u_1^{(0,-2)} \\ u_2^{(1,1)} & u_2^{(-1,1)} & u_2^{(0,-2)} \\ u_3^{(1,1)} & u_3^{(-1,1)} & u_3^{(0,-2)} \end{pmatrix}, \quad \det u = 1\,, \quad u\bar{u} = \bar{u}u = I \tag{5.1}$$

where the couple of superscripts (a,b) stand for the values of two right U(1)-charges. The matrix U transforms as

$$u' = g\,u\,h\,, \quad g \in SU(3)\,, \quad h = e^{i(a_1H_1 + a_2H_2)}$$
$$H_1 = \begin{pmatrix} 1 & 0 & 0 \\ 0 & -1 & 0 \\ 0 & 0 & 0 \end{pmatrix}, \quad H_2 = \begin{pmatrix} 1 & 0 & 0 \\ 0 & 1 & 0 \\ 0 & 0 & -2 \end{pmatrix} \tag{5.2}$$

The harmonic calculus on SU(3)/U(1)×U(1) is analogous to that for the N=2 case. There exist 6 (as many as the dimension of SU(3)/U(1)×U(1)) covariant harmonic derivatives compatible with the conditions (5.1):

$$
\begin{aligned}
D^{(1,3)} &= -u_i^{(1,1)}\frac{\partial}{\partial u_i^{(0,-2)}} + \bar{u}^{(0,2)i}\frac{\partial}{\partial \bar{u}^{(-1,-1)i}} \\
D^{(-1,3)} &= u_i^{(-1,1)}\frac{\partial}{\partial u_i^{(0,-2)}} - \bar{u}^{(0,2)i}\frac{\partial}{\partial \bar{u}^{(1,-1)i}} \\
D^{(2,0)} &= u_i^{(1,1)}\frac{\partial}{\partial u_i^{(-1,1)}} - \bar{u}^{(1,-1)i}\frac{\partial}{\partial \bar{u}^{(-1,-1)i}}
\end{aligned}
\qquad (5.3)
$$

and their conjugates.

An analog of the N=2 analytic SS (2.28) is the following SS

$$\{\zeta_A^M = (x_A^{\alpha\dot\alpha}, \theta^{(1,-1)\alpha}, \theta^{(0,2)\alpha}, \bar\theta^{(1,1)\dot\alpha}, \bar\theta^{(-1,1)\dot\alpha}), u\} \qquad (5.4)$$

$$x_A^{\alpha\dot\alpha} = x^{\alpha\dot\alpha} + 2i\left[\theta^{(0,2)\alpha}\bar\theta^{(0,-2)\dot\alpha} - \theta^{(-1,-1)\alpha}\bar\theta^{(1,1)\dot\alpha}\right] \qquad (5.5)$$

$$\theta^{(a,b)\alpha} = \bar{u}^{(a,b)i}\theta_i^{\alpha}, \quad \bar\theta^{(a,b)\dot\alpha} = u_i^{(a,b)}\bar\theta^{i\dot\alpha} \qquad (5.6)$$

It is closed both under the N=3 SUSY transformations and the corresponding operation $\overline{*}$:

$$
\begin{aligned}
&u_i^{(1,1)} \overset{(\overline{*})}{\longleftrightarrow} u^{(0,2)i}, \quad \theta_\alpha^{(1,-1)} \overset{(\overline{*})}{\longleftrightarrow} -\bar\theta_{\dot\alpha}^{(-1,1)}, \quad x_A^{\alpha\dot\alpha} \overset{\overline{*}}{\longleftrightarrow} x_A^{\alpha\dot\alpha} \\
&u_i^{(0,-2)} \longleftrightarrow u^{(-1,-1)i}, \quad \theta_\alpha^{(0,2)} \longleftrightarrow \bar\theta_{\dot\alpha}^{(1,1)} \\
&u_i^{(-1,1)} \longleftrightarrow -u^{(1,-1)i},
\end{aligned}
\qquad (5.7)
$$

The latter property allows us to consider real analytic N=3 SFs.

5.2. <u>Geometry of N=3 SYM</u>. The N=3 SYM constraints being written in the standard real N=3 SS, are similar to those for the N=2 case[19,20]. An important difference is that these put the theory on-shell. Contracting SU(3)-indices of gauge covariant spinor derivatives $\mathcal{D}_\alpha^i$ with proper harmonics $u_i^{(a,b)}$ one may again represent the N=3 SYM constraints as the commutativity conditions expressing the covariant preservation of analyticity and solve them in terms of a "bridge" $e^{iv(x,\theta,\bar\theta,u)}$:

$$e^{iv'} = e^{i\lambda} e^{iv} e^{-i\tau} \qquad (5.8)$$
$$D^{(a,b)}\tau = 0,\ D_\alpha^{(1,1)}\lambda = \bar D_{\dot\alpha}^{(0,2)}\lambda = 0;\ \left(D_\alpha^{(a,b)} = u_i^{(a,b)} D_\alpha^i\right)$$

Using this bridge one may pass to the λ -frame where the analyticity is explicit and the derivatives $\mathcal{D}_\alpha^{(1,1)}, \bar{\mathcal{D}}_{\dot\alpha}^{(0,2)}$ are short:

$$\mathcal{D}_{\alpha\,(\lambda)}^{(1,1)} = D_\alpha^{(1,1)},\quad \bar{\mathcal{D}}_{\dot\alpha\,(\lambda)}^{(0,2)} = \bar D_{\dot\alpha}^{(0,2)}$$

On the other hand, the notion of u -independence becomes covariant in the λ -frame and the analyticity preserving harmonic derivatives (5.3) lengthen:

$$\mathcal{D}^{(a,b)} = D^{(a,b)} + iV^{(a,b)},\ \left((a,b) = (1,3), (-1,3), (2,0)\right) \qquad (5.9)$$

$$V^{(a,b)} = -ie^{iv} D^{(a,b)} e^{-iv} \qquad (5.10)$$

$$D_\alpha^{(1,1)} V^{(a,b)} = \bar D_{\dot\alpha}^{(0,2)} V^{(a,b)} = 0 \qquad (5.11)$$

Thus, we arrive at the description of N=3 SYM in terms of three analytic gauge connections $V^{(-1,3)}, V^{(2,0)} = \overset{*}{V}{}^{(-1,3)}, V^{(1,3)}$ Owing to the representation (5.10), the covariantized $\mathcal{D}^{(a,b)}$ should obey the algebra of flat derivatives:

$$[\mathcal{D}^{(\pm1,3)}, \mathcal{D}^{(1,3)}] = [\mathcal{D}^{(1,3)}, \mathcal{D}^{(2,0)}] = 0,\ [\mathcal{D}^{(-1,3)}, \mathcal{D}^{(2,0)}] = \mathcal{D}^{(1,3)} \qquad (5.12)$$

All the N=3 SYM dynamics in the λ-frame is concentrated in eqs. (5.12) which play now the rôle of equations of motion properly constraining analytic connections $V^{(a,b)}$ (recall that we have started with the standard constraints in $\mathbb{R}^{4|12}$ which are equivalent to the equations of motion). $V^{(a,b)}$ even in the W.Z.gauge have an infinite number of components but all these except the physical ones (the latter consist of three complex scalars φ^i, gauge field b_μ and four Majorana spinors $\psi^i_\alpha, \chi_\beta$) prove to be auxiliary and are eliminated by eqs. (5.12).

5.3. Action. In fact, to be certain about which fields in $V^{(a,b)}$ are really propagating we should indicate an invariant action resulting in eqs. (5.12). Such an action exists and has an unexpectedly simple form[x)]:

$$S = \int d\zeta_A^{(-2,-6)} du\, tr\{V^{(2,0)}(D^{(1,3)}V^{(-1,3)} - D^{(-1,3)}V^{(1,3)}) - V^{(-1,3)}(D^{(1,3)}V^{(2,0)} - D^{(2,0)}V^{(1,3)}) + V^{(1,3)}(D^{(-1,3)}V^{(2,0)} - D^{(2,0)}V^{(-1,3)}) - (V^{(1,3)})^2 + 2i V^{(1,3)}[V^{(-1,3)}, V^{(2,0)}]\} \tag{5.13}$$

The integrand in (5.13) is not tensor, it changes by total harmonic derivatives under the λ -gauge group, thus ensuring the invariance of action. Note a remarkable similarity of (5.13) to the well-known Chern-Simons terms in the conventional (N=0) field theory[21)].

Nontensorial character of action (5.13) implies, just as in the case of ordinary Chern-Simons terms, that in an appropriate background field formalism this action is not renormalized at all. Thereby, the finiteness of theory becomes manifest. In fact, having a complete quantization procedure for N=3 SYM in the harmonic SS (it is not worked out as yet) one may prove the finiteness in a more straightforward manner following the line of ref.[18)] based merely on the dimensionality arguments. Analogously to the N=2 case all the quantum corrections in the N=3 SYM theory are expected to be representable as integrals over the whole N=3 harmonic SS $\{x^m, \theta^\alpha_i, \bar\theta^{\dot\alpha i}, u\}$. Then, the typical

x) It is written in a "first-order formalism". One may express $V^{(1,3)}$ by means of the third of eqs. (5.12) and substitute it into (5.13), thus going over to a "second-order" form of action.

n -point contribution to the effective action is (schematically):

$$A_n = \int (d^4p)^{n-1} d^{12}\theta \, V(p_1,\theta) \cdots V(p_n,\theta) \, \Gamma(p_1, \cdots p_n, \theta)$$

$$[V(\zeta_A, u)] = m^0 \Rightarrow [V(p,\theta)] = m^{-4}, \; [A_n] = 0$$

Thus, $[\Gamma(p_1 \cdots p_n, \theta)] = m^{-2}$ and there is no room for divergences in the theory.

6. CONCLUSION

The harmonic superspace era is now only starting. A numerous lot of things remain to be understood and done, such as off-shell N=4 SYM and N=2...8 supergravity theories (till now we understand only N=2 SG and merely at the level of its supergroup[1], a differential geometry and tensor apparatus are not worked out as yet), soft breaking in the finite theories[22], and so on. An interesting problem ahead is to apply the harmonic superspace techniques to superstring theories[23].

I end with a comment concerning the relation to popular theories of the Kaluza-Klein type. There are intriguing parallels between the latter theories and the harmonic superspace ones. Both require extra bosonic dimensions, moreover the number of them for N=2 and N=3 SYM in the harmonic superspace approach is, respectively, 2 and 6, i.e. the same as in the dimensional reduction scheme for these theories. However, the status of extra dimensions in the harmonic superspace theories is quite different. Before all, these are not introduced "by hand" as higher coordinates associated with extra components of the translation operator, but are required by the intrinsic geometric structure of theory and correspond to a specific choice of superspace of an ordinary four-dimensional extended SUSY (these appear as parameters of a certain coset of the SUSY automorphism group).

While in the KK -theories higher dimensions lead to an infinite tower of massive physical excitations, the theories in harmonic SS are multidimensional only off-shell, because their extra coordinates manifest themselves merely in appearance of an infinite number of nonpropagating degrees of freedom, gauge and (or) auxiliary. On-shell these formulations yield the familiar four-dimensional component results with a finite number of physical excitations.

REFERENCES

1. Galperin A., Ivanov E., Ogievetsky V. and Sokatchev E. JETP Pisma 40, 155 (1984);
   Galperin A., Ivanov E., Kalitzin S., Ogievetsky V. and Sokatchev E. Class Quant. Grav. 1, 469 (1984).
2. Galperin A., Ivanov E., Kalitzin S., Ogievetsky V. and Sokatchev E. Phys. Lett. 151B, 215 (1985); Class. Quant. Grav. 2, 155 (1985).
3. Galperin A., Ivanov E., Ogievetsky V. and Sokatchev E. Preprints JINR E2-82-127,128, Dubna 1985.
4. Brink L., Lindgren O. and Nilsson B.E.W. Nucl. Phys. B212, 401 (1983); Phys. Lett. 123B, 323 (1983).
   Mandelstam S. Nucl. Phys. B213, 149 (1983).
   Howe P.S., Stelle K.S. and West P.C. Phys. Lett. 124B, 55 (1983);
   Derendinger J.P., Ferrara S. and Masiero A. Phys.Lett. 143B, 133 (1984).
5. a) Mezincescu L. JINR P2-12572, Dubna, 1979.
   b) Howe P.S., Stelle K.S. and Townsend P.K. Nucl. Phys. B236, 125 (1984).
6. Fayet P. Nucl. Phys. B113, 135 (1976);
   Sohnius M.F. Nucl. Phys. B138, 109 (1978).
7. Stelle K.S. California University preprint NSF-ITP-85--01, Santa Barbara 1985;

Howe P.S., Stelle K.S. and West P.C., in preparation.
8. Roček M. and Siegel W., Phys. Lett. 105B, 275 (1981).
Rivelles V. and Taylor J.G., J. Phys. A15, 163(1982).
9. Galperin A., Ivanov E. and Ogievetsky V. JETP Pisma 33, 176 (1981).
10. Bagger J. and Witten E., Nucl. Phys., B222, 1 (1983).
11. Taub A. Ann. Math. 53, 472 (1951).
Newman E., Tamburino L. and Unti T., J. Math. Phys. 4, 915 (1963).
12. Howe P.S., Stelle K.S. and Townsend P.K. Nucl. Phys. B214, 519 (1983).
13. Wess J., Acta Phys. Austriaca 41, 409 (1975).
Siegel W., Nucl. Phys. B173, 51 (1980).
Karlhede A., Lindstrom U. and Roček M.
Phys..Lett. 147B, 297 (1984).
14. Ogievetsky V. and Polubarinov I. Yad. Fiz. 4, 216 (1966).
15. Alvarez-Gaumé L. and Freedman D.Z. Comm. Math. Phys. 80, 443 (1981);
Kogan Ya., Morozov A., Perelomov A., JETP Pisma 40, 38 (1984).
16. Grimm R., Sohnius M. and Wess J. Nucl. Phys. B133, 275 (1978).
17. Rosly A.A., in Proc. Int. Seminar on Group Theoretical Methods in Physics (Zvenigorod 1982). Nauka, Moscow, v. I, p. 263.
18. Gates S.J., Grisaru M.T., Roček M. and Siegel W., "Superspace" (Benjamin/Cummings, Reading (1983)).
19. Sohnius M., Nucl. Phys. B136, 461 (1978).
20. Witten E., Phys. Lett. 77B, 394 (1978).
21. Deser S., Jackiw R. and Templeton S. Ann. Phys. 140, 372 (1982).
Breitenlohner P., Maison D. and Stelle K.S.
Phys. Lett. 134B, 63 (1984).

Townsend P.K. in Proc. 3rd Trieste School on Supersymmetry and Supergravity, April 1984.

22. Parkes A. and West P., Nucl. Phys. B222, 269 (1983); Namazie M.A., Salam A. and Strathdee J. Phys. Rev. D28, 1481 (1983);
Sazdowich B. and Tarasov O., JETP Pisma 37, 602 (1983); Nucl. Phys. B250, 39 (1984).

23. Green M. and Schwarz J. Phys. Lett. 149B; 117 (1984); Phys. Lett. 151B, 21 (1985).
Gross D., Harvey J., Martinec E. and Rohm R. Phys. Rev. Lett. 54, 502 (1985).

SYMMETRIES AND SUPERSYMMETRY IN NUCLEAR PHYSICS

Stanisław Szpikowski

Institute of Physics, M. Curie-Sklodowska University Lublin, POLAND

The seminar is devoted to the symmetries specially suited for nuclear physics that is, for nucleon configurations and their interactions. However, there are many symmetries of interest and we will choose only a special kind of them being guided by the - so far - an unique example of a supersymmetry in nuclear physics given first a couple of years ago by Iachello and Arima [1]. At first we introduce the fermion and boson symmetries relevant for the supersymmetry example given in the end.

1. FERMION CONFIGURATIONS ON THE j-LEVEL

Let us take n-one kind of nucleons, say protons on the j-level. The one-particle vectors

$$a^{+}_{jm}|0\rangle \qquad (1)$$

span the 2j+1 dimension vector space in which we can perform the unitary transformation U(2j+1). A very efficient representation of the generators in the transformation U(2j+1) is furnished by the fermion creation and annihilation operators with the standard anticommutation relations

$$\{a^+_{jm}, a_{j'm'}\} = \delta_{jj'}\delta_{mm'}$$
$$\{a^+_{jm}, a^+_{j'm'}\} = \{a_{jm}, a_{j'm'}\} = 0 \qquad (2)$$

It can be checked that the operators

$$a^+_{jm}; \quad = j, j-1, \ldots, -j$$

and

$$\tilde{a}_{jm} \equiv (-1)^{j-m} a_{j-m} \qquad (3)$$

form two independent tensor operators of the rank j in the group SU(2). Then we construct the tensor product of the single particle operators

$$(a^+_j \tilde{a}_j)^{(J)}_M \equiv \sum_{m_1 m_2} (jm_1 jm_2 | JM) a^+_{jm_1} \tilde{a}_{jm_2} \qquad (4)$$

where the Clebsch-Gordan coefficient is given in the bracket (1). The operators (4) are closed under commutation relations and they generate the transformation U(2j+1). We can select from (4) the subsets of operators also closed under commutation

$$(a^+_j \tilde{a}_j)^{(J)}_M \rightarrow (a^+_j \tilde{a}_j)^{(J_{odd})}_M \rightarrow (a^+_j \tilde{a}_j)^{(J=1)}_M \qquad (5)$$

The operators (5) are known to be the generators of the respective chain of groups

$$U(2j+1) \supset Sp(2j+1) \supset SU_J(2) \qquad (6)$$

The vector basis for irreducible representations of the group chain in (6) reads

$$|n_f; V_f; JM; \alpha\rangle \qquad (7)$$

where n_f is the fermion number, V_f - the seniority number, JM are the angular momentum quantum numbers and α is the set of additional quantum numbers needed for a complete classificiation.

The classification (7) came from the old and classical papers of Flowers [2,3]. There is, however, much more efficient classification of shell model nuclear states given by the so called quasi-spin formalism [4-8]. Namely, let us define

$$S_+ \equiv \frac{1}{2}\sqrt{2j+1}\,(a_j^+ a_j^+)^{(J=0)}$$

$$S_- \equiv (S_+)^+ \qquad (8abc)$$

$$S_0 \equiv -\frac{1}{2}\sqrt{2j+1}\,(a_j^+ \tilde{a}_j)^{J=0} - \frac{1}{4}(2j+1)$$

The operators (8) have exactly the same commutation relations as the angular momentum operators J_+, J_-, and J_0 and hence they generate the $SU_q(2)$ transformations in, so called, the abstract quasi-spin space. The equivalent, to (6), group chain is now

$$SU_q(2)\times Sp(2j+1)\supset SU_q(2)\times SU_J(2) \qquad (9)$$

The group chain (9) has many advantages as compare to the chain (6) but that is another problem.

2. BOSON SYMMETRIES IN THE INTERACTING BOSON MODEL

The pairs of nucleons coupled to J = 0 (8a), like the Cooper pairs, are of great importance in nuclear structure, giving rise to the theory of pairing interaction. However, it has been well recognized thatbesides of the pairs coupled to J = 0, the pairs J = 2 play also the relevant role in nuclear structure. However, the communication relations of such creation and annihilation pairs are rather complicated and, moreover, they create pairs of higher and higher angular momenta. Hence we approximate the fermion pairs with J = 0 and J = 2 by single bosons with the same angular momenta. Moreover, the bosons are considered as properly correlated pairs of nucleons over the relevant set of j-levels. They fulfill the standard commutation relations

$$[b_{l_1 m_1}, b^+_{l_2 m_2}] = \delta_{l_1 l_2}\delta_{m_1 m_2}$$
$$[b^+_{l_1 m_1}, b^+_{l_2 m_2}] = [b_{l_1 m_1}, b_{l_2 m_2}] = 0 \tag{10}$$

The bosons are sitting, by assumption, on two different energy levels

ε_d ———————— d-bosons (l = 2)

ε_s ———————— s-bosons (l = 0)

and the boson interacting Hamiltonian contains, by definition, single particle and interaction terms which however do not change the boson number as bosons are considered as images of properly correlated fermion pairs

$$H_b = \varepsilon_s\hat{n}_s + \varepsilon_d\hat{n}_d + \sum_{\substack{l_1 l_2 l_3 l_4 \\ L}} C^{(L)}_{l_1 l_2 l_3 l_4} (b^+_{l_1} b^+_{l_2})^{(L)} \cdot (\tilde{b}\tilde{b})^{(L)} \tag{11}$$

where a dot means the scalar product and l_i = 0, 2; L = 0, 2, 4. There are six independent one particle boson vector-states

$$b^+_{lm}|0\rangle \tag{12}$$

and they form the basis for the vector irreducible representation of the group U(6) with generators the same in structure as in the fermion case

$$(b^+_{l_1} \tilde{b}_{l_2})^{(L)}_M \tag{13}$$

The chain of boson symmetry groups in so called transitional limit reads

$$U(6) \supset SO(6) \supset SO(5) \supset SO(3) \tag{14}$$

with their generators

$$(b^+_{l_1} \tilde{b}_{l_2})^{(L)}_M \rightarrow (d^+_2 \tilde{d}_2)^{(L_{odd})}_M ;\ (d^+ s + s^+ d)^2_M \rightarrow (d^+_2 \tilde{d}_2)^{(L_{odd})} \rightarrow (d^+_2 d_2)^{(L=1)}_M \tag{15}$$

The state vectors of the n-boson system can be factorized by the quantum numbers related to the irreducible representation factors of the chain (15)

$$|n_b;V_b;V_d;JM;\beta\rangle \tag{16}$$

where n_b is the number of bosons, V_b (V_d) is the seniority number of the group SO(6), (SO(5)), JM are the total angular momentum quantum numbers, and β is to complete the classification. The same remark as in the fermion case can be said to improve the state factorization (14-16) with the help of the quasi spin group SU(1,1) which is, in a boson case, a noncompact unitary group [9]. The full description of the Interacting Boson Model (IBM) is given in the four extensive papers [9-13] and the application of the model together with its extensions is a subject of hundreds and hundreds papers since 1973.

3. FERMION AND BOSON SYSTEM AND ITS SUPERSYMMETRY

It is a trivial matter to construct the supersymmetry groups and its graded Lie algebra for a given fermion and boson groups and algebras. It is not, however, a trivial matter, to find the application for such a supersymmetry.

Keeping the same notation for the boson (b) and fermion (a) operators, we symbolically can write four kinds of generators

$$\left.\begin{matrix} a^+a \\ b^+b \end{matrix}\right\} \rightarrow G$$

$$\left.\begin{matrix} a^+b \\ b^+a \end{matrix}\right\} \rightarrow F \tag{17abcd}$$

which we have divided into two parts, G and F. Then the commutation and anticommutation relations reads

$$[G,G] \rightarrow G$$
$$[G,F] \rightarrow F \tag{18}$$
$$\{F,F\} \rightarrow G$$

which is in accord with the graded Lie algebra definitions. We have assumed in (18) that the fermion operators commute with boson ones. The supersymmetry group generated by (18) has as its subgroup the direct product of a fermion unitary group generated by (17a) and a boson unitary group generated by (17b)

$$U(n/m) \supset U_b(n) \times U_f(m) \tag{19}$$

where n(m) is the dimension of the vector boson (fermion) space.

The irreducible representation (IR) of the group U(n/m) is labelled by the total number of particles N (bosons + fermions) while the IR of the group $U_b(n)$ is represented by a fully symmetric Young diagram of the n_b bosons, and the IR for the group $U_f(m)$ - by a fully antisymmetric Young diagram of the n_f fermions

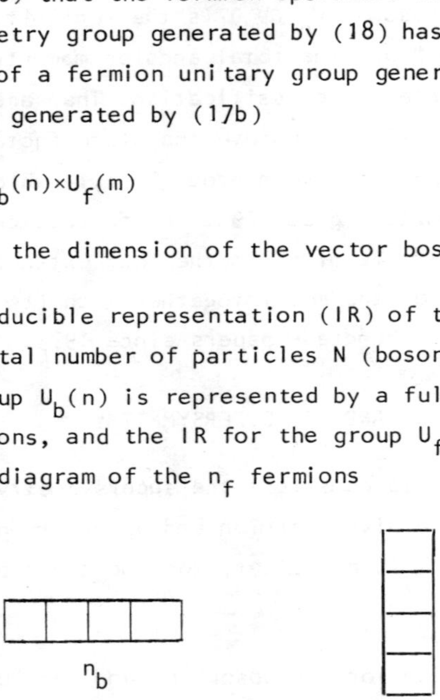

n_b n_f

Such a IR of the group U(n/m) reads, by definition

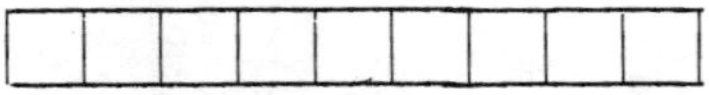

$N = n_b = n_f$

Suppose now that the fermions are sitting on the j = 3/2 level. Then the fermion chain of groups in this case is, by (6)

$$U_f(4) \supset SU_f(4) \supset Sp_f(4) \supset SU_f(2) \tag{20}$$

The important observation for further decomposition of the supergroup (19) is the following isomorphism

$$SO(6) \approx SU(4) \rightarrow Spinor(6)$$

$$SO(5) \approx Sp(4) \rightarrow Spinor(5) \qquad (21)$$

$$SO(3) \approx SU(2) \rightarrow Spinor(3)$$

in which we attach to each pair of groups the spinor group [14]. For example, the generators of the group Spinor(6) are formed from the generators of the boson group $SO_b(6)$

$$\left((d_2^+\tilde{d}_2)_\mu^{(J=1,3)}; (d_2^+s+s^+d_2)_\mu^{(2)} \right) \equiv B_\mu^{(J)} \qquad (22)$$

and from the generators of the fermion group $SU_f(4)$

$$(a_{3/2}^+\tilde{a}_{3/2})_\mu^{(J=1,2,3)} \equiv F_\mu^{(J)} \qquad (23)$$

in the following way [15]

$$S_\mu^{(1)} \equiv B_\mu^{(1)} - \frac{1}{2} F_\mu^{(1)}$$

$$S_\mu^{(2)} \equiv B_\mu^{(2)} + F_\mu^{(2)} \qquad (24)$$

$$S_\mu^{(3)} \equiv B_\mu^{(3)} + \frac{1}{2} F_\mu^{(3)}$$

and respectively for the groups Spinor(5) and Spinor(3). The supersymmetry chain of groups for the system with $N = n_b+n_f$ particles now reads

$$\begin{array}{l} U(6/4) \supset U_b(6) \times U_f(4) \supset SO_b(6) \times SU_f(4) \supset Spin(6) \supset Spin(5) \supset Spin(3) \\ \quad N \qquad\quad n_b \qquad\quad n_f \qquad\quad v_b \qquad\qquad\qquad \sigma_1\sigma_2\sigma_3 \quad \tau_1\tau_2 \qquad\quad L \end{array} \qquad (25)$$

The numbers labelled irreducible representations of the groups are also shown in (25).

The most general second order Hamiltonian, scalar in the group decomposition (25) is built from the second order Casimir operators of those groups

$$H = E_0+E_1\hat{C}_{6/4}+E_2\hat{C}_{U6}+E_3\hat{C}_{U4}+E_4\hat{C}_{06}+E_5\hat{C}_{S6}+E_6\hat{C}_{S5}+E_7\hat{C}_{S3} \qquad (26)$$

The eigenvalues of the Casimir operators in the state

$$|N;n_b;n_f;v_b;(\sigma_1\sigma_2\sigma_3);(\tau_1\tau_2);L> \qquad (27)$$

are well known and hence the eigenenergy of the (26) reads

$$E = E_0+E_1N+E_2n_b+E_3n_f+E_4v(v+4)+E_5[\sigma_1(\sigma_1+4)+\sigma_2(\sigma_2+2)+\sigma_3^2]$$
$$+E_6[\tau_1(\tau_1+3)+\tau_2(\tau_2+1)]+E_7L(L+1) \qquad (28)$$

The lowest excitations for a given number of bosons and fermions are only within the lowest irreducible representations of the groups SO(6) and Spinor(6). Hence, the relevant parts of the (28) for the lowest excited states are only those parts with E_6 and E_7

$$\Delta E = E_6[\tau_1(\tau_1+3)+\tau_2(\tau_2+1)]+E_7L(L+1) \qquad (29)$$

4. EXPERIMENTAL EVIDENCE

As a starting nucleus let us take ${}^{190}_{76}Os_{114}$. The two last proton levels below the magic number 82 are $d_{3/2}$ and $s_{1/2}$ and hence the next protons will occupy the level j = 3/2 in accord with (20). The nucleus ${}^{190}_{76}Os_{114}$ consists of 12 neutron holes equivalent to 6 boson holes (below the neutron magic number 126) and 6 proton holes equivalent to 3 boson holes. There are given below further nuclei in the same supermultiplet.

boson holes	fermions (protons)	nucleus
9	0	${}^{190}_{76}Os_{114}$
8	1	${}^{191}_{77}Ir_{114}$
7	2	${}^{192}_{78}Pt_{114}$
6	3	${}^{193}_{79}Au_{114}$

The excited levels of the first nucleus $^{190}_{76}Os_{114}$ can be well described in the frame of a pure boson model (IBM) with two adjusted parameters E_6 = 40 keV and E_7 = 10 keV [15]. The other members of the supermultiplet should be equally well described with the same parameters if the supersymmetry really exists. Let us show the excited levels for the second member the odd-even nucleus $^{191}_{77}Ir_{114}$ (Fig. 1) [15].

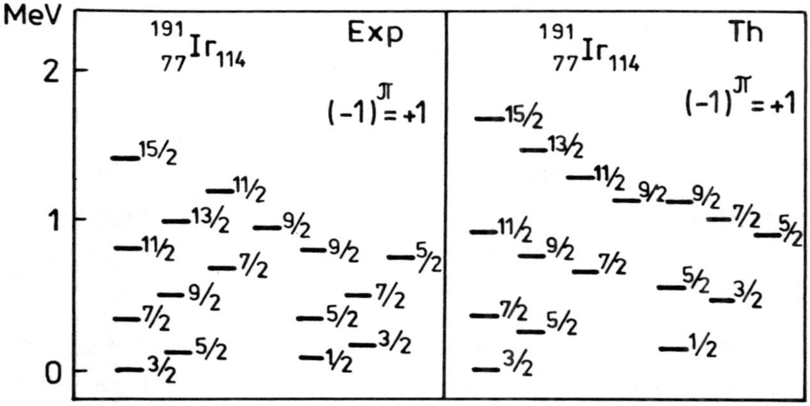

Fig. 1. Comparison of experimental and theoretical spectra of a nucleus $^{191}_{77}Ir_{114}$. The theoretical results has been obtained with E_6 = 40 keV and E_7 = 10 keV in the supersymmetry scheme using eq. 28 (after [15]).

The excellent agreement with experimental data is a good check for the supersymmetry of nuclei under consideration.

We will also show the second example of intimate relation between nuclei of in the supermultiplets [15]. Let us consider the binding energy for three members of the same supermultiplet

$$\Delta \equiv E(N;n_f=0)-2E(N;n_f=1)+E(N;n_f=2) \qquad (30)$$

The total binding energy, Δ, does not depend on the particle number N. Hence, if the boson parameters in (26) do not change appreciably for two neighbouring supermultiplets, the binding energy (30) should be the same. Let us take the nuclei from two supermultiplets N = 9 and N = 8

$$
\begin{aligned}
&{}^{190}_{76}\mathrm{Os}_{114} - 2\,{}^{191}_{77}\mathrm{Ir}_{114} + {}^{192}_{78}\mathrm{Pt}_{114} \equiv \Delta_9 \\
&{}^{192}_{76}\mathrm{Os}_{116} - 2\,{}^{193}_{77}\mathrm{Ir}_{116} + {}^{194}_{78}\mathrm{Pt}_{116} \equiv \Delta_8
\end{aligned}
\qquad (31ab)
$$

for which it has been checked from a mass table that

$$\Delta_9 = (-44\pm28)\ \mathrm{keV}; \qquad \Delta_8 = (-123\pm25)\ \mathrm{keV}$$

It is a very good check of the supermultiplet prediction if one takes into account that each member in (31ab) is of the order of 40 MeV. It is also an interesting feature that each sum in (31ab) is approximately equal to zero.

In 1983-84 there were published several papers [16-19] showing other interesting examples of nuclei supermultiplets. In the same time the theoretical base for a supermultiplet treatment has been also enlarged toward considering several j-levels for fermion configurations [20,21]. It seems that this direction of research will be widely developed.

5. REFERENCES

[1] Iachello, F. and Arima, A., Phys. Lett. 53B, 309 (1974).

[2] Flowers, B.H., Proc. Roy. Soc. A210, 497 (1951).

[3] Flowers, B.H., Proc. Roy. Soc. A212, 248 (1952).

[4] Kerman, A., Ann. Phys. 12, 300 (1961).

[5] Flowers, B.H. and Szpikowski, S., Proc. Phys. Soc. 84, 193 (1964).

[6] Ichimura, M., Progr. Theor. Phys. 32, 757 (1964).

[7] Hecht, K.T., Nucl. Phys. 63, 177 (1965).

[8] Lipkin, H., Lie Groups for Pedestrians (North Holland, Amsterdam 1965).

[9] Szpikowski, S. and Góźdź, A., Nucl. Phys. A344, 76 (1980).

[10] Arima, A. and Iachello, F., Ann. of Phys. 99, 253 (1976).

[11] Arima, A. and Iachello, F., Ann. of Phys. 111, 201 (1978).

[12] Scholten, O., Iachello, F. and Arima, A., Ann. of Phys. 115, 325 (1978).

[13] Arima, A. and Iachello, F., Ann. of Phys. 123, 468 (1979).

[14] Gilmore, R., Lie groups, Lie Algebras and Some of their applications (Wiley, New York, 1974).

[15] Balantekin, A.B., Bars, I. and Iachello, F., Nucl. Phys. A370, 284 (1981).

[16] Balantekin, A.B., Bars, I., Bijker, R. and Iachello, F., Phys. Rev. C27, 1761 (1983).

[17] Sun Hong Zhou, Franck, A. and Van Isacker, P., Phys. Rev. C27, 2430 (1983).

[18] Sun Hong Zhou, Franck, A. and Van Isacker, P., Phys. Lett. 124B, 275 (1983).

[19] Vallieres, M., Hong-Zhou Sun, Da Hsuan Feng and Gilmore, R., Phys. Lett. 135B, 339 (1984).

[20] Hong-Zhou Sun, Vallieres, M., Da Hsuan Feng and Gilmore, R., Phys. Rev. C29, 352 (1984).

[21] Bijker, R. and Iachello F. - preprint, March 1984.

MACROSCOPIC QUANTUM PHENOMENA AS WEAKLY COUPLED SPONTANEOUS SYMMETRY BREAKING*

Alfred Rieckers
Institut für Theoretische Physik
Universität Tübingen, W.-Germany

1. INTRODUCTION

Spontaneous symmetry breaking for systems with many degrees of freedom stands for a global reduction of dynamically allowed symmetries without external influence. For a concise theoretical description of such an effect one deals with second quantized models in the thermodynamic limit both for relativistic and non-relativistic descriptions. The far-reaching consequences of the infinite volume extrapolation are, however, much better understood in non-relativistic many body physics, to which we restrict our discussion. In this locally well established theory the thermodynamic limit is not only a calculational simplification but - what is more important - also a conceptual extension of the traditional quantum mechanical formalism. Only in the extended theory one has a clear discrimination between local microscopic and global macroscopic observables with the possibility of a unified dynamical treatment, and only within this framework one may elaborate all consequences of spontaneous symmetry breaking, the most spectacular of which being perhaps the macroscopic quantum phenomena. We shall derive the basic features of these phenomena from the combination of spontaneous symmetry breaking with weak coupling effects and use for this a systematic representation theory by means of operator

*Elaboration of a lecture given at the XXI Winter School of Theoretical Physics, Karpacz, February 1985

algebraic methods. Concerning the group theoretical aspects our treatment is more general than required by the hitherto observed phenomena, which refer solely to spontaneously broken gauge symmetry. The fact that our theoretical derivation deals in principle with the break down of general internal symmetries may perhaps help to recognize also outside the realm of superfluidity and superconductivity macroscopic quantum features.

In every section the emphasis lies on the general reasoning and on the interpretational aspects, whereas the formal arguments are treated rather selectively. In Section 2 we present a global scheme for a quantum mechanics with macroscopic features including a general definition of spontaneous symmetry breaking. In Section 3 the limiting Gibbs states of simple and strongly coupled lattice systems with phase transitions are treated. In Section 4 the reconstructed quantum mechanics of the mentioned systems with the condensed and quasi-particles structure is worked out. In Section 5 the behaviour of symmetry breaking systems under weak coupling is investigated.

2. QUANTUM MECHANICS WITH MACROSCOPIC FEATURES

In order to deal with quantum mechanical theories which exhibit also macroscopic features we need flexible and general theoretical notions. The (bounded) observables are taken from a general C*-algebra $\mathcal{A}$, that is an associative, normed algebra over $\mathbb{C}$, which is complete in the norm topology and which has a *-operation (an idempotent anti-automorphism) fulfilling $\|A^*A\| = \|A\|^2$ for all $A \in \mathcal{A}$ [1]. The state space $\mathcal{S}(\mathcal{A})$, constisting of all elements φ from the dual Banach space $\mathcal{A}^*$, which are positive ($\langle\varphi;A^*A\rangle \geq 0, \forall A \in \mathcal{A}$) and normalized ($\|\varphi\| = 1$), is in general too large for physical applications. An appropriate subset $\mathcal{F} \subset \mathcal{S}(\mathcal{A})$ should be convex, norm-closed and bi-invariant ($\varphi \in \mathcal{F} \Rightarrow \varphi_C \in \mathcal{F}$ for all $C \in \mathcal{A}$, where $\langle\varphi_C;A\rangle := \langle\varphi;A\rangle$ for $\langle\varphi;C^*C\rangle = 0$ and $\langle\varphi_C;A\rangle := \langle\varphi;C^*AC\rangle/\langle\varphi;C^*C\rangle$, otherwise) and is called "folium". A general theoretical "description" [2] consists then of a triple $(\mathcal{A}, \mathcal{F}, \langle;\rangle)$, where $\mathcal{A}$ is a C*-algebra, $\mathcal{F}$ a folium and $\langle\cdot;\cdot\rangle$ the duality relation for the expectation values. For mathematical convenience let us assume that $\mathcal{A}$ contains a unit $1 \in \mathcal{A}$, which implies that $\mathcal{S}(\mathcal{A})$ is weak-*-compact. It is also a natural assumption that $\mathcal{F}$ is full, that means $<\varphi;A> \geq 0$, for $A \in \mathcal{A}$ and for all $\varphi \in \mathcal{F}$, implies $A \geq 0$. $\mathcal{F}$ is then weak-*-dense in $\mathcal{S}(\mathcal{A})$ [3].

Traditional quantum mechanics is characterized by the description $<\mathcal{B}(\mathcal{H}), \mathcal{T}_{+,1}, tr\{...\}>$, where $\mathcal{H}$ is a separable Hilbert space, $\mathcal{B}(\mathcal{H})$ the set of all bounded linear operators in $\mathcal{H}$, $\mathcal{T}_{+,1}$ the set of all positive normalized traceclass operators and tr the usual trace. The set of physical states is uniform in that is has no non-trivial subfolium.

For the description of many body systems it is by now customary to use a so-called quasilocal C*-algebra $\mathcal{A}$ which is specified by a family $\{\mathcal{A}_\Lambda;\Lambda \in \mathcal{L}\}$ of C*-subalgebras indexed by finite regions Λ (possibly also in momentum space). The set $\mathcal{L}$ of all considered finite regions is directed by set inclusion. If $\Lambda \cap \Lambda' = \emptyset$ the observables in $\mathcal{A}_\Lambda$ and $\mathcal{A}_{\Lambda'}$ are required to be compatible. The union $\mathcal{A}_0 := \cup_{\Lambda \in \mathcal{L}} \mathcal{A}_\Lambda$ is assumed to be normdense in $\mathcal{A}$. The $\mathcal{A}_\Lambda$ are in most cases taken to be isomorphic to $\mathcal{B}(\mathcal{H}_\Lambda)$ for a finite or separable Hilbert space $\mathcal{H}_\Lambda$ describing finitely many particles in Λ.

In order to understand the physical meaning as well as the mathematical structure of a quasilocal algebra $\mathcal{A}$ it is advantageous to consider $\mathcal{A}$ as constructively generated by the traditional algebras $\mathcal{B}(\mathcal{H}_\Lambda), \Lambda \in \mathcal{L}$. The requirement that the $\mathcal{B}(\mathcal{H}_\Lambda)$ belong to a single infinite system leads to the condition that there is an embedding morphism of $\mathcal{B}(\mathcal{H}_\Lambda)$ into $\mathcal{B}(\mathcal{H}_{\Lambda'})$ for all Λ' containing Λ which typically is of the form

$$\mathcal{B}(\mathcal{H}_\Lambda) \ni A \to A \otimes 1_K \in \mathcal{B}(\mathcal{H}_{\Lambda'}) , \tag{2.1}$$

where $K := \Lambda' \setminus \Lambda$ and 1_K is the unit of $\mathcal{B}(\mathcal{H}_K)$. The C*-inductive limit construction [4] leads then to a quasilocal algebra $\mathcal{A}$ the $\mathcal{A}_\Lambda$ of which are given by the injected $\mathcal{B}(\mathcal{H}_\Lambda)$. Since this injection is based on an iteration of (2.1) it maps compact operators of $\mathcal{B}(\mathcal{H}_\Lambda)$ onto elements in $\mathcal{A}_\Lambda \subset \mathcal{A}$ which are not compact in an abstracted algebraic sense (they are not sums of "abelian" elements). This is the reason why a quasilocal algebra $\mathcal{A}$ is completely different (antiliminary [5]) from a traditional algebra $\mathcal{B}(\mathcal{H}_\Lambda)$ in spite of the local algebraic isomorphy to the latter. On the other hand $\mathcal{A}$ has, like $\mathcal{B}(\mathcal{H}_\Lambda)$, a trivial center and there are thus no nontrivial macroscopic classical observables in it.

The antiliminarity property of $\mathcal{A}$ is connected with a very rich structure of sub-folia in the state space $\mathcal{S}(\mathcal{A})$. Denoting by $\mathcal{F}_\varphi$ the smallest folium containing the state $\varphi \in \mathcal{S}(\mathcal{A})$ we call $\psi \in \mathcal{S}(\mathcal{A})$ macroscopically different from φ (disjoint), if $\mathcal{F}_\varphi \wedge \mathcal{F}_\psi = \emptyset$ and call φ macroscopically pure (factorial), if $\mathcal{F}_\varphi$ does not contain a nontrivial sub-folium. If φ and ψ are macroscopically different, no perturbation of the one state by means of a quasilocal element leads to the other state, that means that they differ in at least one property from each other which is stable against (quasi-) local changes of the states. On the other hand, the states in a minimal folium $\mathcal{F}_\varphi$ cannot have different properties, which are invariant under local excitations. Thus the set of minimal folia defines classes of macroscopically equivalent states. As in the foundations of thermodynamics [6] an empirical total order relation in this set constitutes an empirical macroscopic observable, and if this ordered chain is mapped into $\mathbb{R}$ we have a quantified macroscopic observable. Let us emphasize the direct intuitive appeal of this new concept formation. Only in an enlarged algebra of observables we can associate with the quantified macroscopic observable a self-adjoint operator.

Since in the description of traditional quantum mechanics there is only one (minimal) folium of states it follows that there are no nontrivial macroscopic observables. Thus, as has been often deplored, there is no connection between atomic and macroscopic properties in this theoretical frame. It cannot account for the microscopic explanation of macroscopic processes, and since only the latter are directly observable nothing "happens" in traditional quantum mechanics [7].

In the state space of a quasi-local algebra, however, a mathematical theorem [5] tells us, that there are overcountably many different minimal folia. In this way, a natural extrapolation of traditional quantum mechanics to the thermodynamic limit provides us with principally new concepts which are related to the classically macroscopic properties of many body systems.

In the forthcoming interpretations of our model discussions we shall always stick to the point of view, that the description of an infinite system is an extrapolation of local notions. Because of this, we are only interested in those states on the quasilocal algebra $\mathcal{A}$, which locally correspond to a traditional quantum state. These so-called locally normal states, the restrictions of which to every $\mathcal{A}_\Lambda$ are given by a density operator, constitute a subfolium $\mathcal{S}_0$ of $\mathcal{S}(\mathcal{A})$.

2.1. Definition: A description $(\mathcal{A}, \mathfrak{F}, \langle\cdot,\cdot\rangle)$ of an infinite many body system consists of a quasilocal algebra $\mathcal{A}$, a locally normal folium $\mathfrak{F} \subset \mathcal{S}_0$ and the duality relation $\langle\cdot,\cdot\rangle$ of $\mathcal{A}$ and its Banach space dual $\mathcal{A}^*$ restricted to $\mathcal{A}$ and $\mathfrak{F}$.

Such a description $(\mathcal{A}, \mathfrak{F})$ (for which the standard duality relation will be dropped in the notation henceforth) gives the frame for the more specific theoretical attributes. It is well known that the correspondence between local observables and self-adjoint elements in $\mathcal{A}$ is not unique. Any bijective transformation in $\mathcal{A}$ which may be compensated for by a transformation in $\mathfrak{F}$, such that the expectations of the transformed and original quantities are the same, leads to a physically equivalent formulation [8]. Transformations ν in $\mathfrak{F}$, which have a dual linear transformation $\nu^*: \mathcal{A} \to \mathcal{A}$, are $\sigma(\mathfrak{F},\mathcal{A})$-continuous. This is especially for the dynamics too severe a restriction.

2.2. Definition: A structural symmetry of a description $(\mathcal{A},\mathfrak{F})$ is an affine bijection $\nu: \mathfrak{F} \xrightarrow{\text{on-to}} \mathfrak{F}$ (where $\nu(\lambda_1\varphi_1+\lambda_2\varphi_2)=\lambda_1\nu(\varphi_1)+\lambda_2\nu(\varphi_2)$ for $0 \le \lambda_i \le 1$, $\lambda_1+\lambda_2=1$, $\varphi_i \in \mathfrak{F}$, is the affinity property).

Evidently the set of all structural symmetries is a group. The Heisenberg picture of a structural symmetry requires in general an enlargement of $\mathcal{A}$. An important property in the Schrödinger picture is that ν is norm continuous and maps folia onto folia. Symmetry groups are then groups G which act by means of structural symmetries $\nu_g, g \in G$, in $\mathcal{F}$ such that $\nu_{gg'} = \nu_g \circ \nu_{g'}$.

2.3. Definition: A dynamics in $(\mathcal{A}, \mathcal{F})$ is a weak-*-continuous representation of the time-translation group $\mathbb{R}$ by structural symmetries σ_t, $t \in \mathbb{R}$. A dynamical symmetry group G is a group which is represented by structural symmetries ν_g, $g \in G$, such that

$$[\nu_g, \sigma_t]_- = 0 \ , \quad \forall \ t \in \mathbb{R}, \quad g \in G \ .$$

For a statistical analysis various decompositions of a state $\varphi \in \mathcal{F}$ into less mixed states are of interest, especially for the definition of spontaneous symmetry breaking. For physical applications it is sufficient to consider so-called orthogonal probability measures μ on $\mathcal{S}$. Their σ-algebra is generated by all weak-*-open sets (the Borel sets), and for any measurable $\mathcal{K} \subset \mathcal{S}$ it holds, that any positive linear functional q on $\mathcal{A}$ with $q \leq \int_{\mathcal{K}} \varphi \, d\mu(\varphi)$ and $q \leq \int_{\mathcal{S} \setminus \mathcal{K}} \varphi \, d\mu(\varphi)$ is necessarily zero. A barycentric decomposition

$$\psi = \int_{\mathcal{S}} \varphi \, d\mu(\varphi)$$

of a state $\psi \in \mathcal{S}_0$ by means of an orthogonal measure μ remains in $\mathcal{S}_0$, since $\mathcal{S}_0$ is a stable face which also satisfies the so-called separability condition S ([1], p. 346). Because of the latter property of $\mathcal{S}_0$ we can define the support $\mathcal{T}$ of μ as the smallest Borel set with $\mu(\mathcal{T}) = \mu(\mathcal{S}_0)$. If $\psi \in \mathcal{F}$, then in general $\mathcal{T} \not\subset \mathcal{F}$. Physically it is, however, sufficient to know that for every $\mathcal{K}$ with $\mu(\mathcal{K}) > 0$ the state $\int_{\mathcal{K}} \varphi \, d\mu(\varphi)/\mu(\mathcal{K}) \in \mathcal{F}$, since $\mathcal{F}$ is a face.

2.4. Proposition: For every $\psi \in \mathcal{S}_0$ there exists a unique orthogonal measure μ_ψ, the support $\mathcal{T}_\psi$ of which consists of pair-wise disjoint factor states. (Proof not published in the literature?)

2.5. Definition: For given $\psi \in \mathcal{S}_0$ the measure μ_ψ of Prop. 2.4 is called the "central measure of ψ" and the corresponding barycentric decomposition is called the "central decomposition of ψ".

According to our previous argumentation, the central decomposition of a state ψ is the finest de-mixture of ψ into macroscopically different sub-states. This filtering may be achieved by purely macroscopic manipulations. It is now first remarkable, if a "natural" state preparation leads to a macroscopically mixed (non-factorial) state and second if the central decomposition of it produces states of lower symmetries than allowed for by the (Schrödinger) dynamics. This is mostly discussed for time invariant states, but is equally interesting for non-equilibrium states. In the framework of the present formalism the following definition seems to be adequate.

2.6. Definition: A dynamical symmetry group G is spontaneously broken in $\psi \in \mathcal{F}$, if $\nu_g(\psi) = \psi$, for all $g \in G$, but there is a $\mathcal{K} \subset \mathcal{J}_0$ with $\mu_\psi(\mathcal{K}) > 0$, such that $\int_{\mathcal{K}} \varphi \, d\mu_\psi(\varphi)/\mu_\psi(\mathcal{K}) \in \mathcal{F}$ is not invariant under ν_g for a $g \in G$.

If ν_g is defined directly on $\mathcal{J}_\psi$, which is always the case for internal symmetries, then spontaneous symmetry breaking implies the existence of a $\varphi \in \mathcal{J}_\psi$ and of a $g \in G$, such that $\nu_g(\varphi) \neq \varphi$.

All previous notions were introduced without reference to a representation of the quasilocal algebra $\mathcal{A}$. But for structural and physical purposes representation theory is very helpful. With every $\varphi \in \mathcal{J}$ there is canonically associated a triple $(\pi_\varphi, \mathcal{H}_\varphi, \Omega_\varphi)$ of a representation $\pi_\varphi : \mathcal{A} \to \mathcal{B}(\mathcal{H}_\varphi)$, a Hilbert space $\mathcal{H}_\varphi$, and a vector $\Omega_\varphi \in \mathcal{H}_\varphi$, which is uniquely (up to unitary equivalence) characterized by the two relations

$$\langle \varphi ; A \rangle = (\Omega_\varphi, \pi_\varphi(A) \, \Omega_\varphi) \quad , \forall \, A \in \mathcal{A}$$

and

$$\mathcal{H}_\varphi = \overline{\pi_\varphi(\mathcal{A}) \, \Omega_\varphi} \quad \text{(cyclicity)} .$$

For a barycentric decomposition of $\psi \in \mathcal{F}$ by means of an arbitrary orthogonal measure one has a corresponding decomposition of the so-called GNS-triple $(\pi_\psi, \mathcal{H}_\psi, \Omega_\psi)$. Especially for the central decomposition we find

$$(\pi_\psi, \mathcal{H}_\psi, \Omega_\psi) = \int_{\mathcal{J}_\psi}^{\oplus} (\pi_\varphi, \mathcal{H}_\varphi, \Omega_\varphi) \, d\mu_\psi(\varphi)$$

as well as

$$\mathcal{M}^\psi := \pi_\psi(\mathcal{A})'' = \int^{\oplus} \mathcal{M}^\varphi \, d\mu_\psi(\varphi)$$

with

$$\mathcal{M}_\varphi := \pi_\varphi(\mathcal{A})'' \quad .$$

The center $\mathcal{Z}^\psi$ of $\mathcal{M}^\psi$ consists then exactly of the diagonal operators $\int^\oplus_{\mathcal{J}_\psi} f(\varphi)\, 1_\varphi \, d\mu_\psi(\varphi) =: \kappa_\psi(f)$, $f \in \mathcal{L}^\infty(\mathcal{J}, \mu_\psi)$. From our definition it follows that spontaneous symmetry breaking in ψ may occur only if $\mathcal{Z}^\psi$ is non-trivial.

Let us finally mention the partial universally representation $(\pi_\mathcal{F}, \mathcal{H}_\mathcal{F})$ associated with the description $(\mathcal{A}, \mathcal{F})$ where

$$\mathcal{H}_\mathcal{F} := \sum_{\varphi \in \mathcal{F}}^{\oplus} \mathcal{H}_\varphi \, , \quad \pi_\mathcal{F} := \sum_{\varphi \in \mathcal{F}}^{\oplus} \pi_\varphi \, .$$

The von Neumann algebra $\mathcal{M}^\mathcal{F} := \pi_\mathcal{F}(\mathcal{A})''$ has the set of normal (σ-weakly continuous) states $\mathcal{J}_n(\mathcal{M}^\mathcal{F})$, which is affine isomorphic to $\mathcal{F}$. To every sub-folium $\mathcal{F}' \subset \mathcal{F}$ corresponds a central projection in $\mathcal{M}^\mathcal{F}$. Thus this set up covers also the macroscopic classical observables. Furtheron, every structural symmetry ν_g acting in $\mathcal{F}$ has a dual mapping ν_g^* acting as a Jordan automorphism (which is always σ-weakly continuous) in $\mathcal{M}^\mathcal{F}$ [1], [9].

3. SPONTANEOUS SYMMETRY BREAKING FOR SIMPLE AND STRONGLY COUPLED SYSTEMS

In this section we shall demonstrate that even in the specialized situation of thermodynamic equilibrium, where the appropriate description $(\mathcal{A}, \mathcal{F}_\beta)$ involves the smallest folium $\mathcal{F}_\beta$ which contains the limiting Gibbs state ω^β (or limiting ground state for $\beta=\infty$), we have need for the general set up developed in the foregoing section. Our main concern will be the origin of non-trivial macroscopic features by spontaneous symmetry breaking and their physical meaning. We start with the most simple cases.

3.1. Simple Systems

After having specified the quasi-local algebra $\mathcal{A}$ and the family of local Hamiltonians $\{H_\Lambda, \Lambda \in \mathcal{L}\}$, the next step in a systematic model construction is the calculation of the Gibbs state in the thermodynamic limit. This is the most difficult part of the discussion and restricts the class of treatable models considerably. Explicit insights into the symmetry breaking mechanisms in performing the thermodynamic limit are gained only for long-range interacting (quantum) lattice models, and we shall confine the subsequent reasoning to this type of systems.

The lattice is represented by $\mathbb{Z}^d$, $d \in \mathbb{N}$, and for each $i \in \mathbb{Z}^d$ one has a (quasi-spin) observable algebra $\mathcal{A}_i$, which is $*$-isomorphic to a fixed finite-dimensional matrix algebra $\mathcal{M}^{(n)} = \mathcal{B}(\mathbb{C}^n)$. For every $\Lambda \subset \mathbb{Z}^d$, with card $(\Lambda) = |\Lambda| < \infty$, one forms $\mathcal{A}_\Lambda := \bigotimes_{i \in \Lambda} \mathcal{A}_i$ and arrives at the quasi-local algebra

$$\mathcal{A} = \bigotimes_i \mathcal{A}_i = \overline{\bigcup_\Lambda \mathcal{A}_\Lambda}^{\,\|\cdot\|} =: \overline{\mathcal{A}_0}^{\,\|\cdot\|} . \tag{3.1}$$

Having chosen the basis $\{\sigma^\lambda; \lambda=1\ldots n^2\}$ in $\mathcal{M}^{(n)}$, $*$-isomorphy gives the basis $\{\sigma^\lambda{}_i; \lambda=1,\ldots,n^2\}$ in $\mathcal{A}_i$ for all $i \in \mathbb{Z}^d$. Many simple local model Hamiltonians are of the form

$$H_\Lambda = \sum_{i \in \Lambda, \lambda} f_\lambda \sigma^\lambda{}_i + \sum_{i,j \in \Lambda} \sum_{\lambda,\kappa} (g_{\lambda\kappa}/|\Lambda|) \sigma_i^{\lambda *} \sigma_j^\kappa . \tag{3.2}$$

The limiting Gibbs state ω^β is defined by

$$\langle \omega^\beta; A \rangle := \lim_{\Lambda' \to \infty} \mathrm{tr}_{\Lambda'} \{ e^{-\zeta_{\Lambda'} - \beta H_{\Lambda'}} A \} \tag{3.3}$$

if the limit exists for all $A \in \mathcal{A}_0$, where $\zeta_{\Lambda'}$ is the logarithm of the partition function for the finite region Λ' and $\mathrm{tr}_{\Lambda'}$ is the trace in $\mathcal{H}_{\Lambda'} := \bigotimes_{i \in \Lambda'} \mathbb{C}_i^n$. Observe that (3.3) determines ω^β uniquely as a state on $\mathcal{A}$. In (3.3) the trace in Λ' may be performed by first calculating the trace in $\Lambda' \setminus \Lambda =: K$, where $A \in \mathcal{A}_\Lambda$, and then in Λ:

$$\mathrm{tr}_{\Lambda'} \{ e^{-\zeta_{\Lambda'} - \beta H_{\Lambda'}} \} = \mathrm{tr}_\Lambda \{ A \, \mathrm{tr}_K \{ e^{-\zeta_{\Lambda'} - \beta H_{\Lambda'}} \} \} .$$

Up to a normalization constant the relative trace in K produces a density operator in Λ, the limiting form of which for $K \to \infty$, ρ_Λ^β, gives the restriction of ω^β to $\mathcal{A}_\Lambda$. In the part of the exponent of $\exp(-\beta H_{\Lambda'})$ which belongs to K the energy eigenvalues E_K^r are in competition with the degeneration entropy $k_B\, T \log d(E_K^r)$, where $d(E_K^r)$ is the multiplicity of the eigenvalue E^r_K. In the limit $|K| \to \infty$ only the absolute minima of

$$f_K(\beta,r) := (E^r_K - k_B T \log d(E^r_K))/|\Lambda'| \tag{3.4}$$

give a contribution to the form of ρ^β_Λ .

Observe that (3.4) has the form of a free energy density in which the physical value of the label r (or of a related variable) is determined by a minimal principle. In the composite system approach to thermodynamic stability [11][12] one obtains a minimum principle for the subtracted physical free energy with respect to the physical extensive state variables by a combination of the second law and thermodynamical homogeneity. It is interesting to note that the thermodynamic limit provides a supplementary extremum principle for the formal variables connected with an unsubtracted formal free energy. (For the distinction between physical and formal quantities cf. [13]).

The properties of the absolute minima in (3.4) are decisive for the occurrence or non-occurrence of spontaneous symmetry break down. Investigations on the specified class of models have led to the following picture. In case of an attractive interaction the degeneration entropy may be dominated by the energy in (3.4) if T is below a critical (absolute) temperature T_c and the minima move into a family of asymmetric positions. Via the calculable local density operators ρ^β_Λ the limiting Gibbs state ω^β decomposes then into a family of other states with lower symmetry. As comes out from our model studies [14]-[21] there are two essentially different mechanisms to obtain a continuous desintegration. First, the absolute minima of the $f_K(\beta,r)$ may develop into a continuous family for $K \to \infty$ giving a Riemannnian sum approximation for the limiting decomposition integral. Second, a conservation law for particles outside of every finite region Λ may become asymptotically exact. Since this law must be converted from a Kronecker symbol δ_{N_1,N_2} referring to K to a formulation, which is possible in Λ, one has to substitute δ_{N_1,N_2} by $\int_0^{2\pi} \exp[i\nu(N_1-N_2)]d\nu/2\pi$ in calculating ρ^β_Λ. Here, the Goldstone theorem is replaced by a desintegration of ω^β over a gauge angle ν to indicate the occurrence of condensed particles.

Let us illustrate this first by means of the Weiß-Heisenberg-ferromagnet [15] with

$$H_\Lambda = -(I/|\Lambda|) \sum_{i,j\in\Lambda} \vec{\sigma}_i \cdot \vec{\sigma}_j .$$

The absolute minima of $f(\beta,r)$ move below T_c away from the symmetric origin and lead to a spontaneous magnetization per particle with the absolute value $m_s(\beta)$. Their degeneration is signified by the directions of the vector

$$\vec{m}(\beta,z,\nu) = (\sqrt{m_s(\beta)^2-z^2}\cos\nu, \sqrt{\cdot}\sin\nu, z) , \tag{3.5}$$

where $z \in [-m_s(\beta), +m_s(\beta)]$ and ν is the angle around the z-axis. For $A \in \mathcal{A}_\Lambda$ the limiting Gibbs state has the form

$$\langle\omega^\beta;A\rangle = \int_{-m_s}^{m_s} \int_0^{2\pi} \langle\omega^{\beta,z,\nu}; A\rangle \frac{dz}{2m_s} \frac{d\nu}{2\pi} , \tag{3.6}$$

where

$$\langle\omega^{\beta,z,\nu}; A\rangle = tr_\Lambda \{\exp[-\zeta_\Lambda - 2\beta I \vec{m}(\beta,z,\nu)\cdot\vec{\sigma}_\Lambda]A\} \tag{3.7}$$

with $\vec{\sigma}_\Lambda := \sum_{i\in\Lambda} \vec{\sigma}_i$. This is a concise formulation of the well known spontaneous breaking of rotational symmetry by spontaneous magnetization. The limiting Gibbs state is still totally symmetric but decomposes into a continuum of asymmetric states $\omega^{\beta,z,\nu}$. Here only the z-integration is locally approximated by a Riemannian sum. In ω^β, the mean value of the z-component of the spontaneous magnetization is zero. For fixed z, the ν-integration expresses conservation of the angular momentum in z-direction of the condensed spins. "Condensed" means "incorporation into an infinite (that is a macroscopic number) set of aligned spins". If ν is also fixed, then condensed spins (with a definite z-component) may enter or leave the system, since their number is no longer conserved. In this way z labels the different types of condensed particles and the ν-distribution gives the degree of their conservation: total conservation in a state of equipartition, complete non-conservation in a state (as $\omega^{\beta,z,\nu}$) with sharp ν-value.

Formula (3.6) describes a barycentric decomposition of ω^β by means of a probability measure on $\mathcal{J}(\mathcal{A})$, the support of which is $\{\omega^{\beta,z,\nu};\ z,\ \nu\}$. The $\omega^{\beta,z,\nu}$ are factor states because of their cluster property (cf. [1], Ch. 2.6) and pairwise disjoint, since

$$\lim_{\Lambda} <\omega^{\beta,z,\nu};\ \frac{\vec{\sigma}\Lambda}{|\Lambda|}> = \vec{m}(\beta,\nu,z) \tag{3.8}$$

is pairwise different [22]. This identifies (3.6) as the central decomposition of ω^{β} according to Prop. 2.4.

As a second simple but still very important example let us consider the strong coupling BCS-model which is defined by the local Hamiltonians

$$H_{\Lambda} = \sum_{k\in\Lambda,\sigma} \varepsilon C^*_{k,\sigma} C_{k,\sigma} - (g/|\Lambda|) \sum_{k,l\in\Lambda} C^*_{k\uparrow} C^*_{-k\downarrow} C_{-k\downarrow} C_{k\uparrow} \ . \tag{3.9}$$

Here Λ is a finite set of wave vectors taken from a shell around the Fermi surface, σ is the spin index and $C_{k,\sigma}$ is the annihilation operator of an electron in the state (k,σ). The constant coupling energy dominates the kinetic energy, which may be replaced by an average ε. The limiting Gibbs state ω^{β} was first rigorously calculated in [23], but only the calculation in [18], applying the scheme outlined above, revealed the symmetry breaking mechanism. By means of this method ω^{β} is again directly calculated in terms of its central decomposition

$$<\omega^{\beta};A> = \int_0^{2\pi} \frac{d\nu}{2\pi} <\omega^{\beta\nu};A> \ , \tag{3.10}$$

where for $A \in \mathcal{A}_{\Lambda}$

$$<\omega^{\beta,\nu};A> = tr_{\Lambda} \{\rho^{\beta,\nu}_{\Lambda} A\}$$

with

$$\rho^{\beta\ \nu}_{\Lambda} = \exp [-\zeta_{\Lambda} - \beta \sum_{k\in\Lambda,\sigma} \varepsilon C^*_{k,\sigma} C_{k,\sigma} -$$

$$- (gw(\beta)/|\Lambda|) \sum_{k\in\Lambda} (e^{-i\nu} C^*_{+k\uparrow} C_{-k\downarrow} + e^{i\nu} C_{-k\downarrow} C_{k\uparrow})]$$

Here $w(\beta) > 0$ for β greater than the critical β_c. The phase angle ν labels the disjoint pure phase states and has, therefore, macroscopic meaning. The ν-integral is locally not approximated by a Riemannian sum, but expresses conservation of Cooper pairs at infinity. These special particles, which replace the Goldstone particles, constitute the condensate and are conserved only in a state with uniform distribution of the pure phases $\omega^{\beta\nu}$.

If the model is slightly more complicated than the two ones just mentioned the direct calculation procedure for the limiting Gibbs state fails.

3.2. Strongly Coupled Systems

In the following we outline a more indirect method for dealing with the limiting Gibbs state which largely based on [24]. It covers the case of $r \in \mathbb{N}$ different lattice systems in mutual strong interaction. The super lattice is described by $\mathcal{K} := \mathbb{Z}^d \times \{1,\ldots,r\} \ni (i,q) = k$. The one lattice point algebra $\mathcal{A}_k$, $k \in \mathcal{K}$, is isomorphic to

$$\mathcal{B}^q := \mathcal{B}(\mathbb{C}^{n(q)}),\ n(q) \in \mathbb{N} \ . \tag{3.11}$$

Every $\Lambda \subset \mathcal{K}$ is the union of sets (Λ_q, q), $1 \leq q \leq r$, and may be identified with $(\Lambda_1,\ldots,\Lambda_r)$, $\Lambda_q \subset \mathbb{Z}^d$. The local algebras are

$$\mathcal{A}_\Lambda = \bigotimes_{k\in\Lambda} \mathcal{A}_k = \bigotimes_{q=1}^{r} \bigotimes_{i\in\Lambda q} \mathcal{A}_{(i,q)} \tag{3.12}$$

and the quasilocal algebra is

$$\mathcal{A} = \bigotimes_{k\in\mathcal{K}} \mathcal{A}_k = \overline{\cup_\Lambda \mathcal{A}_\Lambda}^{\|\cdot\|} =: \overline{\mathcal{A}_o}^{\|\cdot\|} \ . \tag{3.13}$$

A permutation P in Λ is a r-tupel of usual permutations in the Λ_q, $1 \leq q \leq r$. A (finite) permutation P in $\mathcal{K}$ is one in a set Λ, leaving $\mathcal{K} \setminus \Lambda$ invariant. The groups of all finite permutations is denoted by $\mathcal{P}$. The prescription

$$\alpha_p(\bigotimes_k A_k) := \bigotimes_k A_{p(k)} \tag{3.14}$$

extends to a *-automorphism of $\mathcal{A}$. A slight extension of the usual arguments ([25], cf. also [14]) shows that the C*-dynamical system $(\mathcal{A}, \mathcal{P}, \alpha_p)$ is norm asymptotic abelian, that is there exists a sequence $\{P_n;\ n \in \mathbb{N}\} \subset \mathcal{P}$ such that

$$\lim_n \|[A,\alpha_{p_n}(B)]_-\| = 0 , \forall A,B \in \mathcal{A} . \tag{3.15}$$

Choosing a basis $\{\sigma_k ; 1 \leq \kappa \leq n^2(q)\}$ we introduce the following class of local Hamiltonians (which generalizes (3.2)):

$$H_\Lambda = \sum_{k\in\Lambda} \sum_\kappa f_{\kappa,q}\, \sigma_k + \sum_d v_\Lambda(d) \sum_{k,l\in\Lambda} \sum_{\kappa,\lambda} \sigma_k^{\lambda *} g^{q,p}_{\kappa,\lambda}(d)\sigma_l^\kappa , \tag{3.16}$$

where $k = (i,1)$, $l = (j,p)$, $f_{\kappa,q} \in \mathbb{C}$, $g^{qp}_{\lambda}(d) \in \mathbb{C}$, and

$$v_\Lambda(d) := 1/(\sum_{q=1}^{r} a_q(d)|\Lambda_q|) , a_q(d) \in \mathbb{R} . \tag{3.17}$$

That is, there are finitely many types of interactions between the lattice points k and l (which may or may not belong to the same sublattice q), which are indexed by $d \in \mathbb{N}$ and which have different factors $v_\Lambda(d)$ providing extensivity (thus not all $a_q(d)$ may be zero). We also employ the subsidiary conditions

$$\lim_\Lambda |\Lambda_q|/|\Lambda| = C_q , 1 \leq q \leq r , \tag{3.18}$$

with $\sum_{q=1}^{r} C_q = 1$ and $C_q \geq 0$, in going to the thermodynamic limit. Clearly

$$\alpha_p(H_\Lambda) = H_\Lambda , \tag{3.19}$$

for all permutations P acting in Λ.

If $u = (u^1,\dots,u^r)$ is a r-tupel of unitaries $u^q \in \mathcal{B}^q$, then let $u_k \in \mathcal{A}_k$, with $k = (i,q)$, be the unitary obtained by the $*$-isomorphism from $\mathcal{B}^q$ onto $\mathcal{A}_k$. The prescription

$$\alpha_u(\bigotimes_k A_k) := \bigotimes_k u_k A_k u_k^* \tag{3.20}$$

is extendable to a $*$-automorphism of $\mathcal{A}$.

3.1. <u>Definition:</u> The group of internal symmetries is

$$H := \{u; \alpha_u(H_\Lambda) = H_\Lambda, \forall \Lambda\} . \tag{3.21}$$

The total symmetry group is

$$G := \mathcal{P} \times H . \tag{3.22}$$

From the definition H is compact. Since H and G act in $\mathcal{A}$ via automorphisms, we have also the Schrödinger transformations $\nu_p := \alpha_p^*$ and $\nu_h := \alpha_h^*$, defined by duality. Clearly $\alpha_g = \alpha_p \circ \alpha_u$ and $\nu_g = \nu_p \circ \nu_u$ if $g = (P,u) \in G$. Let us define, furtheron,

$$\mathcal{S}^P := \{\varphi \in \mathcal{S} ; \nu_p(\varphi) = \varphi, \forall P \in \mathcal{P} \} \tag{3.23}$$

$$\mathcal{S}^G := \{\varphi \in \mathcal{S} ; \nu_g(\varphi) = \varphi, \forall g \in G \} . \tag{3.24}$$

$\mathcal{S}^P$ and $\mathcal{S}^G$ are convex, weak-*-compact subsets of $\mathcal{S}$ with non-empty extreme boundary sets $\partial_e \mathcal{S}^P$ and $\partial_e \mathcal{S}^G$. Since P and hence G act in a norm asymptotic abelian manner on $\mathcal{A}$, $\mathcal{S}^P$ and $\mathcal{S}^G$ are simplices ([1], Ch. 4.3), that is: for every state in the convex subset, there is exactly one barycentric decomposition supported by the extremal states of this subset.

In the following we want to use symmetry arguments to obtain some information on the limiting Gibbs state. Quite generally we know from the weak-*-compactness of $\mathcal{S}$ that every net of local thermodynamic (canonical or grand canonical) equilibrium states has accumulation points. Let us choose in the subsequent reasoning of this section a fixed accumulation point ω^β, to which converges a sub-net of local equilibrium states of the temperature β in the weak-*-topology. Then $\omega^\beta \in \mathcal{S}^G \subset \mathcal{S}^P$. Any $\varphi \in \partial_e \mathcal{S}^P$ is a product state $\bigotimes_k \varphi_k$, where $\varphi_k \in \mathcal{S}(\mathcal{A}_k)$, and the components are the same within one sub-lattice q [24]. If $\varphi, \varphi' \in \partial_e \mathcal{S}^P$ differ from each other in one component, then they do in infinitely many of them and they are disjoint [26]. The unique decomposition of ω^β into extremal permutation invariant states is thus supported by disjoint factor states and coincides by Prop. 2.4 with the central decomposition. Thus we have:

3.2. Proposition: The central decomposition of a thermodynamic accumulation point ω^β has a support $\mathcal{T}^\beta$ contained in $\partial_e \mathcal{S}^P$ and permutation invariance is never broken spontaneously.

For the spontaneous symmetry breaking of H the following result is useful [24]:

3.3. Proposition: The following two assertions are equivalent: (i) $\omega^\beta \in \partial_e \mathcal{S}^G$ (ii) $\exists\, \varphi \in \partial_e \mathcal{S}^P$ such that $\mathcal{T}^\beta = \{\nu_h(\varphi);\ \varphi \in H\}$. If these conditions are satisfied the central decomposition reads

$$\omega^\beta = \int_H \nu_h(\varphi) dh \tag{3.25}$$

where dh is the unique left (and right) invariant Haar measure of the compact group H.

As has been illustrated in the previous examples, the states

$$\omega^h := \nu_h(\varphi) \tag{3.26}$$

are the pure phase states. If they form an H-orbit (3.25) tells us, that they are equi-distributed in the sense of the Haar measure, which is of purely group theoretical, and not of dynamical, origin.

A further specification of the thermodynamic accumulation points is gained by means of thermodynamic variational principles. Given a state $\psi \in \mathcal{S}(\mathcal{A})$, its "action" on $A \in \mathcal{A}_\Lambda$ may be written as

$$\langle\psi;A\rangle = \mathrm{tr}_\Lambda\ \{\rho^\psi_\Lambda A\}\quad , \tag{3.27}$$

where ρ^ψ_Λ is a density operator in the finite-dimensional Hilbert space $\mathcal{H}_\Lambda := \bigotimes_{k\in\Lambda} \mathbb{C}^{n(k)}$, $n(k) := n(q)$ for $k = (i,q)$, and $\mathrm{tr}_\Lambda\{\cdot\}$ is the usual trace on $\mathcal{H}_\Lambda$. Again the local form of the states is used to define their physical properties by means of the thermodynamic limit. For the densities of the extensive observables the vector $c := (c_1,\ldots,c_r)$ of the subsidiary conditions (3.18) comes, however, into play, and we denote this restricted limit by "$\text{c-}\lim_\Lambda$".

3.4. Definition: For $\varphi \in \mathcal{S}$ with local density operators ρ^φ_Λ we define the energy, entropy, and free energy densities, respectively, by

(i) $u(c,\varphi) := \text{c-}\lim_\Lambda \langle\varphi;H_\Lambda\rangle/|\Lambda|$

(ii) $s(c,\varphi) := \text{c-}\lim_\Lambda \mathrm{tr}_\Lambda\{-k_B\rho^\varphi_\Lambda \log \rho^\varphi_\Lambda\}/|\Lambda|$

(iii) $f(c,\beta,\varphi) := u(c,\varphi) - Ts(c,\varphi)$,

if these limits exist.

In spite of the long range interactions of our model class, which makes the usual reasoning (cf. [1], Ch. 6.2) inapplicable, the thermodynamic functions are well-defined on $\mathcal{J}^P$ (and also on the smallest folium containing $\mathcal{J}^P$) as shows a slight extension of the results in [24].

3.5. Theorem: The limits in Def. 3.4 exist for all $\varphi \in \mathcal{J}^P$ and define there weak-*-continuous, affine functions. Furtheron, it holds

$$\begin{aligned} f(c,\beta) :&= \text{c-}\lim_{\Lambda} (-1/\beta|\Lambda|)\mathrm{tr}_\Lambda\{\exp(-\beta H_\Lambda)\} \\ &= \inf f(c,\beta,\mathcal{J}^P) = \inf f(c,\beta,\partial_e\mathcal{J}^P) \ . \end{aligned} \tag{3.28}$$

This shows, that the physical (specific) free energy, defined as the limit of canonical free energies, satisfies a variational principle, which may be restricted to pure (non-equilibrium) phase states. Beside other things this justifies a posteriori the usual calculation procedure of the finite temperature free energy of the BCS-model by means of the very restricted set of product states [27].

Let us now discuss the central support $\mathcal{J}^\beta$ of the thermodynamic accumulation state ω^β. Since ν_h, $h \in H$, is the dual of a C*-automorphism it is weak-*-continuous and

$$\begin{aligned} \omega^\beta = \nu_h(\omega^\beta) &= \int_{\mathcal{J}} \nu_h(\varphi)d\mu^\beta(\varphi) \\ &= \int_{\mathcal{J}} \varphi' d\mu^\beta(\nu_h^{-1}(\varphi')) \ . \end{aligned}$$

The support of $\mu^\beta(\nu_h^{-1}\cdot)$ is $\nu_h^{-1}(\mathcal{J}^\beta)$, which again consists of disjoint factor states (of product form) and thus $\mu^\beta(\nu_h^{-1}\cdot)$ is the unique central measure by Prop. 2.4. Then, of course, $\nu_h(\mathcal{J}^\beta) = \mathcal{J}^\beta$, for all $h \in H$.

3.6. Theorem: The central support $\mathcal{J}^\beta$ of the thermodynamic accumulation point ω^β satisfies

$$\mathcal{J}^\beta \subset \{\varphi \in \partial_e\mathcal{J}^P;\ f(c,\beta,\varphi) = \inf f(c,\beta,\mathcal{J}^P)\} \ .$$

The set $\inf f(c,\beta,\partial_e\mathcal{J}^P)$ may be determined by means of the extremal conditions, which often are called self-consistency equations.

If it is an orbit under ν_h, $h \in H$, then also the invariant set $\mathcal{T}^\beta$ is one and coincides with the minimal set. Since this holds for all accumulation points, and since by Prop. 3.3 the pure phase distribution is always given by the Haar measure dh, all accumulation points coincide and the limiting Gibbs states exist for all sub-nets with a fixed subsidiary condition c.

As an example for this indirect reasoning let us consider a "dirty" superconductor, where beside the conduction band electrons the impurity spins constitute a second lattice subsystem. Thus, $\mathcal{K} = \mathbb{Z}^3 \times \{1,2\}$ and

$$H_\Lambda = H^{BCS}_{\Lambda_1} - \frac{w}{(|\Lambda_1||\Lambda_2|)^{1/2}} \sum_{i\in\Lambda_1} \sum_{j\in\Lambda_2} [(c^*_{i\uparrow}c_{i\uparrow} - c^*_{i\downarrow}c_{i\downarrow})\sigma^z_j + c^*_{i\uparrow}c_{i\downarrow}\sigma^-_j + c^*_{i\downarrow}c_{i\uparrow}\sigma^+_j] \; . \tag{3.29}$$

The subsidiary condition $|\Lambda_2|/|\Lambda_1 \cup \Lambda_2| = c$ fixes the concentration of the impurities. The limiting Gibbs states depend thus on β and c. The volume function (3.17) is of the form $v_\Lambda = 1/(a_1|\Lambda_1|)$. The internal symmetries are $H = U(1) \times SO(3)$. The existence and form of the limiting Gibbs states can only be discussed indirectly by the orbit argument and numerical evaluation of the free energy minima [24]. For high temperature and all c-values $\omega^{\beta,c}$ is factorial. For low temperatures one has the onset of ferromagnetism (SO(3) is broken), or of superconductivity (U(1) is broken), or of both depending on c. That the latter case is also a pure phase is ensured by the factor property of the corresponding states in the central decomposition. The experimental verification of the coexistence of superconductivity and ferromagnetism in one pure phase is, e.g., reported in [28]. In the following diagram the lines of critical points are depicted. They intersect in critical points of higher order. Interesting is the (collective) Kondo-effect, where for a fixed c the system becomes ferromagnetic, then superconducting, and then again ferromagnetic with decreasing temperature.

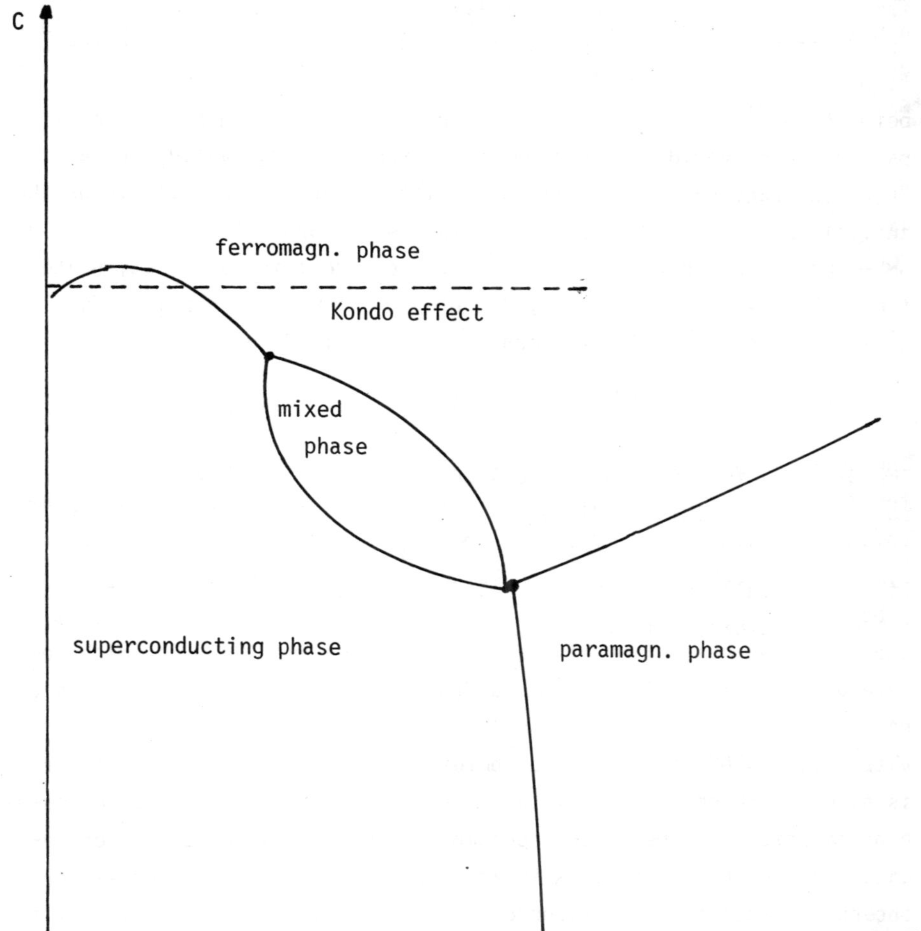

Phase-diagram of a BCS-superconductor coupled with magnetic impurities.

4. CONDENSED AND QUASI-PARTICLES IN SYMMETRY BREAKING SYSTEMS

The central decomposition of the limiting Gibbs state is non-trivial, if there is spontaneous symmetry breaking. In some cases we could interpret it as conservation of particles at infinity by inspection of the local approximation procedure. The most famous connection between spontaneous symmetry breaking and the appearence of special particles is provided by the Goldstone theorem [29], valid for short range interacting systems. If the limiting Gibbs state ω^β is known the analysis of the particle structure may be performed in the associated GNS-representation, irrespectively of the underlying interaction and the broken symmetry group. If, as before, μ^β is the central measure of ω^β with support $\mathcal{J}^\beta \subset \mathcal{S}(\mathcal{A})$ then the GNS-triple decomposes as

$$(\pi_\beta, \mathcal{H}_\beta, \Omega_\beta) = \int_{\mathcal{J}^\beta}^{\oplus} (\pi_\varphi, \mathcal{H}_\varphi, \Omega_\varphi) d\mu^\beta(\varphi) \; . \tag{4.1}$$

The direct integrals over the GNS-triples associated with $\varphi \in \mathcal{J}^\beta$ refer to the representations, Hilbert spaces, and cyclic vectors separately ([1], Ch. 4.4). Every $\pi_\beta(A)$, $A \in \mathcal{A}$, is a decomposable operator in $\mathcal{H}_\beta$. It is physically essential to consider also the weak closure of this set of operators, namely

$$\mathcal{M}^\beta := \pi_\beta(\mathcal{A})'' = \int_{\mathcal{J}^\beta}^{\oplus} \mathcal{M}^\varphi \, d\mu^\beta(\varphi) \tag{4.2}$$

where always $\mathcal{M}^\varphi$ is the double commutant of $\pi_\varphi(\mathcal{A})$. Many features of symmetry breaking are visible only by extending the reconstructed quantum mechanics to this state dependent von Neumann algebra. The center $\mathcal{Z}^\beta := \mathcal{M}^\beta \cap \pi_\beta(\mathcal{A})'$ is non-trivial if μ^β is not a point measure. The internal symmetries are extended by means of the operators

$$W_h^\beta \, \pi_\beta(C)\Omega_\beta := \pi_\beta(\alpha_h(C))\Omega_\beta \; , \quad C \in \mathcal{A}, \; h \in H \; , \tag{4.3}$$

which may be extended to unitaries in $\mathcal{H}_\beta$. Clearly

$$\alpha_h^\beta(M) := W_h^\beta \, M \, W_h^{\beta -1} \; , \quad M \in \mathcal{M}^\beta \; , \tag{4.4}$$

is then an extension of $\alpha_h \in \mathrm{Aut}\,(\mathcal{A})$ to a W*-automorphism of $\mathcal{M}^\beta$. For special elements in $\mathcal{M}^\beta$ of the form

$$M = \int^{\oplus}_{\mathcal{J}^\beta} \pi_\varphi(A^\varphi)d\mu^\beta(\varphi) \ , \ A^\varphi \in \mathcal{A}, \tag{4.5}$$

which constitute the *-subalgebra $\mathcal{M}^\beta_0 \subset \mathcal{M}^\beta$, we find

$$\begin{aligned}
(\pi_\beta(C_1)\Omega_\beta, \ \alpha^\beta_h(M)\pi_\beta(C_2)\Omega_\beta) &= \\
&= (\Omega_\beta, \pi_\beta\ (\alpha_h^{-1}(C_1^*))M\pi_\beta(\alpha_h^{-1}(C_2))\Omega_\beta) \\
&= \int_{\mathcal{J}^\beta} <\varphi;\ \alpha_h^{-1}(C_1^*)A^\varphi\alpha_h^{-1}(C_2))d\pi^\beta(\varphi) \\
&= \int_{\mathcal{J}^\beta} <\nu_h^{-1}(\varphi);\ C_1^*\alpha_h(A^\varphi)C_2 > d\mu^\beta(\varphi) \\
&= \int_{\mathcal{J}^\beta} <\varphi';\ C_1\alpha_h(A^{\varphi'^h})C_2 > d\mu^\beta(\varphi')\ ,
\end{aligned}$$

where

$$\varphi^h := \nu_h(\varphi) \tag{4.6}$$

and C_1, $C_2 \in \mathcal{A}$, since μ^β is invariant under ν_h according to the reasoning before Theorem 3.6. Since $\pi_\beta(\mathcal{A})\Omega_\beta$ is dense in $\mathcal{H}_\beta$ we obtain

4.1. <u>Proposition:</u> On all $M \in \mathcal{M}^\beta_0$ the extended internal symmetry transformations α^β_h, $h \in H$, act as

$$\alpha^\beta_h(M) = \int_{\mathcal{J}^\beta} \pi_\varphi(\alpha_h(A^{\varphi^h}))d\mu^\beta(\varphi)\ . \tag{4.7}$$

Thus only by extension to elements $M \in \mathcal{M}^\beta_0$, with φ-dependent components $\pi_\varphi(A^\varphi)$, the macroscopic transformation property of the internal symmetries, namely the change of the phase index φ to φ^h, is displayed.

Let us introduce the stability groups

$$H_\varphi := \{h \in H;\ \nu_h(\varphi) = \varphi\} \tag{4.8}$$

and the group of unbroken symmetries

$$S := \bigcap_{\varphi \in \mathcal{J}^\beta} H_\varphi\ . \tag{4.9}$$

Observe that by ν_h^{-1} H acts homomorphic in $\mathcal{J}^\beta \subset \mathcal{S}$. For all $s \in S$, all $h \in H$ and all $\varphi \in \mathcal{J}^\beta$ we have

$$\nu^{-1}_{hsh^{-1}}(\varphi) = \nu^{-1}_h \circ \nu^{-1}_s \circ \nu^{-1}_{h^{-1}}(\varphi) = \varphi$$

since $\mathcal{T}^\beta$ is stable under H. Thus S is a normal subgroup of H in any case. If $\mathcal{T}^\beta$ is an orbit for H it is labelled by the quotient group

$$B := H/S \tag{4.10}$$

of broken symmetries. Fix for this one $\varphi =: \varphi e \in \mathcal{T}^\beta$, where e is the unit of H. Every $\psi \in \mathcal{T}^\beta$ has then the form $\psi = \nu_h(\varphi e) =: \varphi h$. If $h' \in hS$ then $\varphi h = \varphi h'$. On the other hand, the latter equality implies $h^{-1}h' \in H_{\varphi e}$ and also

$$\Big(\bigcap_{h_1 \in H} h_1 h H_{\varphi e} h_1^{-1} h^{-1}\Big)h = \Big(\bigcap_{h_1 \in H} h_1 h' H_{\varphi e} h_1^{-1} h'^{-1}\Big)h' .$$

Since $H_{\varphi h} = hH_{\varphi e}h^{-1}$ it follows $Sh = Sh'$ and also $h' \in hS$.

In general B is not a subgroup of H. If the compact group H is also a Lie group it holds locally $H = B \times S$ ([30], p. 17). It is convenient, but for many considerations not essential, to require in the following that

$$H = B \times S \tag{4.11}$$

where B is a (compact) Lie group and to assume that $\mathcal{T}^\beta$ is an H-orbit. Since $\mathcal{T}^\beta$ is then also a B-orbit and since the restriction of the Haar measure dh to B gives the Haar measure db we obtain

$$\omega^\beta = \int_B \varphi b \, db . \tag{4.12}$$

The W*-isomorphism

$$\kappa_\beta : \mathcal{L}^\infty(\mathcal{T}, \mu^\beta) \xrightarrow{\text{onto}} \mathcal{Z}^\beta \tag{4.13}$$

from the general theory of orthogonal measures ([1], Ch. 4.1) modifies then to

$$\kappa_\beta : \mathcal{L}^\infty(B, db) \xrightarrow{\text{onto}} \mathcal{Z}^\beta . \tag{4.13a}$$

Labelling the decompositions (4.1) and (4.2) by $b \in B$ we find

$$(\Omega_b, \pi_b(\alpha_b^{-1}(A))\Omega_b) = (\Omega e, \pi e(A)\Omega e) \tag{4.14}$$

for all $A \in \mathcal{A}$. This gives rise to a famity of unitary mappings

$$V_b : \mathcal{H}_e \longrightarrow \mathcal{H}_b$$

$$V_b\, \pi_e(A)\Omega_e := \pi_b(\alpha_b^{-1}(A))\Omega_b$$

for all $b \in B$. It leads furtheron to the W*-isomorphy of all $\mathcal{M}^b$ with $\mathcal{M}^e$. This, finally, gives the W*-isomorphism

$$\mathcal{M}^\beta \cong \mathcal{L}^\infty(B,db)\ \bar{\otimes} \mathcal{M}^e\ , \tag{4.15}$$

where $\bar{\otimes}$ denotes the W*-tensorproduct [9]. Since $\mathcal{M}^e$ is a factorial pure phase von Neumann algebra it seems at first sight, that we have succeeded by (4.15) in splitting off the macroscopic classical observables. By means of the indicated W*-isomorphism, the quasilocal $\pi_\beta(A)$, $A \in \mathcal{A}$, are, however, not mapped into $\mathcal{M}^e$. In this way, the collective modes in $\mathcal{Z}^\beta$ together with their symmetry structure are not separated from a normal, microscopic background.

A slight improvement of our assumptions, connected with a generalized Bogoliubov transformation, will, however, lead to a perfect two-fluid picture. For motivational purposes only let us mention that every pure phase state φb restricted to $\mathcal{A}_\Lambda$ is given by a density operator which has a state dependent "model Hamiltonian" H_Λ^b in its exponent. (The nomenclature is taken from the BCS-theory [27]. The connection with the generator of the pure phase limiting dynamics is worked out in [24],[31].) A Bogoliubov transformation serves to "diagonalize" the H_Λ^b for all Λ. It should be compatible with the unbroken symmetries. If one inserts the original (not Bogoliubov transformed) field operators into the diagonalized model Hamiltonians the full internal symmetry should be restored. Formulated directly in terms of the pure phase states this leads to

4.2. Assumption: There is a $*$-automorphism χ^e of $\mathcal{A}$ such that

$$[\chi^e, \alpha_s]_- = 0\ , \qquad s \in S \tag{4.16}$$

and

$$\nu_b \circ \chi^{e*}(\varphi e) = \chi^{e*}(\varphi e),\ \forall\, b \in B\ . \tag{4.17}$$

4.3. Definition: Given χ^e of the above assumption we introduce the pure phase Bogoliubov transformations

$$\chi^b := \alpha_b^{-1} \circ \chi^e \circ \alpha_b \tag{4.18}$$

and the multi-phase Bogoliubov transformation

$$\chi : \mathcal{M}_o^\beta \to \mathcal{M}_o^\beta$$

$$\chi[M] := \int_B^{\oplus} \pi_b(\chi^b(A^b))db. \tag{4.19}$$

It is clear that χ^b and χ are $*$-automorphisms of $\mathcal{A}$ and $\mathcal{M}_o^\beta$ respectively. It holds

$$[\chi, \alpha_b^\beta]_- = 0 \ , \ \forall \ b \in B \ . \tag{4.20}$$

We calculate:

$$(\chi[\pi_\beta(A_1)]\Omega_\beta, \chi[\pi_\beta(A_2)]\Omega_\beta) = \tag{4.21}$$

$$= \int_B <\nu_b(\varphi e); \alpha_b^{-1}\chi^e\alpha_b(A_1^* A_2)>db$$

$$= (\pi_b(\chi^b[A_1])\Omega_b, \pi_b(\chi^b[A_2])\Omega_b) \ ,$$

for all $b \in B$. This leads to unitary mappings between

$$\mathcal{H}_\beta^q := \overline{\chi[\pi_\beta(\mathcal{A})]\Omega_\beta} \subset \mathcal{H}_\beta \tag{4.22}$$

and the pure phase spaces $\mathcal{H}_b$. And this gives then W*-isomorphisms between

$$\mathcal{M}_q^\beta := \chi[\pi_\beta(\mathcal{A})]'' \subset \mathcal{M}^\beta \tag{4.23}$$

and the $\mathcal{M}^b$, associating $\chi[\pi_\beta(A)]$ with $\pi_b(\chi^b[A])$ for all $b \in B$. The latter fact implies that χ^b and χ are not σ-continuous in the respective Hilbert spaces, since the pure phase GNS-representations are mutually disjoint.

Let us observe, that any vector $\psi \in \mathcal{H}_\beta$ of the form $\kappa_\beta(f)\chi[\pi_\beta(A)]\Omega_\beta$ determines $f \in \mathcal{L}^\infty(B,db)$ and $A \in \mathcal{A}$ uniquely up to a c-number, and that the linear hull of these vectors is dense in $\mathcal{H}_\beta$. ($\pi_b(\chi^b[\mathcal{A}])\Omega_b$ is dense in $\mathcal{H}_b$ and $\mathcal{L}^\infty(B,db)$ is dense in $\mathcal{L}^2(B,db)$.) We find

$$(\kappa_\beta(f_1)\chi[\pi_\beta(A_1)]\Omega_\beta, \kappa_\beta(f_2)\chi[\pi_\beta(A_2)]\Omega_\beta) \tag{4.24}$$
$$= \int_B \bar{f}_1(b) f_2(b) db (\chi \circ \pi_\beta(A_1)\Omega_\beta, \chi \circ \pi_\beta(A_2)\Omega_\beta) \ .$$

4.4. Theorem: The mapping

$$U \ \kappa_\beta(f)(\chi \circ \pi_\beta)(A)\Omega_\beta := f \otimes (\chi \circ \pi_\beta)(A)\Omega_\beta$$

for all $f \in \mathcal{L}^\infty(B,db)$ and all $A \in \mathcal{A}$, is well defined and extends to a unitary mapping

$$U : \mathcal{H}_\beta \to \mathcal{L}^2(B,db) \otimes \mathcal{H}_\beta^q \ . \tag{4.25}$$

It holds

$$U \mathcal{M}^\beta U^{-1} = \mathcal{L}^\infty(B,db) \otimes \mathcal{M}_q^\beta . \tag{4.26}$$

Since for all $h \in H$

$$U \ W_h^\beta \ \kappa_\beta(f)(\chi \circ \pi_\beta)(A)\Omega_\beta = {}_hf \otimes (\chi \circ \pi_\beta)(\alpha_h(A))\Omega_\beta$$

we have the decomposition

$$U \ W_h^\beta U^{-1} = W_h^c \otimes W_h^q \tag{4.27}$$

where W_h^c is the left-translation in $\mathcal{L}^2(B,db)$, which transforms $f(h')$ into ${}_hf(h') := f(hh')$, and W_h^q is given by the quasilocal action of α_h. From its original definition the α_h are strongly continuous in h as operators on $\mathcal{A}$ and the W_h are strongly continuous in h as operators on $\mathcal{H}_\beta$. Thus the one-parameter subgroups have self-adjoint generators X_l in $U\mathcal{H}_\beta$, $1 \leq l \leq n$, if n is the dimension of H. In the tensor-product decomposition, which we call the "two-fluid picture", the multiplication operators in $\mathcal{L}^2(B,db)$ by $f \in \mathcal{L}^\infty(B,db)$ are still denoted by $\kappa_\beta(f)$. From (4.27) follows an additive decomposition of important observables.

4.5. Theorem: A symmetry breaking system satisfying assumption 4.2 has a two-fluid picture where the represented generators of the internal symmetry group H decompose as

$$X_l = X_l^c \otimes 1_q + 1_c \otimes X_l^q \ , \ 1 \leq l \leq n \ . \tag{4.28}$$

into a condensed and quasi-particle part. In the condensed part, the basis elements of the Lie group of S vanish and we have there the group contraction form H to B. The spectral projections of the X_l, $1 \le l \le m$, where $m \le n$ is the dimension of B, generate together with the $\kappa_\beta(f)$, $f \in \mathcal{L}^\infty(B,db)$, the algebra of macroscopic quantum observables $\mathcal{B}(\mathcal{L}^2(B,db))$.

Since we are dealing here with a representation of a quantum theory (in $U\mathcal{H}_\beta$), the generators of the unitarily represented internal symmetries are directly interpretable observables, which satisfy conservation laws. Thus we are forced from physical reasons to supplement our algebra of observables $\mathcal{M}^\beta$ (resp. $U\mathcal{M}^\beta U^{-1}$), relevant for the system in thermodynamic equilibrium, by the (spectral projections of the) X_l. These self-adjoint operators are not affiliated with $U\mathcal{M}^\beta U^{-1}$ for $0 < \beta < +\infty$, and we have to perform a real extension of the formalism. Since the X_l^c , $1 \le l \le m$, are parts of the total generators they necessarily supplement the classical macroscopic observables of $\kappa_\beta(\mathcal{L}^\infty(B,db))$. Since these differential operators on $\mathcal{L}^2(B,db)$ are not affected by local perturbations, they are also "macroscopic". This shows, that there arises a macroscopic quantum theory with $\mathcal{B}(\mathcal{L}^2(B,db))$ as its algebra of observables, which commutes with all quasi-particle operators in $\mathcal{M}_q^\beta$. The lowering and rising operators of B, represented as operators in $\mathcal{L}^2(B,db)$, define a condensed particle structure. We claim that these particles replace the particles from the Goldstone theorem, which is not applicable to the considered long range interacting systems [33]. These condensed particles are, however, measurable only at weakly coupled systems.

5. SYMMETRY BREAKING AND PARTICLE STRUCTURE OF WEAKLY COUPLED SYSTEMS

We consider a composition of two systems 1 and 2 of the type discussed in Section 4. The local regions are then $\Lambda := (\Lambda_1, \Lambda_2) \in \mathcal{L}_1 \times \mathcal{L}_2$ and the local algebras are $\mathcal{A}_\Lambda = \mathcal{A}_{\Lambda_1} \otimes \mathcal{A}_{\Lambda_2}$. The quasilocal algebra has the form

$$\mathcal{A} = \mathcal{A}_1 \otimes \mathcal{A}_2 , \tag{5.1}$$

where here " $\otimes$ " signifies the abstract C*-algebraic tensor product with minimal cross norm [4].

A weak coupling Hamiltonian is by definition of the form

$$H'_\Lambda = H_{\Lambda_1} + H_{\Lambda_2} + h_\Lambda = H_\Lambda + h_\Lambda \tag{5.2}$$

with

$$h_\Lambda = P(\sigma^\kappa_{k,1}, \sigma^\lambda_{l,2}, m^\kappa_{\Lambda_1}, m^\lambda_{\Lambda_2}) \tag{5.3}$$

being a fixed polynomial in the operators of the argument and their Hermitian adjoints. The interaction consists thus in an exchange of finitely many normal and condensed particles, the latter being described by the representation dependent weak limits of the averages

$$m^\kappa_{\Lambda_1} := \sum_{k\in\Lambda_1} \sigma^\kappa_{k,1}/|\Lambda_1| \ . \tag{5.4}$$

If the families $\{H_{\Lambda i}; \Lambda_i \in \mathcal{L}_i\}$, i=1,2, define the internal symmetries H_i, then $H = H_1 \times H_2$ is the internal symmetry group of $\{H_\Lambda; \Lambda \in \mathcal{L}\}$. The internal symmetries of the coupled system are given by

$$H' := \{h \in H;\ \alpha_h(h_\Lambda) = h_\Lambda, \forall\, \Lambda \in \mathcal{L}\} \quad , \tag{5.5}$$

where $\alpha_h = \alpha_{h_1} \otimes \alpha_{h_2} \in \mathrm{Aut}\,(\mathcal{A})$. H' is a strict subgroup of H.

Clearly the net of local canonical equilibrium states of the uncoupled composite system converges to the limiting Gibbs state

$$\omega^\beta = \int_{B_1 \times B_2} \omega^{b_1} \otimes \omega^{b_2}\, db_1 db_2 \ , \tag{5.6}$$

if we know this for the separate systems (perhaps under subsidiary conditions for the net of local regions), and if the separate orbit conditions are satisfied. Then the support $\mathcal{T}^\beta \subset \mathcal{S}(\mathcal{A})$ of the central measure associated with ω^β is an orbit under the direct product of the broken symmetries B_1 and B_2, which justifies (5.6). By means of a Dyson series expansion with respect to the general weak coupling h_Λ and by an interchange of limits one can treat also the coupled equilibrium states [32].

5.1. Theorem: The net of local canonical states of the weakly coupled model converges in the weak-$*$-topology to the limiting Gibbs state $\omega'^\beta \in \mathcal{S}(\mathcal{A})$, the central decomposition of which has the form

$$\omega'^{\beta} = \int_{B_1 x B_2} \omega'^{b_1 b_2} v(b_1,b_2) db_1 db_2 \ , \tag{5.7}$$

where $v(b_1,b_2)$ is a strictly positive, normalized weight function and $db_1 db_2$ is the Haar measure on $B_1 \times B_2$.

In general $\omega'^{b_1 b_2} \neq \omega^{b_1} \otimes \omega^{b_2}$ and the central support $\mathcal{T}'^{\beta} \neq \mathcal{T}^{\beta}$. Since the uncoupled broken symmetries have direct product form $\alpha_b = \alpha_{b_1} \otimes \alpha_{b_2}$, $b \in B_1 x B_2$, they do not, in general, leave h_Λ invariant, because the latter property would imply a certain connection between b_1 and b_2. Thus, for a coupling with normal and/or condensed particle exchange

$$B_1 \times B_2 \not\subset H'. \tag{5.8}$$

The reduction of the internal symmetries leads to the loss of their transitive action on $\mathcal{T}'^{\beta}$, which still has, as a manifold, the dimensionality of $\mathcal{T}^{\beta}$. This provides the possibility that the desintegration measure of ω'^{β} is not a Haar measure. One finds, in fact, that $v(b_1,b_2) \neq 1$, if there is an exchange of condensed particles. But $v(b_1,b_2) db_1 db_2$ is, as a measure, still equivalent to $db_1 db_2$, which gives the following result.

5.2. Proposition: The GNS-representation of ω'^{β} is unitarily equivalent to that of ω^{β}.

This assertion contrasts with the strong coupling case, where the interacting theory requires a representation, which is disjoint to that of the uncoupled one. In the weak coupling case we have still the whole condensed and quasi-particle structure of the constituing subsystems, since the representation von Neumann algebra still has the uncoupled form

$$\begin{aligned} \mathcal{M}^{\beta} &= \mathcal{M}_1^{\beta} \bar{\otimes} \mathcal{M}_2^{\beta} \\ &= \mathcal{L}^{\infty}(B_1 x B_2, db_1 db_2) \bar{\otimes} \mathcal{M}_{q_1}^{\beta} \bar{\otimes} \mathcal{M}_{q_2}^{\beta} \ . \end{aligned} \tag{5.9}$$

By our previous analysis in Section 3 we conclude from the weighted central measure, that the condensed particles are not conserved in the subsystems.

Let us illustrate the general reasoning by means of a model for the Josephson junction, which consists of two weakly coupled BCS-superconductors with the separate Hamiltonians H_{Λ_i}, i=1,2, from (3.9). The total system is defined by a family of local Hamiltonians of the form

$$H'_\Lambda = H_{\Lambda_1} + H_{\Lambda_2} + P(C^1_{k\sigma}, C^2_{l\rho}, m_{\Lambda_1}, m_{\Lambda_2}) , \tag{5.10}$$

where in the fixed polynomial $P(\cdot)$ $m_{\Lambda_i} = \sum_{k\in\Lambda_i} C^i_{-k\downarrow} C^i_{k\uparrow}/|\Lambda_i|$, i=1,2. In a representation these averages converge in the weak operator topology to central elements describing condensed Cooper pairs. From the stability of the condensed and quasi-particle structure against weak interactions one may argue that during the tunneling their status (to be condensed or not) does not change. Thus, the interaction polynomial in (5.10) should decompose into a sum of h^c and h^q, where h^q converges to a polynomial in the quasi-particle operators (see below) and

$$h^c_\Lambda = \tau(m^*_{\Lambda_1} m_{\Lambda_2} + m_{\Lambda_1} m^*_{\Lambda_2}) \tag{5.11}$$

converges to the leading exchange term for condensed Cooper pairs. The coupling constant τ is proportional to the pair tunnel frequency.

The internal symmetry group for the uncoupled composite system $H_1 x H_2 = U(1)xU(1)$ ($= B_1 x B_2$ for $\beta > \beta^1_c, \beta^2_c$) is reduced by the interaction to

$$H' = U(1) \tag{5.12}$$

describing identical gauge transformations in both subsystems.

The limiting Gibbs state of the grand canonical ensemble is of the form [34]

$$\omega'^\beta = \iint_0^{2\pi} \omega'^\theta c(\beta)\exp[-2\beta w_1 w_2 \cos(\theta_1 - \theta_2)]d\theta , \tag{5.13}$$

where $c(\beta)$ is a positive normalization constant, $\theta = (\theta_1, \theta_2)$ labels the elements of $B_1 x B_2$, the normalized Haar measure is $d\theta := d\theta_1 d\theta_2/(2\pi)^2$, and the real parameters $w_i = w_i(\beta)$ are the same as in (3.10). For $\beta > \beta^i_c$,

$i=1,2$, both w_i (and the gaps) are greater than zero and the decomposition in (5.13) is nontrivial, exhibiting a weight function with the Josephson cosine potential. This special interaction potential does not show up in the specific free energy (as in [38]) but, nevertheless, brings about an exchange of condensed Cooper pairs.

According to Prop. 5.2 we may use for the coupled theory the uncoupled GNS-representation $(\pi_\beta, \mathcal{H}_\beta, \Omega_\beta)$. Here the quasi-particle defining Bogoliubov transformation is determined by the special applications

$$\chi(\pi_\beta(C^i_{k\uparrow})) := u^i_k \pi_\beta(C^i_{k\uparrow}) - r^i_k m_i(\beta) \pi_\beta(C^{*i}_{-k\downarrow}) =: \gamma^i_{ko} \quad , \ i=1,2 \qquad (5.14)$$

$$\chi(\pi_\beta(C^i_{-k\downarrow})) := u^i_k \pi_\beta(C^i_{-k\downarrow}) + r^i_k m_i(\beta) \pi_\beta(C^{*i}_{k\uparrow}) =: \gamma^i_{k1}$$

where u^i_k, v^i_k are the usual constants [27], $r^i_k := |v^i_k|/w_i(\beta)$, and

$$m_i(\beta) := \text{weak-}\lim_\Lambda m_{\Lambda_i} \qquad (5.15)$$
$$= w_i(\beta) \exp(-i2\Theta^\beta_i) \quad ,$$

with

$$2\Theta^\beta_i = \int\!\!\int^\oplus \theta_i \pi_\theta(1) d\theta \ . \qquad (5.16)$$

The condensed Cooper pair annihilation operators $m_i(\beta)$, $i=1,2$, and the macroscopic phase operators Θ^β_i, $i=1,2$, are elements of the center of $\mathcal{M}^\beta = \pi_\beta(\mathcal{A}_1 \otimes \mathcal{A}_2)''$ and decompose according to the pure phase representations, labelled by θ. The $\gamma^i_{k\sigma}$ generate the von Neumann algebra $\mathcal{M}^\beta_q$, and again $\mathcal{M}^\beta \cong \mathcal{L}^\infty([0,2\pi)^2, d\theta) \otimes \mathcal{M}^\beta_q$. According to (4.7) the extended gauge transformations $\alpha^\beta_{\theta_i}$, $\theta_i \in [0,2\pi)$, act non-trivially on the $m_i(\beta)$ and hence covariantly on the quasi-particle fields $\gamma^i_{k\sigma}$, $\sigma=0,1$. The generators of the unitaries, which implement the gauge automorphisms and leave Ω_β invariant, are the appropriate, renormalized particle number operators of the representation and decompose as

$$N^\beta_i = N^c_i \otimes 1_q + 1_c \otimes N^q_i \ , \qquad (5.17)$$

where N^c_i is essentially the differentiation to θ_i. It holds

$$[N_i^\beta, \gamma_{k\sigma}^j] = [1_c \otimes N_i^q, \gamma_{k\sigma}^j] = -\delta_{ij}\gamma_{k\sigma}^j \tag{5.18}$$

$$[N_i^\beta, m_j(\beta)] = [N_i^c \otimes 1_q, m_j(\beta)] = -2\delta_{ij}m_j(\beta)$$

as well as

$$[N_i^\beta, \Theta_j^\beta] = [N_i^c \otimes 1_q, \Theta_j^\beta] = -i\pi_\beta(1)\delta_{ij} . \tag{5.19}$$

The last relation is typical for macroscopic quantum mechanics. Observable effects are, up to now, only dynamical ones. As is elaborated in [34] the limiting Heisenberg dynamics acts as well defined W*-automorphisms in $\mathcal{M}^\beta$. If the weak interaction decomposes as described before the limiting dynamics leaves both $\mathcal{Z}^\beta$ and $\mathcal{M}_q^\beta$ invariant and acts in $\mathcal{Z}^\beta$ in a nontrivial manner. For the macroscopic phase difference it holds in particular

$$\Theta_1^\beta(t) - \Theta_2^\beta(t) = \Theta_1^\beta - \Theta_2^\beta + (\mu_1 - \mu_2)t\pi_\beta(1) . \tag{5.20}$$

In order to calculate the current we introduce also a time dependence for the N_i^β, which are not affiliated with $\mathcal{M}^\beta$ and thus not locally approximable in the usual sense. For a, so to say, generalized local approximation we have need for the modular *-antiautomorphisms

$$j : \mathcal{M}^\beta \to \mathcal{M}^{\beta'} \tag{5.21}$$

of the Tomita-Takesaki-theory [1], [39]. In the strong resolvent sense it holds then in $\mathcal{H}_\beta$

$$N_i^\beta = \lim_\Lambda(\pi_\beta(N_{\Lambda i}) - j(\pi_\beta(N_{\Lambda i}))) , \tag{5.22}$$

where $N_{\Lambda i} = \sum_\sigma \sum_{k\in\Lambda_i} C_{k\sigma}^{i*}C_{k\sigma}^i$. We consider the subtraction terms in (5.22) as operator-valued regularization and renormalization quantities, required for the thermodynamic limit, which should be time-independent in a fixed representation. Locally we have

$$N_{\Lambda_i}(t) = \exp(itH_\Lambda')N_{\Lambda_i}\exp(-itH_\Lambda') , \tag{5.23}$$

and for the associated current operator

$$J_{\Lambda_1}(t) = dN_{\Lambda_1}(t)/dt . \tag{5.24}$$

5.3. Theorem: For weakly coupled BCS-models with interactions of the form (5.10), (5.11) the following limits exist in the strong resolvent sense of $\mathcal{H}_\beta$ for all $t \in \mathbb{R}$

$$N_i^\beta(t) := \lim_\Lambda (\pi_\beta(N_{\Lambda_i}(t)) - j(\pi_\beta(N_{\Lambda_i}))) \tag{5.25}$$

and define a family of unbounded self-adjoint operators on the common dense domain $\mathcal{D}(N_i^\beta)$, i=1,2. There it holds

$$dN_1^\beta(t)/dt = \lim_\Lambda \pi_\beta(J_{\Lambda_1}(t) \tag{5.26}$$

$$=: J_\beta(t) = J_\beta^c(t) + J_\beta^q(t) ,$$

where $J_\beta^q(t) \in \mathcal{M}_q^\beta$ and

$$J_\beta^c(t) = 4\tau w_1 w_2 \sin(2\Theta_1^\beta(t) - 2\Theta_2^\beta(t)) . \tag{5.27}$$

Equations (5.20) and (5.27), which involve only central observables, constitute the Josephson relations in operator form [36]. If $\mu_1 = \mu_2$, relation (5.27) gives a directed current, which is typical for weak coupling theories. In strong coupling theories with ([40]) or without ([41]) reservoirs an alternating current is predicted also for vanishing voltage difference. A direct measurement of the dc-Josephson relation seems to be outside the present experimental possibilities [42]. Our treatment confirms a statement of [43] that the Josephson relations for themselves are formulated in terms of classical macroscopic observables only. This is, however, quite different for two Josephson junctions combined in a SQUID ("superconducting quantum interference device"), where the incompatibility of phase and particle number differences seems to produce macroscopic quantum effects [44],[45].

In our treatment the envisaged macroscopic quantum phenomena are part of a general theory of condensed and quasi-particles. This structure arises from a twofold contraction, where the internal symmetry reduction $H \to S$ is the prerequisite for the algebraic contraction $\mathcal{A} \to \mathcal{L}^\infty(B,db)$. In the latter commutative algebra the creation operators of the condensed particles are located. In non-relativistic many body physics this type of particles is absolutely real. It is only the way they enter the theory, where only their collective behaviour is described, which gives them a slightly exotic touch. The direct observation of their properties seems to be possible only in those coupled systems with spontaneous symmetry breaking, where the interaction is strong enough to induce exchange processes but weak enough not to destroy the internal condensed and quasi-particle structures.

Acknowledgements

This work belongs to a research project supported by the Deutsche Forschungsgemeinschaft. Discussions with E. Duffner, W. Fleig, W. Hauser, and G. Raggio are gratefully acknowledged.

References

[1] O. Bratteli and D.W. Robinson, Operator Algebras and Quantum Statistical Mechanics I, II (Springer, New York, 1979,1981)

[2] G.G. Emch, Algebraic Methods in Statistical Mechanics and Quantum Field Theory (Wiley-Interscience, New York, 1972)

[3] R.V. Kadison, Topology 3, Suppl. 2, 177 (1965)

[4] S. Sakai, C*-Algebras and W*-Algebras (Springer, Berlin, 1971)

[5] G.K. Pedersen, C*-Algebras and their Automorphism Groups (Academic Press, London, 1979)

[6] G. Falk and H. Jung, Hdb. Physik III/2 (Springer, Berlin, 1959)

[7] H. Primas, Chemistry, Quantum Mechanics and Reductionism, Lecture Notes in Chemistry 24 (Springer, Berlin, 1981)

[8] A. Rieckers, in: Groups, Systems, and Many-Body Physics, p. 69, ed. P. Kramer and M. Dal Cin (Vieweg & Sohn, Braunschweig, 1980)

[9] M. Takesaki, Theory of Operator Algebras I (Springer, New York, 1979)

[10] R. Haag, R.V. Kadison, and D. Kastler, Commun. Math. Phys. 16, 81 (1970)

[11] A. Rieckers, Z. Naturforsch. 33a, 1406 (1978)

[12] H. Stumpf and A. Rieckers, Thermodynamik I (Vieweg & Sohn, Braunschweig, 1976)

[13] L. Tisza, Evolution of the Concepts of Thermodynamics (MIT Press, Cambridge, Mass., 1966)

[14] A. Rieckers, Physica 108A, 107 (1981)

[15] H. Hellmich, Diplom Thesis (Tübingen, 1982)

[16] E. Duffner, Diplom Thesis (Tübingen, 1982)

[17] E. Widmann, Diplom Thesis (Tübingen, 1982)

[18] W. Fleig, Acta Phys. Austr. 55, 135 (1983)

[19] A. Rieckers and M. Ullrich, Acta Phys. Austr. 56, 131 (1985)

[20] A. Rieckers and M. Ullrich, Acta Phys. Austr. 56, 259 (1985)

[21] H.-J. Volkert, Preprint (Tübingen 1985)

[22] K. Hepp, Helv. Phys. Acta 45, 237 (1972)

[23] W. Thirring, Commun. Math. Phys. 7, 181 (1968)

[24] W. Fleig, Ph. D. Thesis (Tübingen, 1985)

[25] E. Størmer, J. Funct. Anal. 3, 48 (1969)

[26] loc. cit. [5], Ch. 6.5.

[27] M. Tinkham, Introduction to Superconductivity (McGraw-Hill, Tokyo, 1975)

[28] M. Wilhelm and B. Hillenbrand, Z. Naturforsch. 26a, 141 (1971)

[29] H. Ezawa and J. Swieca, Commun. Math. Phys. 5, 330 (1967) and references therein.

[30] A.O. Barut and R. Rączka, Theory of Group Representations and Applications, (PWN, Warszawa, 1977)

[31] loc. cit. [20], p. 268

[32] M. Ullrich, to appear in Rep. Math. Phys. (1985)

[33] G. Morchio and F. Strocchi, Commun. Math. Phys. 99, 153 (1985)

[34] A. Rieckers and M. Ullrich, On the Microscopic Derivation of the Finite Temperature Josephson Relation in Operator Form, Preprint (Tübingen, 1985)

[35] E.B. Davies, One-Parameter Semigroups (Academic Press, London, 1980)

[36] B.D. Josephson, Phys. Lett. 1, 251 (1962)

[37] A. Barone and G. Paterno, Physics and Applications of the Josephson Effect (J. Wiley & Sons, New York, 1982)

[38] P.W. Anderson, in: Lectures on the Many-Body Problem, Vol. 2, ed. E.R. Caianiello (Academic Press, London, 1964)

[39] M. Takesaki, Tomita's Theory of Modular Hilbert Algebras and its Applications (Springer, Berlin, 1970)

[40] K. Hepp and E.H. Lieb, Helv. Phys. Acta 46, 573 (1973)

[41] E. Duffner, The Macroscopic Pure Phase Dynamics of Two Strongly Coupled Superconductors, Preprint (Tübingen, 1985)

[42] Ch. Nöldeke, private communication

[43] A. Leggett, Proc. Int. Symp. Foundations of Quantum Mechanics, p. 74 (Tokyo, 1983)

[44] J.E. Mercereau, in: Superconductivity, ed. R.D. Parks (Dekker, New York, 1969)

[45] A. Rieckers, On Macroscopic Quantum Coherence at the Josephson Junction, in preparation.

Works related to the presented material are also:

[46] W. Thirring, Quantenmechanik großer Systeme (Springer, Wien, 1980)

[47] L. van Hemmen, Fortschr. Phys. 26, 397 (1978)

[48] H. Roos, Physica 100A, 183 (1980)

[49] M. Fannes, H. Spohn, and A. Verbeure, J. Math. Phys. 21, 355 (1980)

